Optimization Over Time

Optimization Over Time

Dynamic Programming and Stochastic Control

Volume I

PETER WHITTLE F.R.S.

Churchill Professor of the Mathematics of Operational Research, University of Cambridge
Fellow of Churchill College, Cambridge

175 YEARS OF PUBLISHING
1807 1982

JOHN WILEY & SONS
Chichester · New York · Brisbane · Toronto · Singapore

Library of Congress Cataloging in Publication Data:

Whittle, Peter.
Optimization over time.

(Wiley series in probability and mathematical statistics. Applied probability and statistics section)
Includes index.
Contents: v. 1. Dynamic programming and stochastic control
1. Mathematical optimization. 2. Dynamic programming. I. Title. II. Series.
QA402.5.W44 519.7′03 81–16401

ISBN 0 471 10120 6 AACR2

British Library Cataloguing in Publication Data:

Whittle, Peter
Optimization over time.—(Wiley series in probability and mathematical statistics (applied section)
Vol. 1
1. Dynamic programming
I. Title II. Series
519.7′03 T57.83

ISBN 0 471 10120 6

Photosetting by Thomson Press (India) Limited., New Delhi
Printed by Page Bros. (Norwich) Ltd.

Even a journey of one thousand li *begins with a single step.*

LAO TZE

千里之行始于足下

Contents of Volume I

Preface

By 'optimization over time' I mean the optimization of decisions to be made for the running of a dynamic system. As a subject this has been developing vigorously for some decades under the names of dynamic programming and of stochastic control.

In the face of this onrush of development any pretensions to encyclopaedism had to be abandoned fairly early, and survive only in that the text grew to two volumes, against my earnest wish. Within this compass I would claim, however, to give a unified account of the method of dynamic programming as an analytic tool for the solution of temporal optimization problems, to consider applications of this and related methods to major areas such as control and inference and to investigate a number of interesting particular cases in detail. Much of the material is recent. That having been said, it must be admitted that a text could be written on the subject of every single chapter, and others on subjects which find no coverage at all: e.g. large-scale systems, multi-person dynamic games and the optimization of simultaneous inference and control.

Most classifying dichotomies of the subject that I visualized when starting proved to be either false or grossly asymmetric. For example, the deterministic/stochastic split is not a dichotomy but a contrast between less and more general cases. Moreover, as argued in Chapter 9, the real contrast is not between deterministic and stochastic models but between models with complete and incomplete information. Another possible contrast is that between finite and infinite horizon models. Some recent texts develop an almost complete preoccupation with the quirks of the infinite horizon case. I have taken this topic as being important, but not dominating. Another recent preoccupation of theorists has been a painstaking analysis of the regularity properties of cost and control functions. I applaud the intention, but cannot see that anything natural or attractive has emerged. This is presumably a point which will find its natural resolution some time in the future, and as such I have left it.

This work is companion to my earlier book *Optimization under Constraints* (Whittle, 1971), although no more dependent upon that than upon other references quoted. It has been written during my tenure of the Churchill Chair, founded by the Esso Petroleum Company, to whom I am glad to acknowledge my gratitude.

Cambridge P. WHITTLE
June 1981

Structure

It is at least recommended that the reader, before tackling a given chapter, should acquaint himself with those preceding it in the directed graph of the diagram.

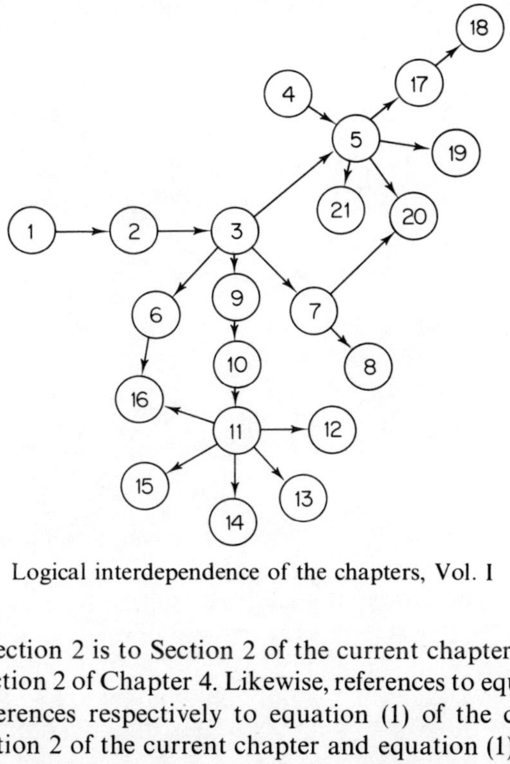

Logical interdependence of the chapters, Vol. I

A reference to Section 2 is to Section 2 of the current chapter; a reference to Section 4.2 is to Section 2 of Chapter 4. Likewise, references to equations (1), (2.1) and (4.2.1) are references respectively to equation (1) of the current section, equation (1) of Section 2 of the current chapter and equation (1) of Section 4.2. Theorems and figures always carry section identification as well. Therefore Theorem 2.1 refers to Theorem 1 of Section 2 of the current chapter; theorem 4.2.1 refers to Theorem 2.1 of Chapter 4. A similar convention is followed for figures: e.g. Figure 2.1 and Figure 4.2.1.

CHAPTER 1

Introduction

1. OPTIMIZATION OVER TIME

At both the individual and the corporate level we are all optimizers over time, in that we seek to choose our actions and decisions moment by moment with some consideration of their future consequences. We also know, then, how difficult it can be to make that choice and how difficult to assess those future consequences. Even if one had unlimited powers of logic and computation (which one has not) the future is usually imperfectly predictable; one's decisions are also contingent on decisions taken at other times, or by other people. Furthermore, criteria for judgement are often lacking, in that it is not even clear which outcomes are to be regarded as 'good' or 'right'. Under these circumstances 'optimization' is meaningless, and most of us take actions which seem reasonable, judged against a foreseeable future.

However, there are situations which can be formalized sufficiently that the optimization problem is well posed, and it can be maintained that study of these generates understanding and concepts that are useful more generally. Such situations (restricted to the case, already difficult enough, of a single decision-taker) constitute the compass of this work. The special feature of a temporal optimization problem is that variables, observations and decisions can be associated with moments in time, which have not only a natural ordering but also a natural direction. The principle of causality requires process dynamics to be such that present actions can affect only future outcomes. (However, process *optimization* will usually entail the consequence that future *policy* affects present *policy*—a recurrent theme). Another indicator of the direction of flight of the 'arrow of time' is that one's knowledge, upon which decisions are based, generally increases with the passage of time. (That is unless one forgets, see Section 43.4, although even then there is an asymmetry between past and future.)

The evolutionary character of a temporal problem makes it natural to express the optimization recursively as a choice of one's current action in the light of an already optimized action policy for the future. This is the *dynamic programming principle*, or *Bellman optimality principle.* It is much more than the computational reduction of the problem it is sometimes represented as being; it formulates the natural structural evolution (*backwards* through time) of policy optimization, and is basic.

A variant of this recursive formulation yields the *Pontryagin Maximum Principle* (Chapters 6 to 8). Other approaches to temporal optimization rely very much on special features of the problem. For example, if process-dynamics are linear and time-invariant, in senses to be clarified, then the variables of the process can be expressed as sums of terms exponential in time. By considering what exponents may occur one is led to a study of *stability* (Chapter 4). If such a process is stable, but excited by random inputs of time-invariant statistical structure, then it is natural to represent variables by linear combinations of sinusoids in time and to tackle optimization by so-called *spectral methods* (see Chapter 21).

This introductory chapter will briefly illustrate and develop these ideas.

2. DISCRETE AND CONTINUOUS TIME

We shall denote time consistently by t. In physical contexts it is supposed that t can take any real value, i.e. time is *continuous*. In such cases, we shall write $t \in R$, where R is the real axis. In some cases we envisage a situation starting at $t = 0$, and are interested only in non-negative values of t. In such cases we write $t \in R_+$.

However, there are contexts where observations and decisions are made only at regularly separated moments of time. For instance, it may be that production and demand figures are made available in a factory on a daily basis, with production decisions also taken daily. If one considers events only at these instants of time then it is said that one is working in *discrete time*. If the time interval between such consecutive instants is constant, then it is conventional to take this interval as the unit of time and to let t take the values $\ldots$, $-2, 1, 0, 1, 2, \ldots$. We then write $t \in Z$, where Z is the set of signed integers. If we are interested only in the t values $0, 1, 2, \ldots$, then we write $t \in Z_+$ where Z_+ is the set of non-negative integers.

Results of genuine structural significance in discrete time generally have an analogue in continuous time, and conversely. One advantage of discrete-time formulation is that the notion of a recursive relation is then simple and explicit (whereas in continuous time it must be expressed in differential or integral form). Also, by excluding the possibility of arbitrarily rapid variation with time, one excludes a class of pathologies.

However, at the other extreme, it is only in continuous time that one can achieve continuous paths, and it is the possibility of continuous-path variation which gives the continuous-time maximum principle (Chapter 7), the diffusion process (Chapter 38) and related boundary conditions (Sections 37.4, 37.5) their simplifying special features.

In general we shall work in discrete time, transferring to continuous time only when convenience or realism prompts, but indicating the analogues between the two cases wherever these are direct.

3. POLICIES

A *policy* (or *strategy*) is a rule for a choice of action under all possible circumstances as a function of perceived circumstances. We shall give more formal definitions in Chapters 2 and 10. Note that, whereas actions are taken in real time, a policy will usually be formulated in advance. That is, the decisions are really taken in advance, rather than at the time. To say this is not to say that the decision one will take at a given time is formulated in advance, but rather that the decision one will take under given circumstances is formulated in advance (the distinction between open- and closed-loop control; see Section 4.7).

Of course, in human life decisions are, very often, taken at the time. This is because life is, mercifully, too complex to admit the advance formulation of an all-inclusive policy for dealing with it.

4. RECURSIVE OPTIMIZATION; MINIMAL-COST PATHS ON A GRAPH

Consider a motorist wishing to reach one given point on a road-net from another by the shortest route. If he has a road-map with distances, etc., marked then he can certainly plan his route optimally in advance. However, he traces out his route in continuous time and could in principle change his mind at every point; this is therefore a temporal optimization problem.

It may be presumed that the only points on the route at which any decision is called for are at road junctions. We can thus imagine the road-net as a graph on which road junctions, initial position and desired final position define nodes. Two given nodes are *adjacent* if there is a direct road joining them, i.e. a route without intermediate nodes.

Let us label nodes by a variable x and let $c(x, x')$ denote the distance between a pair of adjacent nodes x, x'. If x, x' are joined by more than one direct route then let $c(x, x')$ be the distance of the shortest such route. If x, x' are not adjacent then let us set $c(x, x') = +\infty$. By pretending that the direct route (non-existent) between x and x' has infinite distance, we in effect forbid the motorist to consider it.

Actually, $c(x, x')$ can be generalized from the concept of distance to that of a *cost* of direct passage from x to x'. The motorist's preferences may well be a function of other factors as well as distance; they may be affected by whether the route sector is hilly, rough, motorway, urban, scenic or familiar, for example. The cost $c(x, x')$ reflects what he would pay not to have to make the direct passage x to x', and the planning of his route becomes a matter of cost-minimization rather than necessarily of distance-minimization. One will normally presume that $c(x, x') \geq 0$, with inequality for $x \neq x'$. The assumption $c(x, x') > 0$ may not be unreasonable if dawdling is to be penalized.

Suppose that the motorist wishes to go from node x to node y and suppose that he does so through the sequence of distinct nodes $x_0, x_1, x_2, \ldots, x_T$, where $x_0 = x$ and $x_T = y$. Assuming additive costs, the total cost of this route is

$$C = \sum_{t=0}^{T-1} c(x_t, x_{t+1}). \tag{1}$$

We have used t to indicate the stage on the route in the sense of the number of nodes traversed after the first. So, although the problem is actually in continuous time, the fact that the journey breaks naturally into stages means that we can formally regard it as a discrete-time problem. The total number of stages on the route, T, is of course not specified; the only specification is

$$x_0 = x, \qquad x_T = y. \tag{2}$$

The optimization problem is, then, the minimization of expression (1) over paths $(x_0, x_1, x_2, \ldots, x_T)$, subject to (2), and with T unspecified.

One will test a route for optimality by varying it. Variation of the *next* step on one's route yields in fact a recursive reduction of the problem. Let $F(x; y)$ denote the minimum of C subject to (2), the minimal cost which need be incurred in passage from x to y. If one moves first from x to x', and optimally thereafter, the cost incurred will be $c(x, x') + F(x', y)$. The optimal choice of x' will presumably be that which minimizes this quantity, and so we conjecture the relation

$$F(x; y) = \min_{x' \neq x} [c(x, x') + F(x'; y)], \tag{3}$$

To this is to be adjoined, as a matter of definition,

$$F(y; y) = 0 \tag{4}$$

(although see Exercise 1).

We shall formally prove the validity of (3) under appropriate conditions in Chapters 2 and 10. Note that, if valid, it provides not merely a functional equation for the minimal cost function F but also determines the optimal action at x—the choice of which node x' to aim for first. Relation (3) is genuinely a recursion in the sense that x' is 'nearer' to y than x (see Exercise 2). Relation (3) is the *dynamic programming equation*, or the *Bellman optimality equation*, for this particular problem. We shall in general simply use the term 'optimality equation'. It can sometimes be solved analytically; it must often be solved computationally. Quite possibly, understanding will come from direct study of the equation rather than from either solution.

The example is more general than it seems. The minimal-cost routing through a *physical* space could just as well have been a minimal-cost routing through a *conceptual* space. That is, x could represent a 'state', $c(x, x')$ the cost of changing the state from the current value x to x' at the next step and y a desired target state. The next two chapters will bring out these aspects. We take advantage of the exercises to make some technical points on the argument of this section.

Exercise and comments

1. If $c(x, x') < 0$ then the trip from x to x' is 'pleasurable', i.e. one would pay to make it. If $c(x, x') + c(x', x) < 0$ then the round trip (x, x', x) is pleasurable, and there is an incentive to repeat it indefinitely. That is, $F(x; x)$ has the value $-\infty$ rather than 0. In general, if there is any round trip (i.e. cycle) which is pleasurable, and this cycle is accessible from x, then $F(x; y) = -\infty$, even for $y = x$. That is, cost-minimization does not provide an incentive to reach y, but rather to reach the pleasure trip and embark on it for ever.
2. Show that cycling is excluded if $c(x, x') \geq 0$ with inequality for all $x' \neq x$. Note that then the minimizing x' of equation (3) is 'nearer' to y than x, in that $F(x'; y) < F(x; y)$.
3. Suppose $c(x, x') \geq 0$ with inequality for all $x' \neq x$. A minimal-cost path from x to y defines a *geodesic* on the graph between those points. Note that

$$F(x; y) \leq F(x; z) + F(z; x)$$

with equality if and only if z is an intermediate point on an (x, y) geodesic. This statement may be regarded as a generalization of the optimality equation (3).

5. SOLUTION OF THE OPTIMALITY EQUATION

It is not obvious how one would go about solving the optimality equation (4.3) numerically, even though this is a recursion, in the sense that movement from x to x' brings one nearer the target point (see Exercise 4.2). A way to make the recursivity evident is to keep an explicit tally of the number of stages required to reach the target. This is achieved by considering the modified problem, in which at most s stages are permitted, and a cost $K(x; y)$ is incurred if one is in state x when time has run out. That is, a path $x_0, x_1, \ldots, x_s$ over s stages carries a total cost of

$$C_s = \sum_{t=0}^{s-1} c(x_t, x_{t+1}) + K(x_s; y). \tag{1}$$

We shall assume that $c(y, y) = 0$ and $K(y, y) = 0$, so that, if one reaches target y before time has run out, then no further cost accumulates.

The severest prescription of the non-arrival penalty K would be

$$K(x; y) = \begin{cases} 0 & (x = y), \\ +\infty & (x \neq y), \end{cases} \tag{2}$$

although one may work with a milder measure of 'cost-distance' of x from y.

Let us denote the infimum of expression (1) over paths for which $x_0 = x$ by $F_s(x; y)$. This is the minimal cost incurred if at most s moves are permitted ('at most' s rather than 'exactly' s, because an early arrival at y counts as fewer than s moves). Then certainly

$$F_0(x;y)=K(x;y) \tag{3}$$

and, plausibly,

$$F_s(x;y)=\min_{x'}[c(x,x')+F_{s-1}(x';y)] \qquad (s>0), \tag{4}$$

where the minimizing x' is the best node to aim for from x if at most s moves are permitted. Relation (4) is clearly recursive in s, and (3), (4) can be solved for increasing s. Presumably also $F_s(x;y)\to F(x;y)$ as s increases, at least if the penalty $K(x;y)$ for non-arrival is severe enough (see Exercises 2 and 3).

The variable s is often spoken of as 'time-to-go', or the 'horizon'. In the formulation of Section 4 the horizon was indefinite. The optimal policy may well find the target in a bounded number of moves, but no explicit restriction was imposed. The placing of such a restriction allows one to establish a recursion on horizon length.

Exercises

1. Assume $K(x;y)\geq 0$, $c(x,x')\geq 0$. Show that $F_s(x;y)$ is non-increasing in s and has a limit $F_\infty(x;y)$.
2. Assume prescription (2) and that the graph contains n nodes. Show that $F_n(x;y)=F(x;y)$.
3. Suppose that $c(y,y)=0$, but that otherwise $c(x,x')\geq\delta>0$ and that $K(x;y)\geq 0$. Show then that $F_s(x;y)=F(x;y)$ for

   $$s\geq s(x,y)=\frac{F(x;y)}{\delta}.$$

   The hypotheses of Exercises 2 and 3 are designed to enforce a preference for arrival at y.
4. Modify the problem by assuming that $K(x;y)$ is the cost of 'terminating' at x, where 'termination' is an option open at any time. Equation (3) then modifies to

   $$F_s(x;y)=\min\left[K(x;y),\min_{x'}[c(x,x')+F_{s-1}(x';y)]\right].$$

   Note the equivalence of the two versions if $c(x,x)=0$ for all x.

6. LINEAR SYSTEMS; STABILITY

Let $x(t)$ be a scalar function of continuous time whose value is required to be kept as near zero as possible (the *regulation* problem). It might, for example, be the deviation of greenhouse temperature from an ideal value. Suppose that the values of $x(t)$ are generated by a dynamic equation

$$Ax(t) = 0 \tag{1}$$

where A is an operator relating $x(t)$ to its past (i.e. to $x(\sigma)$ for $\sigma < t$). In practice, of course, one will have many variables simultaneously interacting, since the temperature will vary over the greenhouse and be affected by many factors. The ideas we shall sketch generalize to this situation, but for the moment we consider the scalar case.

Under common simplifying assumptions the operator A may be regarded as linear and time-invariant. This is equivalent to the statement that

$$Ae^{\lambda t} = a(\lambda)e^{\lambda t} \tag{2}$$

where $a(\lambda)$ is a scalar function and (2) holds for all t and for all λ for which $a(\lambda)$ is properly defined. Indeed, the solutions of (1) are arbitrary linear combinations

$$x(t) = \sum_j c_j e^{\lambda_j t} \tag{3}$$

of the exponential functions $\exp(\lambda_j t)$, where the λ_j are the roots of

$$a(\lambda) = 0. \tag{4}$$

The values of the coefficients c_j in (3) are determined by initial conditions. If we require that $x(t) \to 0$ as t increases, for arbitrary initial conditions, then it is necessary that

$$\mathrm{Re}(\lambda_j) < 0$$

for all j. This is the requirement of *stability* of the system.

The system will in fact commonly be a controlled one, so that (1) should be replaced by a relation of the type

$$Ax(t) = u(t) \tag{5}$$

where $u(t)$ is the control variable to be chosen by the optimizer. In the case of the greenhouse, $u(t)$ would represent the effects that could be achieved by deliberate heating, cooling, ventilation, etc. Usually one will choose the correcting control as a function of recently observed values of x itself (*feedback*) so that

$$u(t) = Dx(t) \tag{6}$$

where D is an operator such that $Dx(t)$ is a function of those values of x regarded as observable at time t, usually $x(\sigma)$ for $\sigma < t$. (Again, in the multivariate situation $u(t)$ will be chosen as a function of any observable that is helpful.) Note that the operator A is prescribed by the physics of the situation; the operator D can be chosen by the optimizer, subject to conditions of physical realizability.

If D is also linear and time-invariant, with $De^{\lambda t} = d(\lambda)e^{\lambda t}$, then we see from (5) that, for the controlled system to be stable, it is necessary that $a(\lambda) - d(\lambda)$ have all its zeros in the negative half of the complex plane: $\mathrm{Re}(\lambda) < 0$.

This was very much the way of traditional control theory—to find operators D corresponding to easily realizable devices which assured stability. The more recent optimization approach rather sets a cost on deviations of the controlled variable from target value, and also on any other features one would wish to discourage (e.g. heavy control, rapidity of change, frequent re-setting). One then chooses an operator D in the available set to minimize total cost. The expectation is that the optimized system will automatically be stable, if stability is possible and costs effectively penalize instability. The optimization can be formulated recursively, as in the last two sections.

This approach is, of course, equally applicable to non-linear systems, and the concept of stability can also be given a meaning for such systems. However, the particular rôle of exponential solutions (3) for the linear case continues to penetrate thinking.

A yet more realistic model than (5) is

$$Ax(t) = u(t) + \alpha(t), \tag{7}$$

where $\alpha(t)$ represents exogenous disturbances, i.e. disturbances injected into the system from outside. In the case of the greenhouse, this might represent effects of sun and weather, for example. Assuming the future course of α known (an assumption seldom fully justified; see Exercise 17.4.2) one will probably replace the control rule (6) by something like

$$u(t) = D_1 x(t) + D_2 \alpha(t)$$

Usually $D_1 x(t)$ will be a function only of past x (because that is all that is observable) and $D_2 \alpha(t)$ a function only of future α (because past α has already wrought its effect on x). These two components of control are referred to as *feedback* and *feedforward* components respectively.

Exercises and comments

1. Note the examples

$$A = \sum a_k \left(\frac{\mathrm{d}}{\mathrm{d}t}\right)^k$$

$$Ax(t) = \int_0^\infty \mathrm{e}^{-\alpha s} x(t-s)\,\mathrm{d}s$$

for which $a(\lambda)$ has the evaluations $\sum a_k \lambda^k$ and $(\lambda + \alpha)^{-1}$ respectively.

7. IGNORANCE; STOCHASTIC EFFECTS

The contrasting terms 'stochastic' and 'deterministic' are used to describe processes respectively with and without some random element. This may give the impression that the two classes are disjoint. In fact, one should probably regard the class of deterministic models as a subset of the class of stochastic models. Our examples, of route-finding and greenhouse control, were both formulated deterministically, but realism would demand some stochastic element.

The greenhouse example illustrates three of the principal ways in which randomness can enter. The first is that the disturbing variable $\alpha(t)$ of (6.2) may be random to some extent, and so not perfectly predictable, or even not at all predictable. In such a case one speaks of it as 'process noise'. The second is that relevant variables such as $x(t)$, regarded as observable at least to some extent, may perhaps now be observable only with some error, usually assumed random. This error is referred to as 'observation noise'.

The third possibility is that the operator A, which describes the intrinsic dynamics of the greenhouse, may not be known completely, and its nature may have to be inferred in the course of time. Any attempt at rational control requires, of course, knowledge of A. Properly speaking, this is not an example of randomness, but of lack of knowledge ('ignorance'). However, the two concepts lie close.

One way of formulating ignorance is to regard the unknown quantity as a random variable, whether it be structural parameter, process/observation noise or any other quantity of interest. This is what we shall call the *stochastic* approach. A general rationale for optimization based upon it is formulated in Chapter 10 and is developed in the rest of the book.

Another way of coping with ignorance is to assume the 'worst case', and pessimistically assign unknown quantities their least favourable values. This is the *minimax* approach, dealt with in Chapter 9. Despite the rather extreme temperamental bias of such a philosophy, it leads to an attractive mathematical structure in strong analogue to that derived from the stochastic formulation.

A formulation, stochastic or minimax, which concedes ignorance of some relevant variables clearly sets the direction of the arrow of time. This is because, if new knowledge is acquired at all, it is acquired in the course of time. Consider, for example, the route-finding problem of Section 4. Once one's policy had been decided, then the route which would be followed was perfectly predictable, and there was no need to make contingency plans for the unexpected. Indeed, the problem of a cost-minimization from x to y could just as well have been posed as the problem of cost-minimization from y to x, with the modified stage-cost function $c^*(x, x') = c(x', x)$. However, suppose random events (floods, accidents, road-works) disturb one's progress. One's route is then unpredictable, and one may have to revise it as one goes along (although by a policy decided on in advance). One's knowledge increases as one's uncertainty about the state of the route is progressively resolved by observation. This monotonicity fixes the direction of time.

For the stochastic case also there are alternatives to a recursive formulation of the optimization problem in special cases. Consider, for example, the greenhouse control problem, assuming a time-invariant linear structure and random process noise. If the system is intrinsically stable, so that variables can settle down to a statistically time-invariant pattern of variation ('stationarity'), then the relation $x(t) = 0$ to which (6.3) would converge for large enough t becomes, rather,

$$x(t) = \int e^{i\omega t}\, d\zeta(\omega) \tag{1}$$

Here the integral is over real values of *frequency* ω, and $\zeta(\omega)$ is a *random* amplitude function. The characteristics of ζ are determined by the corresponding characteristics of process noise and by the form of the process and control operators A and D. The cost can be expressed in terms of these characteristics and optimized with respect to available D by Fourier methods (see Chapter 21). However, this is a technique available only for rather particular cases.

8. A SIMPLE EXAMPLE OF THE COST OF INFORMATION

If one goes along observing and coping with unpredictables as they occur (e.g. encountering the floods, accidents and road-works as they happen) then in a sense this new knowledge costs nothing; it simply comes. In some cases it can be deliberately acquired, at a cost which may be worth while. For example, in the route-finding case, it may be that advance enquiry would have given helpful information about the state of the road-net.

Another example is the case of the greenhouse with unknown dynamics A. One may have to experiment somewhat in order to determine these dynamics. That is, one may have to operate the greenhouse for a while in a different way than if one had known A, in order to provide the extreme conditions which exhibit the nature of A. The information thus yielded will have been obtained at a cost—the cost of less-than-ideal operation of the greenhouse during the experimental period.

This example illustrates one of the most interesting and least satisfactorily treated problems of temporal optimization: the use of control actions both to optimize control itself and to acquire information. This problem occurs recurrently in Part VI, although without general solution. Here we consider an example drastically simplified in that the phases of 'experimentation' and 'operation' are completely separated.

Suppose that an industrial operation requires the setting of a parameter θ (pressure, say) whose ideal value θ_0 is unknown. If one takes the setting θ then there is a loss of production of value $(\theta - \theta_0)^2$ per unit time. If one stops to experiment then one incurs a cost of c per unit time. The process goes on in discrete time; at each instant of time one has a choice of either operating the process (which, it is supposed, yields no information on θ_0) or of stopping the process and obtaining a reading $\theta_0 + \eta$, where the η are independent observation errors of zero mean and unit variance. If one has taken r observations on θ_0 one then has effectively an unbiassed estimate $\hat{\theta}_0$ of θ_0 with $E(\hat{\theta}_0 - \theta_0)^2 = 1/r$. This would also be the expected cost of production loss if one operated for unit time at setting $\hat{\theta}_0$.

The variable r may be taken as indicating one's 'state of knowledge' about θ_0. Let $F_s(r)$ denote the minimal expected cost over a horizon s if one starts in the state of knowledge r. If one allows the two alternatives of either operating with setting equal to the current best estimate $\hat{\theta}_0$ or of taking another observation then, plausibly, the optimality equation takes the form

$$F_s(r) = \min\left[\frac{1}{r} + F_{s-1}(r), c + F_{s-1}(r+1)\right] \qquad (s > 0) \tag{1}$$

with

$$F_0(r) = 0 \tag{2}$$

In fact, as the problem is formulated, one acquires no new information which would change one's mind about the number of observations one intends to take. It is also clear that one should take all these observations at the beginning of the period when they will be of most benefit. That is,

$$F_s(r) = \min_u\left[cu + \frac{s-u}{r+u}\right] \tag{3}$$

where u is the number of further observations taken and is restricted to being an integer in the range $[0, s]$. If one neglects the fact that u must be an integer then one derives the optimizing value

$$u = \left[\sqrt{\left(\frac{r+s}{c}\right)} - r\right]_+ \tag{4}$$

For large s this is of order $\sqrt{s}$. That is, if the horizon is long then it is worth while to make the period of initial experimentation also long, because an improvement in the estimate of θ_0, even if small, brings large integrated benefits. In practice, one will be less concerned about costs in the distant future and will accord them less weight in the cost function (see Exercise 2).

For simplicity we have totally separated the phases of experimentation and operation. In practice, one might well combine them to some degree, so that one varies θ during operation and discovers the effect on production. Deviation from the value $\hat{\theta}_0$, currently believed best, will cost production but yield information. Determination of the optimal compromise remains a difficult matter.

Exercises

1. Verify that (3) solves (1), (2). Verification proves that all the observations taken should indeed be taken at the beginning of the period.
2. Let g_t be the cost incurred at time t, so that we are minimizing the value of

$\sum_0^{s-1} g_t$. In a *discounted* formulation one rather minimizes $\sum_0^{s-1} \beta^t g_t$, where β is a *discount factor* with a value between 0 and 1. The smaller the value or β, the less one weights future costs relative to present ones. Show that, in the infinite horizon case $s = +\infty$, the optimal value of u is

$$u = \left[\frac{1 + \sqrt{(1 + 4c/\alpha)}}{2c} - r\right]_+$$

if one neglects the fact that u must be integral. Here $\alpha = -\log \beta$.

9. SOME GENERAL POINTS ON OPTIMIZATION

Even in the one-stage case an optimum need not exist. Consider, for example, the minimization of $C(u) = u^{-1}$ for u positive and finite. We have $C \geq 0$ so $C = 0$ would certainly be optimal if attainable. It cannot be attained for any finite u, although it can be approached arbitrarily closely by taking u large enough.

In general, if $C(u)$ is a scalar function of u in $\mathscr{U}$, one defines the *infimum* of C, written $\inf C(u)$, as the largest value of λ such that $C(u) \geq \lambda$ for all u in $\mathscr{U}$. If there exists a value u_0 in $\mathscr{U}$ such that

$$C(u_0) = \inf C(u)$$

then the infimum is *attainable*. Then one says also that a minimum exists, or that an optimum exists, and u_0 is optimal (if minimization is the criterion for optimality). However, even if an optimum does not exist, for any prescribed positive ε one can always find a u_0 in $\mathscr{U}$ such that

$$C(u_0) \leq \inf C(u) + \varepsilon.$$

Such a value is said to be *ε-optimal.*

One can produce problems of some physical plausibility for which an optimum does not exist, although a more careful formulation of the problem will usually remove the pathology (see Exercise 1).

Often an optimum fails to exist because the optimizing value is a limit point of the values of the variable in the permitted set $\mathscr{U}$ (i.e. lies in the closure of $\mathscr{U}$, but not in $\mathscr{U}$). This was the case in the example above. For another, $\mathscr{U}$ might be a class of continuous functions and the optimizing value u_0 may be discontinuous; then u_0 lies outside $\mathscr{U}$, but is arbitrarily closely approximated by elements of $\mathscr{U}$.

Restriction on the class of solutions can often create problems, sometimes artificial. For example, any action rule must certainly express action u as a function $u(W)$ of observables W: this condition of *realizability* is a natural structural constraint. However, sometimes, for technical reasons, the class of permitted functions is restricted, e.g. to Borel-measurable functions. The theory generated by preoccupation with this point is so heavy as to lead one to believe that the natural approach has yet to be found. For this reason, and because these matters take care of themselves in a properly formulated problem, we shall not concern ourselves with questions of technical measurability.

Exercise and comments

1. Suppose one arrives to catch a train leaving at $t=0$. If one arrives before it leaves one incurs a cost equal to the waiting time; if one arrives at or after the departure time then one misses the train and incurs unit cost. That is, if $-u$ is the time of arrival then the cost is

$$C(u)=\begin{cases} u & (u>0), \\ 1 & (u\le 0). \end{cases}$$

The infimal value is $C=0$, which is not attainable.

The reason for the trouble is assignment of the value 1 rather than 0 to $C(0)$. That is, one makes the fine (and unrealistic) distinction that one misses (rather than catches) the train if one arrives at the exact departure time. The distinction would be blurred and lost by a number of factors in practice. For example, the time τ of departure of the train may be a random variable, distributed continuously about zero. The expected cost if one arrived at time $-u$ would then be $EC(u+\tau)$. This is a continuous function of u, with a minimum.

10. NOTATION AND USAGE

Unless stated otherwise, a vector x is considered to be a column vector. The transpose of a matrix A is denoted A' and the complex conjugate of its transpose is denoted $A^{\dagger}$.

The real axis is denoted R and the set of signed integers Z, with n-dimensional analogues R^n and Z^n. If t denotes time then the cases of continuous time and discrete time correspond to t's adopting values in R or Z respectively. If x is explicitly a function of time then it will be written $x(t)$ or x_t in continuous and discrete time respectively.

However, the subscript notation is also used to indicate partial differentiation. If H is a scalar and x a column vector then H_x is the row vector of differentials of H with respect to the components of x. Correspondingly, if p is a row vector then H_p is the column vector of the differentials of H with respect to the components of p. If x and a are column vectors of dimensions n and r respectively, then a_x is the $r\times n$ matrix of differentials of components of a with respect to components of x.

A relation such as

$$F(x)=K(x) \qquad (x\in D)$$

is often written simply

$$F=K \qquad (D).$$

The abbreviations max, min, sup, inf denote maximum, minimum, supremum and infimum respectively. There are some symbols which have a constant connotation throughout the work: e.g. x for state variable (or, more generally, process-describing variable), u for control or action variable and W for observation history. However, limitations of the alphabet lead one to overwork certain symbols, and the principal ambiguities (resolved by the context, one hopes) are best listed here.

The symbol C can denote cost (see equation (4.1)), the matrix relating observations and state variables (see equation (4.1.4)) or a continuation region (see Section 16.2). It also occasionally denotes capacity (see Sections 3.3 and 6.4).

The symbol K can denote terminal cost (see equation (5.3)). It can also denote a matrix relating control and state variables (see the end of Section 4.2).

The symbol M does multiple duty as a maximizing operator (Chapter 9), as denoting terminal reward (Chapter 14), as denoting the covariance matrix of observation noise (Chapter 18), as denoting a bound on control magnitude (Section 7.4) or as a matrix occurring in the context of controllability (Section 4.3).

The symbol ξ is also used in several connotations, the principal ones being as the combined variable (x, t) in Chapters 7 and 20 and as the vector of exogenous inputs in Chapters 17 and 18.

If P, Q are symmetric matrices of equal order then the inequalities $P > Q$ and $P \geq Q$ will usually mean that $P - Q$ is positive definite and non-negative definite respectively.

The sign $:=$ between two quantities defines the first in terms of the second.

The symbol ■ denotes 'end of proof'.

If a vector random variable x is considered to be distributed normally with mean μ and covariance matrix V then we shall write 'x is distributed $N(\mu, V)$' or '$x \sim N(\mu, V)$'.

A set of references to general texts on the subject is given in the first section of the Bibliography.

Part I

Deterministic problems

CHAPTER 2

The optimality equation and state structure in the deterministic case

We give some preliminary theory for the deterministic case, to be illustrated by examples in the next chapter. There are themes which are not developed because they are best left until Part II, which covers the more general stochastic case. Also, although some infinite-horizon examples occur in Chapter 3, the general theory of such problems is left until Part IV.

1. THE OPTIMALITY EQUATION

Assume the problem posed in discrete time, starting at $t=0$. The decision at time t is denoted by u_t, and the partial sequence of decisions $u_0, u_1, u_2, \ldots, u_t$ by U_t. If we consider events over a horizon T then the cost function can be considered as a function of the decisions made over this period:

$$C = G(u_0, u_1, \ldots, u_{T-1}) = G(U_{T-1}). \tag{1}$$

We suppose all relevant quantities known, which is to assume rather more than determinism. No other variable need then be explicitly indicated in the cost function. Variables are either exogenous to the system, in that they are unaffected by the u_t, or endogenous, in that they are. In the first case they are known, and can be regarded as fixed. In the second case they can themselves be regarded as functions of the u_t. There may be structural point later in indicating the dependence of C upon significant exogenous and endogenous variables, but for the moment there is no necessity.

The value of U_{T-1} will presumably be restricted to some permitted set $\mathscr{U}_{T-1}$. We shall in general not indicate this restriction explicitly, but merely assume that all minimizations, etc., observe it.

Theorem 1.1 (The optimality equation). *Define the functions*

$$G(U_{t-1}, t) = \inf_{u_t, u_{t+1}, \ldots, u_{T-1}} G(U_{T-1}). \tag{2}$$

Then these obey the recursion

$$G(U_{t-1}, t) = \inf_{u_t} G(U_t, t+1) \qquad (0 \le t < T) \tag{3}$$

with terminal evaluation

$$G(U_{T-1}, T) = G(U_{T-1}). \tag{4}$$

Furthermore, the minimizing value of u_t in (3) is the optimal value of u_t for prescribed U_{t-1}.

That is, given a history U_{t-1} of past decisions, which may or may not have been optimal, the minimizing value of u_t in (3) is optimal. The phrase 'the minimizing value of u_t is optimal' is to be regarded as shorthand for 'the minimizing value of u_t, if it exists, is optimal, and a value of u_t bringing the function to within ε of its infimum is ε-optimal'.

Proof. Relation (4) is an immediate consequence of definition. Relation (3) is a simple consequence of the fact that the value of $\inf_{u_t} \inf_{u_{t+1}} \cdots \inf_{u_{T-1}} G(U_{T-1})$ is independent of the order in which the infima are taken and that the jointly minimizing value of $(u_t, u_{t+1}, \ldots, u_{T-1})$ is optimal for prescribed U_{t-1}. The set of recursions (3) for $t, t+1, \ldots, T-1$ simply corresponds to the application of the infimalizing operations in a particular order. The quantity $G(U_{t-1}, t)$ is the infimal cost conditional on the fact that decisions U_{t-1} have already been taken. ∎

Relation (3) *is* the optimality equation in this context. It may appear signally trivial in that the recursive reduction is applicable to the minimization of any function of several arguments and the temporal ordering of these arguments appears no more significant than any other ordering. However, the reduction begins to acquire meaning when we take account of temporal structure. This can manifest itself in two ways.

Firstly, there may be natural endogenous variables $\{x_t\}$ which evolve with time, depending upon past decisions in a manner dictated by the prescribed dynamics of the system. The variable x_t may even describe the *state* of the system at time t, in the sense that optimization of $u_t, \ldots, u_{T-1}$ for prescribed U_{t-1} is equivalent to optimization of $u_t, \ldots, u_{T-1}$ for prescribed x_t. Relation (3) then reduces, as we shall see in the next section.

The other clear sign of temporal structure appears in the case of incomplete knowledge, when the set of observables on which decisions are based increases non-trivially with t (see Chapters 9 and 10). The order of action-optimization is then dictated—backwards in time.

Exercises

1. *Optimal allocation.* This is an example of a problem which does not have temporal structure, but which is nicely decomposable. Let $\sum_{j=1}^{n} g(u_j, j)$ be the total utility gained if a resource vector u_j is allocated to activity j, and let $F_n(x)$ be the maximum of this quantity over the u_j subject to $u_j \geq 0$ $(j = 1, 2, \ldots, n)$

and $\sum u_j \leq x$. That is, $F_n(x)$ is the maximal utility obtainable from the n activities with vector x of total resources. Show that

$$F_n(x) = \max_{0 \leq u \leq x} [g(u, n) + F_{n-1}(x - u)] \qquad (n > 0)$$

and that the maximizing u is the optimal allocation to the nth activity.

2. Consider Exercise 1 with the constraint $\sum u_j \leq x$ modified to $\sum_j c_j u_j \leq x$.
3. A total amount of water x is to be released over time T in such a way as to minimize the maximal flow-rate. That is, non-negative scalars u_t subject to $\sum_0^{T-1} u_t = x$ are to be chosen to minimize $\max_{0 \leq t < T} u_t$. Fairly obviously the solution is $u_t = x/T$; confirm this by solution of the optimality equation for this problem.

2. STATE STRUCTURE (ADDITIVE COSTS)

The optimal value of u_t will generally depend upon the whole decision history U_{t-1}, as we see from (1.3). Since past decisions need not have been optimal, one must be prepared to calculate the optimal u_t for all possible U_{t-1}.

Suppose there is a variable ω_t which effectively summarizes past history, in that, given t and the value of ω_t, one can calculate the optimal u_t and also ω_{t+1} without knowledge of U_{t-1}, for all t. Such a variable is termed *sufficient*.

Of course, U_{t-1} is itself sufficient, but the question is whether there is a more economical description which is also sufficient. The history U_{t-1} takes values in as increasing space in t increases; one seeks a describing variable which takes values in a fixed space—indeed a space which is as small as possible. A sufficient variable ω_t taking values in a space independent of t is called a *state variable*. For a variable to be a state variable is a property of both the dynamics and the cost structure of the process.

In the path-minimization problem of Section 1.4 the current position x_t was a state variable. Knowledge of current position made previous history irrelevant as far as future costs were concerned. We seek a more general formulation of that situation.

Assume then that there is a variable x taking values in a space $\mathscr{X}$, its value at time t denoted by x_t, which enjoys the following properties:

(a) The value of x_t is observable at time t; i.e. calculable from known quantities.

(b) The sequence $\{x_t\}$ obeys a known first-order forward recursion in time

$$x_{t+1} = a(x_t, u_t, t). \qquad (1)$$

(c) The set of values u_t may adopt depends on x_t and t at most: $\mathscr{U}(x_t, t)$, say.

(d) The cost function to be minimized over horizon T has the form

$$C = \sum_0^{T-1} g(x_t, u_t, t) + K(x_T). \qquad (2)$$

We shall prove in Theorem 2.1 that x_t is indeed a state variable. Conditions (a) to (c) are virtually necessary, condition (d) can be greatly relaxed.

Relation (1) is termed the *plant equation* (occasionally the *law of motion*). It describes how x_t evolves; the property demanded in (b) is that the plant equation should have the simply recursive form (1). In the case of a dynamic process without decisions or costs a state variable would be characterized exactly by the fact that it obeyed a simple recursion

$$x_{t+1} = a(x_t, t)$$

and that any quantity of interest at time t could be calculated from x_t. When decisions and costs are involved then hypotheses involving these must be added.

Condition (d) gives an additive decomposition of cost. The term $g(x_t, u_t, t)$ may be regarded as the component of cost incurred at time t, the *instantaneous cost*. The variable x_t remains a state variable under considerable relaxations of this cost hypothesis (see Section 3). Nevertheless, hypothesis (2) remains a favourite, as it so clearly assigns cost over time. If g is zero, so that $C = K(x_T)$, the problem is said to be a *terminal* problem.

Denote the partial sequence $x_0, x_1, \ldots, x_t$ by X_t and define

$$F(X_t, U_{t-1}, t) = \inf_{u_t, u_{t+1}, \ldots, u_{T-1}} \left[\sum_{j=t}^{T-1} g(x_j, u_j, j) + K(x_T) \right]. \tag{3}$$

That is, F is the *minimal future cost* at time t, given the x/u history up to time t.

Theorem 2.1 (Optimization with state structure).

(i) *Under hypotheses (a) to (d) x is a state variable.*

(ii) *More specifically, $F(X_t, U_{t-1}, t)$ and the optimal u_t are functions of x_t and t alone: $F(x_t, t)$ and $u(x_t, t)$, say.*

(iii) *The optimality equation can be written*

$$F(x, t) = \inf_u [g(x, u, t) + F(a(x, u, t), t + 1)] \qquad (0 \le t < T) \tag{4}$$

and

$$F(x, T) = K(x) \tag{5}$$

where x is a generic value of x_t. The minimizing value of u in (4) is the optimal decision $u(x, t)$.

Proof. The proof is inductive. Since $F(X_T, U_{T-1}, T) = K(x_T)$, then $F(X_j, U_{j-1}, j)$ reduces to $F(x_j, j)$ for $j = T$, and (5) holds. Suppose the reduction is valid for $t < j \le T$. Then the optimality equation (1.3) becomes at t

$$F(X_t, U_{t-1}, t) = \inf_{u_t} [g(x_t, u_t, t) + F(a(x_t, u_t, t), t + 1)]. \tag{6}$$

The bracket in (6) is a function of x_t, u_t and t alone; both the minimizing and minimized values are then functions of x_t and t alone. Assertions (ii) and (iii) then follow. Assertion (i) follows, because the optimal u_t is calculable from (x_t, t) alone, and x_{t+1} is then calculable by (1). ■

It was assumed in (a) that x_t is known. All that is necessary for this is that the initial state x_0 be known and (as can be taken henceforth for granted unless the contrary is stated) that plant dynamics as specified by $a(x, u, t)$ are known. The ensuing values $x_1, x_2, \ldots$ can then be calculated from the plant equation. If these assumptions fail, then optimization will also imply an attempt at estimation. The minimax version of this situation is considered briefly in Chapter 9 and the stochastic version extensively in Parts III and VI.

Relation (4), the optimality equation, can be viewed as a simple backward recursion for the sequence of functions $F(x, t)$, which simultaneously determines optimal control in feedback form $u(x, t)$ (see Section 4.7). Much of the rest of the book reduces to a study of this equation, its solution and the nature of its solution.

For the path-minimization problem the decision u_t is essentially a choice of the next node x_{t+1} from the set $\mathscr{U}(x_t, t)$ of nodes adjacent to x_t. The plant equation (1) reduces to

$$x_{t+1} = u_t$$

and the optimality equation to

$$F(x, t) = \inf_u [c(x, u) + F(u, t+1)] \qquad (t < T).$$

We consider further examples in detail in the next chapter.

The set $\mathscr{X}$ of values that the state variable x may assume is termed the *state space*.

Exercises

1. *Discounting*. Denote $g(x_t, u_t, t)$ by g_t for simplicity. Consider the discounted cost function $\sum_0^{T-1} \beta^t g_t + \beta^T K(x_T)$ and define the *minimal discounted future cost* $\inf_{u_t, \ldots, u_{T-1}} [\sum_{j=t}^{T-1} \beta^{j-t} g_j + \beta^{T-t} K(x_T)]$. Show that, with this redefinition of C and F, Theorem 2.1 holds with the modified optimality equation
$$F(x, t) = \inf_u [g(x, u, t) + \beta F(a(x, u, t), t+1)]. \qquad (7)$$
2. Suppose the plant equation is of the form
$$x_{t+1} = a(x_t, x_{t-1}, u_t, t).$$
Show that one has state structure with (x_t, x_{t-1}) as the state variable.
3. Generalize the conclusion of Exercise 2 to the case when the plant equation expresses x_{t+1} in terms of $x_t, x_{t-1}, \ldots, x_{t-h}$; $u_t, u_{t-1}, \ldots, u_{t-k}$ and t.

3. RECURSIVELY GENERATED COSTS

The additive cost structure postulated in (2.2) identifies the cost g incurred *at* a certain time. If one can simply identify the cost $C^{(t)}$ incurred *from* a certain time t, and this obeys a recursion in t, then existence of the state structure follows just as in Theorem 2.1. The fact that $C^{(t)}$ is a *future* cost means that the recursion must go *backwards* in time, which motivates the following variation of assumption (d):

(d′) The cost $C^{(t)}$ incurred from time t obeys the recursion

$$C^{(t)} = \psi(x_t, x_{t+1}, u_t, t, C^{(t+1)}) \tag{1}$$

where ψ is strictly increasing in its last argument, $C^{(t+1)}$, for all values of remaining arguments.

The total cost to be minimized is thus $C = C^{(0)}$. Special assumptions must be added if one wishes to indicate that only events before a horizon T are of any import. Essentially, one would wish $C^{(0)}$ to be independent of policy from time T onwards. The simplest way to assure this is to prescribe $C^{(T)}$ as a function $K(x_T)$ of x_T alone. The most general case is normalizable to this prescription (see Section 9.3).

Theorem 3.1 (State structure with recursively generated costs). *Adopt the hypotheses of Theorem* 2.1, *except that* (d) *is replaced by* (d′) *and* $C^{(T)}$ *is prescribed as* $K(x_T)$. *Then the assertions of that theorem hold, with the modifications that* F *is defined by*

$$F(X_t, U_{t-1}, t) = \inf_{u_t, u_{t+1}, \ldots, u_{T-1}} C^{(t)} \tag{2}$$

and the optimality equation (2.4) *becomes*

$$F(x, t) = \inf_u [\psi(x, a(x, u, t), u, t, F(a(x, u, t), t+1))] \qquad (0 \le t < T). \tag{3}$$

Proof. The proof is the obvious slight generalization of that of Theorem 2.1. ∎

As examples, the recursion

$$C^{(t)} = g(x_t, x_{t+1}, u_t, t) + \beta(x_t, x_{t+1}, u_t, t)C^{(t+1)} \tag{4}$$

would correspond to the additive form (2.2), but with addition of the extra argument x_{t+1} in g and incorporation also of an (x_t, x_{t+1}, u_t, t)-dependent discount factor β. Actually, incorporation of x_{t+1} achieves no extra generality, since this is a function of (x_t, u_t, t). However, the generalization is more significant in the cases to be considered in Chapters 9 to 11 of a 'fuzzy' or stochastic plant equation. The monotonicity demand of (d′) implies that $\beta > 0$.

Another case of real generalization is supplied by the cost function

$$C = \max_{0 \le t < T} [g(x_t, u_t, t)] \tag{5}$$

which can be identified as $C^{(0)}$ for the recursion

$$C^{(t)} = \max\,[g(x_t, u_t, t), C^{(t+1)}]. \tag{6}$$

4. VARIATIONS AND REDUCTIONS

Conditions (a) to (c) of Section 2 and (d′) of Section 3 jointly constitute our standard specification of a deterministic temporal decision process with state structure. This is the deterministic case of what is later termed a *Markov decision process* (see Chapter 11). There are other specifications, seemingly more general or less general, which in fact turn out to be formally equivalent.

A problem is said to be *terminal* if termination at $t = T$ is prescribed and total cost is a function only of the terminal state value, i.e. $C = K(x_T)$. Any problem for which the cost function obeys a *forward* recursion can be transformed to terminal form, at the expense of augmenting the state variable by an additional component. Suppose the alternative hypotheses (d) or
(d′) of the preceding two sections are replaced by the following:
(d″) The cost C_t incurred before time t obeys a forward recursion

$$C_{t+1} = \psi(x_t, x_{t+1}, u_t, t, C_t). \tag{1}$$

The total cost incurred if horizon T is specified is then just C_{T+1}.

The cost over horizon T is then a function purely of the terminal value of the augmented state variable (x_t, C_t), obeying the combined plant equation (2.1) and (1).

So, for the additive case (2.2) the recursion would be

$$C_{t+1} = g(x_t, u_t, t) + C_t \tag{2}$$

with $C_0 = 0$ and $g(x_T, u_T, T) = K(x_T)$.

We see then that, although the additive cost function (2.2) might seem more general than a purely terminal cost function, this latter is in fact as general as the class of cost structures postulated in (d″).

We saw in Exercises 1.2 and 1.3 that the occurrence of lagged values of x and u in the plant equations could be dealt with by definition of an enlarged state variable. The same goes for the occurrence of lagged values in the cost function.

The T-horizon formulation apparently allows termination only at $t = T$. In fact it rather specifies termination *by* time T, and one can represent general stopping rules and costs before that time by appropriate specification. For example, one can incorporate the option of termination at state x and time t (more briefly, at (x, t)) at cost $K(x, t)$ by introducing the *terminating decision* $\bar{u}$ and the *absorbing state* $\bar{x}$ for which

$$g(x, \bar{u}, t) = K(x, t),$$

$$a(x, \bar{u}, t) = \bar{x}$$

and $\bar{x}$ is such that the process remains in this state, once entered, without further cost. If termination at (x, t) is compulsory, then $\mathscr{U}(x, t)$ contains only $\bar{u}$.

Suppose dynamics and costs are independent of time to the extent that the plant equation (2.1) and the cost recursion (3.1) do not involve t as an explicit argument (i.e. the third and fourth arguments are missing from a and ψ respectively). The problem then is said to be *time-homogeneous*. Time-homogeneity can always be formally achieved by inclusion of t in the state variable, so that specification of x also implies specification of t. The normalization is an artificial one, however, because the motion of x_t in state space is then transient in the strong sense that x_t can take a given value at most once.

Time-homogeneity is consistent with a finite horizon. The stronger property of time-invariance, which demands that the whole structure of the process be invariant under a translation of the time-axis, is not. One can expect time-invariance only in the infinite-horizon case. The extent to which the results of Sections 2 and 3 survive passage to an infinite horizon is something we shall investigate systematically first in Part IV.

In the time-homogeneous case we shall usually write $F(x,t)$ as $F_s(x)$, where $s = T - t$ is the *time-to-go*. The optimality equation in its various forms, (2.4), (2.7) and (3.3), can then be written

$$F_s = \mathscr{L} F_{s-1} \qquad (s > 0) \tag{3}$$

with the terminal condition

$$F_0 = K. \tag{4}$$

The argument x is suppressed in (3) and (4); the functions F_s are understood to belong to the space of functions $\mathscr{F}$ from $\mathscr{X}$ to R. The operator $\mathscr{L}$ is implicitly defined by the various versions of the optimality equation, and maps from $\mathscr{F}$ to $\mathscr{F}$. We shall investigate the properties of $\mathscr{L}$ systematically later (Section 22.6) in the more general stochastic context. The exercises below list some useful immediate results.

Exercises

1. Show that $\mathscr{L}$ is *monotonic*. That is, if $\phi, \psi \in \mathscr{F}$ and $\phi \geq \psi$ then $\mathscr{L}\phi \geq \mathscr{L}\psi$.
2. Show that if $F_1 \geq F_0$ $(F_1 \leq F_0)$ then F_s increases (decreases) monotonically with s.
3. Show that the operator $\mathscr{L}$ is *concave*, in that $\mathscr{L}(\alpha\phi + \beta\psi) \geq \alpha\mathscr{L}\phi + \beta\mathscr{L}\psi$, where ϕ and ψ are elements of $\mathscr{F}$ and α, β are non-negative scalars such that $\alpha + \beta = 1$. Show that the same is true of $\mathscr{L}^r$, for positive integer r.
4. Suppose that $F = \mathscr{L}F$ has a non-negative solution F and that $\beta F \leq \mathscr{L}^r(0) \leq F$ for some positive integer r and positive scalar β. Show then that $0 \leq F - \mathscr{L}^{rs}(0) \leq \alpha^s F$, where $\alpha = 1 - \beta$. Show also that $0 \leq F - \mathscr{L}^{rs}K \leq \alpha^s F$ if $0 \leq K \leq F$.

C H A P T E R 3

Deterministic examples

This chapter collects some problems of interest which have a natural state formulation and a somewhat tractable optimality equation. We use $F(x, t)$ uniformly to denote minimal loss (or, when appropriate, maximal reward) incurred from time t with $x_t = x$; similarly, $u(x, t)$ denotes an optimal decision at (x, t). For time-homogeneous problems $F_s(x)$ is used to denote $F(x, T-s)$, where T is the horizon point and so $s = T-t$ is 'time-to-go'. We consider the infinite horizon limit in cases where this has a clear interpretation, although the validity of doing so will not be established before Part IV. Some general structural conclusions emerge in Sections 5 and 6.

1. OPTIMAL ALLOCATION BETWEEN CONSUMPTION AND INVESTMENT

Consider a situation in which one can either invest one's money, when it grows by compound interest, or spend it, deriving some satisfaction from doing so. One problem is to determine a consumption/investment pattern which maximizes one's total satisfaction (or *utility*) over a given period.

When such a model is scaled from the individual up to the national level, then the single commodity *money* is replaced by all the many types of resources, raw materials and goods, and the simple consumption/investment decision by a complex of allocation and production decisions. One has thus a model of *economic development*, for which we shall develop a few ideas in Chapters 6 and 36, but which presents daunting conceptual and practical problems in realistic cases.

The state variable is the amount of capital one possesses currently and is assumed to follow the plant equation

$$x_{t+1} = \alpha(x_t - u_t) \tag{1}$$

where u_t is the amount diverted for consumption at time t and α is the factor by which an amount invested increases over unit time.

The optimality equation is

$$F_s(x) = \max_{0 \le u \le x} [g(u) + \beta F_{s-1}(\alpha(x-u))] \tag{2}$$

if $g(u)$ is the utility derived from consumption u and total utility is the sum of discounted instantaneous utilities. If debt is not permitted then one must have $x \geqq 0$ and so $u_t \leqq x_t$; if consumption cannot be reversed to become productive, then $u \geqq 0$.

We can easily solve (2) with the assumptions

$$g(u) = u^{1-\nu}$$

$$F_0(x) = A_0 x^{1-\nu}$$

where ν lies in $(0, 1)$. The form for g is plausible; there are economic reasons for assuming g concave and non-decreasing. That the terminal reward F_0 should have the same form is perhaps debatable, but the assumption is a convenient one, as we shall see. In fact, we leave the reader to verify the evaluations

$$F_s(x) = A_s x^{1-\nu}, \tag{3}$$

$$u(x, T-s) = \frac{x}{A_s^{1/\nu}} \tag{4}$$

for maximal utility and optimal u at horizon s. Here $\{A_s\}$ is a sequence of constants obeying

$$A_s^{1/\nu} = 1 + \gamma A_{s-1}^{1/\nu} \tag{5}$$

with

$$\gamma = (\beta\alpha^{1-\nu})^{1/\nu}, \tag{6}$$

so that

$$A_s^{1/\nu} = A_0^{1/\nu}\gamma^s + \frac{1-\gamma^s}{1-\gamma}. \tag{7}$$

The constant γ can be viewed as the rate at which utility grows, so that sacrifice of an amount δ of current utility can yield utility $\delta\gamma^s$ after time s (for small δ and with optimal decisions; see Exercise 2).

If $\gamma > 1$, so that there is an incentive to delay consumption, then we see from (3), (4) and (7) that

$$F_s(x) = O(\gamma^{\nu s} x^{1-\nu}),$$

$$u(x, T-s) = O(\gamma^{-s} x)$$

for large s. The latter relation implies that, if utility growth is possible, then one delays consumption to the end of the programme.

If $\gamma < 1$ then expressions (3) and (4) have limits as s increases:

$$\begin{aligned} F_s(x) &\to \frac{x^{1-\nu}}{(1-\gamma)^\nu}, \\ u(x, T-s) &\to (1-\gamma)x. \end{aligned} \tag{8}$$

That is, if to delay consumption is unrewarding, then maximal utility over the finite horizon is finite. Furthermore, optimal policy then has the stationary form that one reinvests a fraction γ of one's capital at all times. However, the stationarity of this rule does not imply a steady-state economy (see Exercise 1).

This kind of model in general implies the conclusion that one sacrifices consumption to investment (i.e. to growth) if economic growth is possible. This is a conclusion which would at least be modified if certain of the basic assumptions were varied: e.g. the form of the utility function or the physical possibility of indefinite growth.

If there are several types of investment, whose rates of return may vary in time, then the problem becomes one of *portfolio selection* (see Section 12.8).

Exercises

1. With the policy (8) capital grows at each stage by a factor $\theta = \alpha\gamma = (\alpha\beta)^{1/\nu}$. This can be greater than or less than unity, consistently with $\gamma < 1$.
2. Show that γ is indeed the growth rate of utility, in that if one forgoes a small amount δ of current utility, one can realize an additional amount $\beta^s\theta^{s(1-\nu)}\delta = \gamma^s\delta$ after time s.
3. Adapt the treatment of this section to the case where there are several types of consumption, and

$$x_{t+1} = \alpha\left(x_t - \sum_j u_{jt}\right),$$

$$g(u_t) = \sum_j c_j u_{jt}^{1-\nu}$$

where u_{jt} is the amount of capital committed to consumption of type j at time t.
4. Solve the linear case $\nu = 0$.
5. Solve the case in which $F_0(x)$ and $g(u)$ are proportional to $\log(x - \bar{x})$ and log $(u - \bar{u})$ respectively. Here $\bar{x}, \bar{u}$ are constants and the logarithms are interpreted as having the value $-\infty$ if the argument is non-positive. The constant $\bar{u}$ thus represents the consumption necessary for subsistence and $\bar{x}$ the amount of capital which it is imperative to have accumulated by the horizon point.

2. PRODUCTION SCHEDULING

This is a version of the inventory problems to be considered in Chapter 13. A factory produces a single commodity and wishes to schedule its production so as to meet a known irregular pattern of demand in the most economical fashion.

Let us define x_t as the stock (inventory) of the commodity held at the beginning of the tth day, u_t as production during the tth day and d_t as demand during the tth day. In such problems one must indicate relative timing within a stage if one is to be clear about the sequence of events.

Assuming stock is to be the state variable, we than have the plant equation

$$x_{t+1} = x_t + u_t - d_t \tag{1}$$

and shall assume the optimality equation for the cost

$$F(x,t) = \min_u [g(u) + c(x + u - d_t) + F(x + u - d_t, t + 1)]. \tag{2}$$

The terms $g(u)$ and $c(x)$ must then be interpretable respectively as the cost of producing amount u (per day) and the cost of carrying a stock x from one day to the next. Plausible simple forms are

$$g(u) = \begin{cases} 0 & (u \leq 0) \\ a + bu & (u > 0), \end{cases} \tag{3}$$

$$c(x) = \begin{cases} +\infty & (x < 0), \\ cx & (x \geq 0). \end{cases} \tag{4}$$

Assumption (3) implies that one can dispose of stock without cost, but to go into production on a given day implies a base cost a and a unit cost b. Assumption (4) implies that negative stock ('backlogging') is forbidden and that to carry positive stock implies a unit cost of c per day. We have not discounted, but assume horizon T with

$$F(x, T) = \begin{cases} +\infty & (x < 0), \\ 0 & (x \geq 0). \end{cases}$$

It is the occurrence of the terms a and c which make the problem non-trivial. If the base cost a were zero then one would produce only enough to satisfy current demand. If the storage cost c were zero then one would produce enough to satisfy total future demand in a single production run. With both costs present then one must steer between these two extremes.

The simplifying features of the problem, as we have formulated it, are best summarized as a theorem.

Theorem 2.1. *In an optimal production programme one produces on a given day if and only if stocks are insufficient to meet that day's demand. When one produces one brings stocks up to such a level that they exactly satisfy demand for an integral number of days.*

Proof. Suppose that under a production plan $\{u_t^*\}$ one has $u_t^* > 0$ for a given t. If $x_t \geq d_t$ and one transferred the production u_t^* to day $t + 1$ then one would save a cost cu_t^* or $a + cu_t^*$ according as $u_{t+1}^* = 0$ or $u_{t+1}^* > 0$. Hence the first assertion.

For the second point, suppose that on day t one produces enough to satisfy demand before time τ, but not enough to satisfy demand at time τ as well. That is,

$$u_t = \sum_{j=t}^{\tau-1} d_j - x_t + \delta$$

where $0 \le \delta < d_\tau$. Then one must produce on day τ in any case. If one decreased u_t^* by δ and increased u_τ^* by δ then this would save a storage cost of $c(\tau - t)\delta$, with no other effect on costs. Hence $\delta = 0$ in an optimal policy. ■

Thus, if $x_t \ge d_t$ one does not produce and lets stocks run down, either naturally or by active disposal (see Exercise 1). If $x_t < d_t$ then one produces so that the demand is met up to time $\tau - 1$ but $x_\tau = 0$, where τ is an integer exceeding t. The optimal τ will be determined by the optimality equation

$$F(x, t) = \min_{t < \tau \le T} [G(t, \tau) - bx + F(0, \tau)] \qquad (x < d_t) \tag{5}$$

where

$$G(t, \tau) = a + b \sum_{t}^{\tau - 1} d_j + c \sum_{t}^{\tau - 1} (j - t) d_j. \tag{6}$$

So, after a first production or disposal decision stocks come to zero level at some time, and from that time one operates between recurrent instants of zero stock. These instants are determined by the reduced form of the optimality equation

$$\phi_t = \begin{cases} \phi_{t+1} & (d_t = 0), \\ \min\limits_{t < \tau \le T} [G(t, \tau) + \phi_\tau] & (d_t > 0; t < T), \end{cases} \tag{7}$$

with $\phi_T = 0$, where $\phi_t = F(0, t)$. The cost function (6) is easily calculated recursively; by solving (7) recursively one then calculates the optimal instants at which to produce if initial stocks are zero. If one begins with positive but inadequate stocks one produces according to (5). If one begins with more than adequate stocks one either lets them run down or disposes of some (see Exercise 1).

The model is, of course, excessively simple. In practice one will wish to utilize plant and manpower at a constant rate. The term 'a' could be regarded as representing embryonically the cost of switching between different types of production (see Exercise 3).

Exercises

1. Show that, if x_t is large enough, then one should dispose of stock so that enough is left to cover demand for an integral number of days.
2. Consider the effect of imposition of a bound upon production rate:

$$0 \le u_t \le M.$$

3. Consider the case of several types of commodity in demand, of which only one can be produced at a given time; let a then be the cost of changing the commodity being produced.

3. THE WAREHOUSING PROBLEM

This, as usually formulated, is a linear programme over time. A warehouseman deals in a single commodity, whose buying and selling prices over the future are known, and wishes to operate so as to maximize his profits over a finite horizon.

Again, the order of events within a stage must be specified. It is assumed that he begins day t with stock x_t, of which he sells an amount u_t at price b_t. He then buys an amount v_t at price c_t, making his stock for the next day

$$x_{t+1} = x_t - u_t + v_t \tag{1}$$

and his net profit for the day $b_t u_t - c_t v_t$. Choice of u_t and v_t is subject to the constraints

$$0 \le u_t \le x_t,$$
$$0 \le v_t \le C - x_t + u_t,$$

expressing the facts that the warehouseman cannot sell more than he stocks and that he cannot then buy in more than would bring his stock x_{t+1} above the capacity C of the warehouse. Total profit $\sum_0^{T-1}(b_t u_t - c_t v_t)$ is to be maximized.

We can evidently take x_t as the state variable and (u_t, v_t) as the joint decision variable, u_t being effective before v_t. The optimality equation is

$$F(x, t) = \max_{0 \le u \le x} \max_{0 \le v \le C - x + u} [b_t u - c_t v + F(x - u + v, t + 1)] \quad (0 \le t < T) \tag{2}$$

with

$$F(x, T) = 0.$$

In relation (2) the v-maximization is taken before the u-maximization, following the general principle that decisions are optimized in reverse time order.

We express the principal feature of the optimal solution as

Theorem 3.1. *$F(x, t)$ is linear in x and some extreme solution is always optimal: to sell everything or nothing; to buy to capacity or nothing.*

Proof. The proof is inductive, but the steps of the induction provide the algorithm for the complete determination of the optimal solution.

The assertion is that F has the form

$$F(x, t) = \alpha_t + \beta_t x. \tag{3}$$

If so, then β_t must be regarded as an effective unit value for present stock and α_t as the value of future trading if one had zero stock. Assertion (3) holds at T; assume that it holds at $t + 1$. Then

$$\max_v [-c_t v + \alpha_{t+1} + \beta_{t+1}(x - u + v)] = \alpha_t + \beta'_{t+1}(x - u),$$

say, where the optimal value of v is $C - x + u$ or 0 according to whether $\beta_{t+1} > c_t$ or $\beta_{t+1} < c_t$, and

$$\begin{aligned} \alpha_t &= \alpha_{t+1} + C(\beta_{t+1} - c_t)_+, \\ \beta'_{t+1} &= \min(\beta_{t+1}, c_t). \end{aligned} \tag{4}$$

That is, one buys to capacity or not at all according to whether the effective unit price β_{t+1} exceeds the buying price c_t or not. In the transitional case $\beta_{t+1} = c_t$ any admissible decision is optimal.

Correspondingly,

$$\max_{0 \le u \le x} [b_t u + \alpha_t + \beta'_{t+1}(x - u)] = \alpha_t + \beta_t x$$

with

$$\beta_t = \max(\beta'_{t+1}, b_t) \tag{5}$$

so that the induction is complete. The optimal value of u is x or 0 according to whether b_t is greater or less than β'_{t+1}. ■

By solving the backward recursions (4) and (5) one determines the coefficients α_t, β_t, the nature of the optimal decision for each t and the future return that can be realized from a given initial stock. Solving the plant equation (1) forwards, one then determines the actual sequence of values $\{x_t, u_t, v_t\}$.

It is the linearity and the certainty of the problem which give the solution its extreme character. If future prices were uncertain, or depended upon the actions of traders in general, then conclusions would be considerably modified.

4. COST-MINIMIZING PATHS IN A CONTINUUM; THE EIKONAL EQUATION

In Section 1.4 we considered the optimal path for a car on a road-net or, more abstractly, for a point on a discrete graph. Let us now consider the same problem for motion in a continuum. This has relevance for the calculation of the path of light rays in a medium of variable refractive index. According to the principles of geometrical optics, the path between assigned points should be such as to minimize the time of passage (actually, to make time of passage stationary under local variations of trajectory).

We shall assume motion in a physical space of d dimensions and in continuous time. The position at time t, denoted $x(t)$, is thus a point in R^d. This will be our state variable.

In a 'minimal-cost' version we assume the dynamics to be such that x moves at unit velocity in an arbitrarily chosen direction, so that the plant equation can be taken as

$$\dot{x} = u$$

where $\dot{x} = \mathrm{d}x/\mathrm{d}t$ (a uniform convention) and u is a unit vector, the chosen direction of movement.

We shall suppose that to stay at x for unit time costs $c(x)$, but that costs cease as soon as x reaches a target set D. The cost function to be minimized is thus

$$C = \int_0^\tau c(x)\,\mathrm{d}t \tag{1}$$

where $x(0) = x$ is prescribed and τ is the smallest t for which $x(t) \in D$.

We shall use $x_1, x_2, \ldots, x_d$ to denote the Cartesian components of x. If $\phi(x)$ is a scalar function of x then $\nabla\phi$ is the gradient of ϕ: the column vector with components $\partial\phi/\partial x_j$.

Theorem 4.1. *Let $F(x)$ denote the minimal cost incurred in passage from x to D. In D, $F(x)$ has a prescribed value zero. Almost everywhere in its complement F satisfies the eikonal equation*

$$|\nabla F(x)| = c(x) \tag{2}$$

and the optimal direction of movement is in the direction of $-\nabla F$.

Proof. The proof will be a heuristic one; we assume F differentiable where necessary and assume that the absence of a horizon causes no trouble. The problem is formulated in a time-invariant fashion, and we assume F to be a function $F(x)$ of x alone.

An infinitesimal form of the optimality equation is

$$F(x) = \inf_u [c(x)\,\delta t + F(x + u\,\delta t) + o(\delta t)]$$

where the infimum is over unit vectors: $|u| = 1$. The formal limit of this equation is

$$\inf_u [c + u'\nabla F] = 0. \tag{3}$$

By Cauchy's inequality the infimal value of $u'\nabla F$ is $-|\nabla F|$, achieved when u is in the direction of $-\nabla F$:

$$u = -\frac{\nabla F}{|\nabla F|} \tag{4}$$

This is the second assertion of the theorem. Substituting (4) into (3) we deduce the first. ■

In the optical context one is actually minimizing the time of passage for a point whose speed $|\dot{x}|$ depends upon position x. The conclusions of Theorem 3.1 still hold if $c(x)$ is interpreted as the reciprocal of the speed, the *local refractive index*.

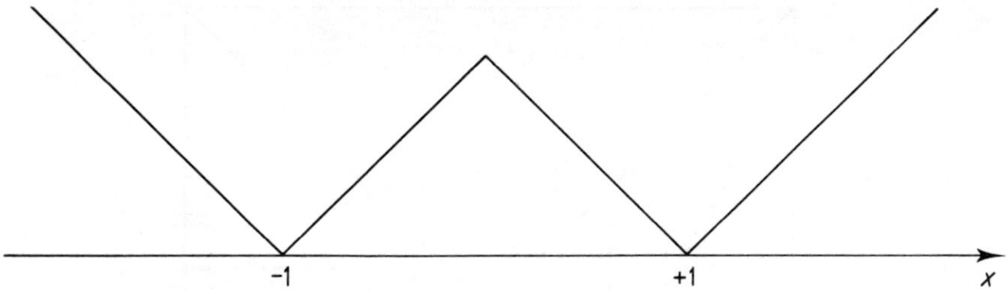

Figure 4.1 Minimal time for passage to one of the target points + 1 or − 1

The quantity δt then represents an infinitesimal element of distance rather than of time.

The eikonal equation (2) is generally written

$$\sum_j \left[\frac{\partial F}{\partial x_j}\right]^2 = c(x)^2$$

when its character as a short-wavelength version of the wave equation becomes more apparent. Note that the characterization (4) of the optimal direction implies that the optimal trajectories will be orthogonal to the level lines of $F(x)$ almost everywhere.

The characterization 'almost everywhere' is necessary, because ∇F may not exist at a point at which an infinitesimal change in x causes a discontinuous change in optimal trajectory. Consider, for example, the case $c(x) \equiv 1$. Plainly, then, optimal trajectories are straight and

$$F(x) = \min_{\bar{x} \in D} |x - \bar{x}|$$

where the minimizing $\bar{x}$ is the optimal target point in D from x. For the case $d = 1$ and $D = (-1, 1)$, for example, $F(x)$ is as illustrated in Figure 4.1. There is a discontinuity in gradient at $x = 0$, because there the target point switches between $+1$ and -1 as x changes sign. There are also discontinuities at the target points ± 1, as these are approached from different directions.

Correspondingly, for $d = 2$ and D the complement of the square $|x_1| < 1$, $|x_2| < 1$, we have

$$F(x) = \min(1 - |x_1|, 1 - |x_2|)$$

The policy is to aim for the nearest side of the square, so there is a discontinuity in policy and in gradient on the diagonals $|x_1| = |x_2|$. The level lines of F are also squares and the optimal trajectories cut these orthogonally, except at the corners (see Figure 4.2).

If one assigns a varying terminal cost $K(x)$ in D, then there can be yet another complication, in that the ideal target point $\bar{x}$ in D may not be reachable from x by

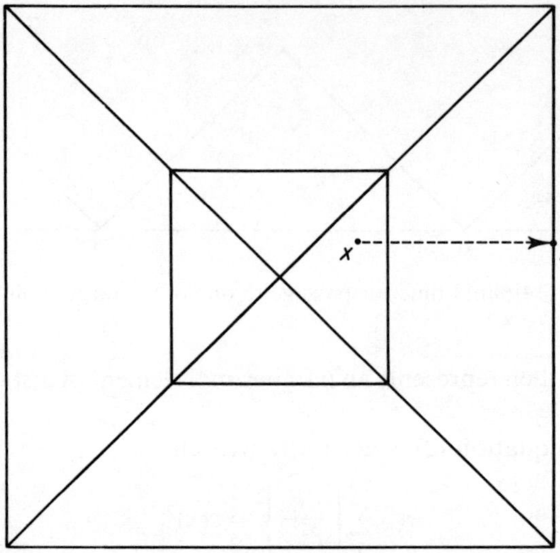

Figure **4.2** Minimal-time passage to the sides of a square. The level lines of constant time are also squares, and the diagonals constitute the loci on which there is a break in policy (i.e. a discontinuous change in ideal target point)

a straight line path (or by a 'free orbit' in the case of variable c). In this case $\mathscr{X}$ breaks up into regions $\mathscr{X}_i$, within each of which the optimal orbit is a free one and F satisfies (2). As one crosses regions the action rule changes, and there is a discontinuity in ∇F, although not in F. For examples see Section 7.4.

Exercises

1. Show that $F(x)$ is always continuous if $c(x)$ is bounded and always has a derivative in a direction of optimal motion if $c(x)$ is continuous.
2. Note an 'analogue solution' of (2) for the case $d = 2$ and $c(x)$ constant. Cut out a card to the shape of the complement of D, hold it horizontal and pile on as much firm, dry sand as the card will hold. The height of the sand is proportional to $F(x)$.

   The case of variable $K(x)$ is also representable if a wall of variable height is placed around the boundary.
3. Suppose c is constant in x. Show that the eikonal equation has a solution

$$F(x) = \max_{\lambda} \left[\sum_j \lambda_j x_j - \theta(\lambda) \right]$$

   where λ is subject to $\sum \lambda_j^2 = c^2$. That is, the level contours of $F(x)$ can be seen as the envelope of a family of planes of slope c.

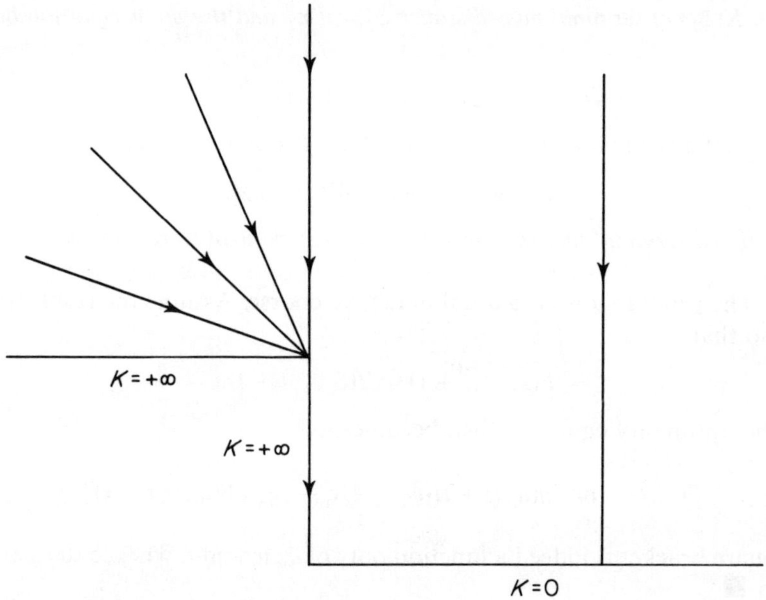

Figure **4.3** The effect of prescription of a variable terminal cost. The minimal-cost paths break into linear segments

4. In fact, it is often easier to determine F from an optimal path, rather than conversely. Consider the two-dimensional situation with constant c and boundary costs as shown in Figure 4.3.

   By appealing to the fact that there are just two types of optimal path, as shown, determine F. What is the locus of discontinuity in ∇F? Verify with the sand model!

5. Suppose the model of this section modified in that, instead of the constraint $|u| \leq 1$, one imposes a cost $\frac{1}{2}|u|^2$ per unit time on travel at velocity u. Show that one then has the relations

$$u = -\nabla F.$$

$$\tfrac{1}{2}|\nabla F|^2 = c.$$

5. TRAJECTORIES AND TERMINAL MISS-DISTANCE

Problems in the guidance of a physical object through physical space are described as 'trajectory problems'. We shall consider a variety of such problems in Chapters 5, 20 and 38, but shall confine ourselves at the moment to making a simple point of some interest.

Theorem 5.1. *Consider a problem with additive cost function* (2.2.2) *for which the instantaneous cost is a function $g(u_t, t)$ independent of x_t, the terminal cost is a* *function* $K(\xi_T)$ *of terminal miss-distance* $\xi_T = d'x_T$ *and the plant equation has the form*

$$x_{t+1} = A_t x_t + B(u_t, t). \tag{1}$$

Then $F(x, t)$ *and the optimal control* $u(x, t)$ *are functions only of* t *and of*

$$\xi_t = d'A_{T-1}A_T \cdots A_t x_t = \phi_t x_t, \tag{2}$$

the predicted terminal miss-distance if no further control were exerted.

Proof. The proof follows the usual inductive course. Assume the result true for $t + 1$, so that

$$F(x_{t+1}, t+1) = H(\xi_{t+1}, t+1),$$

say. The optimality equation then becomes

$$F(x_t, t) = \inf_{u_t} [g(u_t, t) + H(\phi_{t+1} A_t x_t + \phi_{t+1} B(u_t, t), t+1)].$$

The square bracket is indeed a function only of ξ_t, u_t and t, whence the assertion follows. ■

The point of the theorem is that the dimension of ξ_t is the dimension of ξ_T, which may be smaller than that of x_t. Further, the notion of a 'predicted miss-distance for the uncontrolled trajectory' as a measure of deviation of current state from the ideal is an appealing one, which will prove useful in the sequel.

6. AN INTERCHANGE ARGUMENT

The argument of Theorem 5.1 was a structural one. Although it did not solve the optimality equation, it used the equation to demonstrate that the optimal solution had a particular structural property. We now consider another such property.

Consider the general optimality equation in the state-structured case and write it as

$$F(x, t) = \inf_u L_u F(x, t) \tag{1}$$

where

$$L_u \phi(x, t) = \psi(x, a(x, u, t), u, t, \phi(a(x, u, t), t+1)). \tag{2}$$

Suppose that the problem has the following properties:

(a) An optimal policy exists.

(b) Any action $u \in \mathscr{U}(x_t, t)$ will be taken finitely soon after time t in an optimal policy.

(c) If for some u, v, t the choice $u_t = v, u_{t+1} = u$ is permissible, then the choice $u_t = u, u_{t+1} = v$ is also permissible.
(d) For some prescribed x_t, t and $u \in \mathscr{U}(x_t, t)$,

$$L_u L_v F(x_r, r) \leq L_v L_u F(x_r, r)$$

for all $v \in \mathscr{U}(x_t, t)$, all $r \geq t$ and all x_r consistent with the prescribed value of x_t and $u_j \neq u$ for $t \leq j < r$.

Theorem 6.1. *Under conditions* (*a*) *to* (*d*) *it is optimal to take action u at* (x_t, t).

Proof. Suppose one is using an optimal policy. By hypothesis (b) we may assume action u used at some time $r \geq t$. Suppose the policy such that u is first used at $r > t$ and action v is used at time $r - 1$. By hypothesis (c) we may interchange these actions; the cost incurred from time $r - 1$ would then be changed from $L_v L_u F(x_{r-1}, r-1)$ to $L_u L_v F(x_{r-1}, r-1)$. By hypothesis (d) this does not increase costs, so the policy remains optimal. Repeating this procedure, we ultimately achieve $u_t = u$. ■

One will require the problem to have some rather special features if (d) is to be satisfied. We consider two such, for the additive time-homogeneous case when

$$\begin{aligned} L_u \phi(x, t) &= g(x, u) + \beta\phi(a(x, u), t + 1) \\ &= g(x, u) + \beta T_u \phi(x, t), \end{aligned} \tag{4}$$

say. Define the two-stage cost

$$g_{uv}(x) = g(x, u) + \beta T_u g(x, v).$$

Theorem 6.2. *Hypothesis* (*d*) *is implied by either of the conditions* (*d′*) *or* (*d″*) *now defined*:

(*d′*) *The operators* T_u *commute; the inequality*

$$g_{uv}(x_t) \leq g_{vu}(x_t)$$

holds for some prescribed x_t, t, u *and all* $v \in \mathscr{U}(x_t)$ and implies that

$$g_{uv}(x_r) \leq g_{vu}(x_r)$$

for all v, r *and* x_r *characterized in* (*d*).

(*d″*) *The relation* $g_{uv}(x) = g_{vu}(x)$ *holds for all* x, u, v; *the inequality*

$$T_u T_v F(x_t, t) \leq T_v T_u F(x_t, t)$$

holds for some prescribed x_t, t *and* u *and all* $v \in \mathscr{U}(x_t)$, and implies that

$$T_u T_v F(x_r, r) \leq T_v T_u F(x_r, r)$$

for all v, r *and* x_r *characterized in* (*d*).

Proof. As for most theorems with lengthy statement, proof is immediate. ■

Note that if conditions (b) to (d) of Theorem 6.1 are demanded of an arbitrary policy rather than of the optimal policy, then that policy is at least not made worse if actions are re-ordered as in the theorem.

7. A SCHEDULING PROBLEM

By 'scheduling' one means the arranging of tasks in order, subject to physical feasibility, availability of plant, staff, resources, etc., in such a way as to secure a performance which is economic in some sense. Such problems increase rapidly in complexity as numbers of tasks, units and constraints increase. For example, one might consider the scheduling of trains over a rail network; the scheduling of operations in the construction of a building; the scheduling of production in an oil refinery. In each case there will be special constraints and structural features. In the case of an oil refinery, for instance, one must work from given crude products and plant to produce the desired mix of end-products as economically as possible, given constraints on the way the components of the plant can be used and given the amounts of intermediate and final products that each operation will produce from a given input.

What must be about the simplest non-trivial scheduling problem is the case of two machines, A and B, such that B accepts the output of A as input. Suppose that n articles are to be fed through the pair of machines, article j taking times a_j and b_j respectively ($j = 1, 2, \ldots, n$). The aim is to take the articles in such an order that the total time taken for processing is minimized. The moments of action are those at which machine A becomes clear. Suppose that at such a moment machine B requires a time τ before it clears (where the value of τ will depend upon circumstances) and that a set of designated articles S remains to be processed. Then S, τ may be taken jointly as a state variable and the optimality equation is

$$F(S,\tau) = \min_{j \in S} [a_j + F(S - \{j\}, b_j + (\tau - a_j)_+)] \tag{1}$$

with the terminal condition

$$F(\varnothing, \tau) = \tau.$$

If one first processes article j, then article k and proceeds optimally thereafter, one incurs a total cost of

$$a_j + a_k + F(s - \{j\} - \{k\}, \tau_{jk}) \tag{2}$$

where

$$\tau_{jk} = b_j + b_k - a_j - a_k + \max(\tau, a_j, a_j + a_k - b_j). \tag{3}$$

Now, the terms $a_j + a_k$ and $S - \{j\} - \{k\}$ are independent of the order of choice of

j and k, and $F(S, \tau)$ is plainly increasing in τ for all S. Thus, if

$$\tau_{jk} \leq \tau_{kj} \tag{4}$$

for all $k \neq j$ in S, we can assert that condition (d″) of Section 6 is satisfied and apply Theorem 6.1 to deduce that it will be optimal to choose article j first for processing.

Relation (4) will hold for all τ if

$$\max(a_j, a_j + a_k - b_j) < \max(a_k, a_j + a_k - b_k)$$

which is equivalent to

$$\min(a_j, b_k) < \min(a_k, b_j). \tag{5}$$

One can determine an ordering of articles such that (5) holds at any stage for the current article (j) against all remaining articles (k) by the following procedure:

(a) Determine $\min_j \min(a_j, b_j)$.

(b) If the minimizing value is an a, take this article first; if a b, take this article last.

(c) Repeat the procedure on the reduced set of articles.

Ties can be resolved in a number of fairly evident ways.

Sometimes the optimal action rule is of the *index form*: choose the available j for which some function of j, an index I_j, is maximal (see Chapter 14). It is noteworthy that rule (5) is not of this form.

8. CRITICAL PATH ANALYSIS

Dynamic programming arguments can be useful for several scheduling problems (Section 7 and Chapter 14). We shall give an example of a simple and important case for which it is salutorily true, however, that direct arguments lead straight to a solution, which is not obviously derivable by dynamic programming arguments. However, this does not mean that it is pointless to consider a dynamic programming treatment of the problem. It may be that the dynamic programming treatment, once understood, will generalize to cases where the simple treatment would fail. The example we shall consider is the well-known critical path method (CPM). This is appropriate for situations in which one has a complex operation consisting of various component activities, some of which must be completed before others, the duration required to complete each activity being known in advance. This might be the case for a building project, for example, in which conditions were standard enough that the time taken by each phase is well defined and known.

The structure of such a problem can be represented by a graph (see figure 8.1), in which the various phases of intermediate completion are represented by the nodes and the activities by directed arcs between the nodes. For the example illustrated, one starts at node 1 and completion is represented by arrival at node 8. One cannot be said to have arrived at node 5, for example, until activities 25, 35

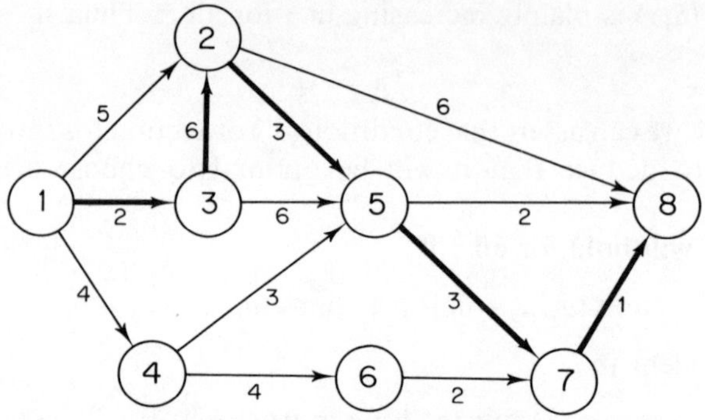

***Figure* 8.1** An activity network represented by a graph. Nodes representing completion of some phase are labelled. Arcs representing activities carry a direction and a duration. The critical path for this particular network is indicated by the heavy lines

and 45 have all been completed. These take 3, 6 and 3 units of time respectively, as indicated. Activities 52 and 53 can be started only once one has arrived at node 5. The aim is to complete the graph from node 1 to node 8 in minimal time.

Note that one completes the graph, not in the sense of moving through it, but in the sense of 'colouring-in' nodes as the corresponding phase of completion is reached. The graph is completed when all nodes are coloured.

In the conventional CP model it is supposed that arbitrarily many activities can be tackled simultaneously, with the reservation that an activity cannot be started until all prerequisite activities have been completed. The constraints on optimization are thus sequencing constraints rather than resource limitations.

This fact makes the derivation of an optimal policy trivial: one will plainly complete the operation in minimal time if all currently permissible activities are driven at full rate, at all instants of time until completion. This will not be the only optimal solution, however. There will usually be activities which can be postponed to some extent without delaying the total project. The aim of a CP analysis is to determine which activities are critical in that they can *not* be postponed without prejudice. This tells one where the urgency lies at any given moment, and gives some feeling for optimal sequences in cases where there are secondary constraints on the choice of activities (i.e. setting-up costs, avoidance of site congestion, spoilage of prematurely completed work, etc.).

If activity jk exists, then we shall say that j is a *predecessor* of k and k a *successor* of j. Let t_{jk} be the time required to complete this activity. Let the starting and terminal nodes be denoted by $j = 1$ and $j = n$ respectively (so that node 1 has no predecessors and node n no successors). Let the set of predecessors and the set of successors for node j be denoted by P_j and S_j respectively.

Let t_j denote the *early completion time* for node j. This is the earliest time after starting by which node j could be reached, using the 'all-out' optimal strategy first proposed. Then the t_j satisfy

$$t_1 = 0, \tag{1}$$

$$t_j = \max_{k \in P_j} (t_k + t_{kj}). \tag{2}$$

Implicit in (2) is the condition that all activities precedent to j must be completed for the phase j to be reached.

The quantity t_n thus determined is the minimal time in which the total operation can be completed. Define now the *late completion time* t_j^* as the latest time by which node j can be reached consistent with total completion by time t_n. These are determined by the backwards recursion.

$$t_n^* = t_n, \tag{3}$$

$$t_j^* = \min_{k \in S_j} (t_k^* - t_{jk}). \tag{4}$$

Equation (4) expresses the fact that, if the late completion times for nodes k in S_j are all to be met, then the late completion time for j is determined by the activity jk which must be started earliest.

The critical path is then determined by the set of j for which $t_j = t_j^*$. In the 'early' solution all activities are begun as early as possible; in the 'late' solution all activities are begun as late as possible, consistent with total completion by time t_n, and the quantity $t_j^* - t_j$ represents the 'slack' at node j. On the critical path there is no slack. That is, if k is a successor to j, and both j and k are on the critical path, then activity jk must be started as soon as node j is reached.

For the numerical example expressed in the graph, one derives the solution

j	1	2	3	4	5	6	7	8
t_j	0	8	2	4	11	8	14	15
t_j^*	0	8	2	8	11	12	14	15

so that 132578 is the critical path. Actually, there may be more than one critical path. In this case, the sequence of critical activities is 13, 32, 25, 57 and 78. However, if t_{35} had been 9 instead of 6, then 35 would also have been a critical activity.

The optimization is trivial in that the 'early' solution is trivially optimal—the art is rather to determine the other solutions and so determine the slack in the optimal solution.

Exercises

1. Consider a dynamic programming formulation. Let $F(x)$ be the minimal time to completion from a situation in which time x_{jk} has been spent on activity jk

$(j, k = 1, 2, \ldots, n)$. Thus $F(t) = 0$, where t is the vector of t_{jk}. Show that the optimality equation is

$$1 + \min_u \sum_j \sum_k u_{jk} \frac{\partial F}{\partial x_{jk}} = 0$$

where $0 \le u_{jk} \le 1$ and u_{jk} may be positive only if node j has been reached.

2. Let $F(0) = \Phi(t)$ be the minimal time needed for completion of the whole task for prescribed $t = \{t_{jk}\}$. Note then that (a) $F(x) = \Phi(t - x)$ and (b) $\Phi(t)$ is piecewise linear. Note that these facts imply optimality of the 'all-out' policy: to take $u_{jk} = 1$ for all j, k such that mode j has been reached and $x_{jk} < t_{jk}$.

9. CONTROL OF PEAKING

We consider a problem in continuous space and time, in which x denotes position, not necessarily a state variable. In fact, let x denote the scalar distance of a vehicle from a point destination D. We assume the equations

$$\dot{x} = -v,$$

$$\dot{v} = u,$$

so that v and u denote velocity and acceleration in the direction of D respectively. We suppose that u can be chosen arbitrarily subject to a physical limitation.

$$|u| \le M.$$

However, it is supposed that there is a velocity-dependent heating effect: the temperature θ of the vehicle obeys the equation

$$\dot{\theta} = \alpha v^2 - \theta.$$

If it is desired to avoid overheating of the vehicle then one might choose a thrust programme $\{u(t)\}$ to minimize $\tilde{\theta}$, where $\tilde{\theta}$ is the peak temperature experienced during passage to D. However, the character of an ε-optimal solution is then clear and unacceptable: one travels towards D at an infinitesimal velocity. A criterion that instils some urgency and ensures existence of an optimum is to minimize $\tilde{\theta} + \lambda\tau$, where λ is a positive scalar and τ the passage-time to D.

This seems like a state-structured problem of the type envisaged in Section 2.3 with the state-variable (x, v, θ). Such is not the case, however. While it is true that the two components $\tilde{\theta}$ and τ of the criterion can be generated by simple scalar recursions, the same is not true of the combination $\tilde{\theta} + \lambda\tau$. One could develop a vector optimality equation, but the simplest course is to rephrase the problem as one of minimizing τ subject to

$$\theta(t) \le \bar{\theta}$$

for all t. This is a problem with state structure. Let $\tau(\bar{\theta})$ be the minimal value of τ

achievable for it. For $\bar{\theta}$ small enough one certainly has $\tilde{\theta} = \bar{\theta}$ and can choose $\bar{\theta}$ to minimize $\bar{\theta} + \lambda\tau(\bar{\theta})$.

The transformed problem can be given a dynamic programming formulation, but solution is easiest by application of the maximum principle (see Section 7.7). We note the form of the optimal policy. Define $\bar{v}$ by

$$\alpha\bar{v}^2 = \bar{\theta},$$

so that $\bar{v}$ is the velocity at which the vehicle would have temperature $\bar{\theta}$ in the steady state. If one starts with a small enough velocity and temperature, and is far enough from the destination, then the optimal policy falls into three phases:

(a) Apply maximal acceleration and let the velocity run up past $\bar{v}$.

(b) Apply maximal deceleration from a moment chosen so that v falls to $\bar{v}$ and θ rises to $\bar{\theta}$ at the same time, $\bar{t}$.

(c) Continue coasting at constant velocity $\bar{v}$ from time $\bar{t}$ until arrival at D. The peak value $\bar{\theta}$ is thus held throughout this coasting phase.

CHAPTER 4

Deterministic dynamic models

There is an immense literature on dynamic models, state-space models and on classical control theory, much of it largely concerned with the stability properties of such models under prescribed feedback control rules. It is a literature scarcely capable of summary; this chapter concentrates on aspects relevant to the optimization of control. We consider principally the discrete-time case; continuous-time analogues are summarized in Section 6.

1. STATE-SPACE MODELS

State structure was defined in Chapter 2. The essential property of a state variable x is its summary quality: the course of relevant aspects of the process after time t is as well determined from knowledge of x_t as it is from that of x_t and of the earlier course of the process. State structure is then a property both of the dynamics of the process and of those aspects of the process considered 'relevant', specified in Chapter 2 by the cost structure.

As far as dynamics are concerned the essential assumption is that x_t should obey a plant equation of the simple recursive form

$$x_{t+1} = a(x_t, u_t, t). \tag{1}$$

Cases not obviously of this form can often be reduced to it (see Exercises 2.2.2 and 2.2.3).

Condition (a) of Section 2.2 required the state variable x_t to be observable at time t (i.e. calculable from observations then available). This is indeed necessary if other quantities are to be calculated from it. However, even if x_t is not completely observable one may still regard it as a conceptual or *process* state variable, which would summarize the physics of the situation if only one could measure it, and this concept is useful. For example, x_t may describe completely the temperature distribution in a furnace at time t, but temperature will be observable only at points where sensors have been placed. The concept of a complete temperature distribution, obeying simply recursive dynamics, is nevertheless fundamental.

In general it must be recognized that at time t one may observe only some aspect of current state:

$$y_t = b(x_t, t).$$

The control u_t is then to be calculated in terms of $(y_t, y_{t-1}, \ldots; u_{t-1}, u_{t-2}, \ldots)$. At least, this would be the *closed-loop* or *feedback* form of control (see Section 7).

By virtue of the plant equation (1) we can write the observation relation as $y_{t+1} = b(a(x_t, u_t, t), t)$, or

$$y_{t+1} = c(x_t, u_t, t), \tag{2}$$

say. The jointly recursive form of (1), (2) turns out to be the natural way to specify the joint relation of the state variable x, action u and observation y.

Equations (1) and (2) constitute a *state-space model* with the state variable x. The variables u and y are often termed the input and output variables of the system respectively, in that it is u whose values can be fixed by the human operator of the system and y whose values are subsequently observed. The state variable constitutes an intermediate—a shadowy entity which can be manipulated or observed only partially. It describes the state of the interior of the black box relating u and y; an interior accessible only to the extent that u can be manipulated and visible only to the extent that y can be observed. Indeed, the state variable of a given system is usually not defined uniquely.

A case of particular importance is that of the *linear system.* One assumes that x, u, y belong to finite-dimensional Euclidean spaces, of dimensions n, m, r respectively, and that relations (1), (2) take the forms

$$x_{t+1} = Ax_t + Bu_t, \tag{3}$$

$$y_{t+1} = Cx_t. \tag{4}$$

This is often referred to as the system $[A, B, C]$, since these three matrices specify it. The *dimension* of the system is n, the dimension of x. The linear system is relatively tractable, which explains much of its popularity. However, for all its relative simplicity, the $[A, B, C]$ system is far from fully understood or analysed, and its study continues to generate a large literature. Description (3), (4) is a time-homogeneous one: more generally, one can study the time-varying linear system $[A_t, B_t, C_t]$. We shall largely keep to time-homogeneous systems, although many results have analogues, more or less immediate, for the time-varying case.

Exercises and comments

1. The linear analogue of (2) might seem to be the slightly generalized form of (4): $y_{t+1} = Cx_t + Du_t$. However, since past actions are presumed known, the form Du_t can always be effectively subtracted off the observation.
2. Suppose one has a model for which the describing variable is a scalar ξ_t obeying the plant equation

$$\sum_{j=0}^{n} a_j \xi_{t-j} = bu_t \tag{*}$$

with $a_0 = 1$. Show that, by definition of an appropriate state vector x_t, one can write equation (*) in the form (3) with

$$A = \begin{bmatrix} -a_1 & -a_2 & \cdots & \cdot & -a_n \\ 1 & \cdot & \cdots & \cdot & \cdot \\ \cdot & 1 & \cdots & \cdot & \cdot \\ \cdots & \cdots & \cdots & \cdots & \cdots \\ \cdot & \cdot & \cdots & 1 & \cdot \end{bmatrix}, \qquad B = \begin{bmatrix} b \\ \cdot \\ \cdot \\ \vdots \\ \cdot \end{bmatrix}.$$

This matrix A is often termed the *companion matrix* of the system. (*).

3. The construction of Exercise 2 has a converse. Consider the uncontrolled model $x_{t+1} = Ax_t$ and define the polynomial $D(\lambda) = |\lambda I - A| = \sum_{j=0}^{n} a_j \lambda^{n-j}$. Show that the elements x_{it} of x_t all satisfy the difference equation $\sum_j a_j x_{i,t-j} = 0$.
4. *Non-uniqueness of the state variable.* Consider the linear system (3), (4) and consider a transformed state variable, $\tilde{x} = Hx$, where H is a non-singular square matrix. Show that $\tilde{x}, u, y$ obey a set of relations (3), (4) with A, B, C transformed to HAH^{-1}, HB and CH^{-1} respectively. Deduced an analogous result for the non-linear model (1), (2).

2. STABILITY

Consider a time-homogeneous model without an explicit control variable

$$x_{t+1} = a(x_t). \tag{1}$$

This may be the plant equation of the system with a prescribed control rule; if state were observable and one considered a control rule $u_t = h(x_t)$ in (1.1) then x would obey the recursion

$$x_{t+1} = a(x_t, h(x_t)). \tag{2}$$

With a redefined this amounts to (1).

The value $\bar{x}$ is an *equilibrium point* of system (1) if it satisfies

$$\bar{x} = a(\bar{x}). \tag{3}$$

Relation (3) ensures that, if x attains the value $\bar{x}$, then it retains the value $\bar{x}$ thereafter. However, the relation does not imply that, if x is near to $\bar{x}$ at a given time, it then continues to approach $\bar{x}$, or even to stay near $\bar{x}$. If a stronger property of this type holds then $\bar{x}$ is said to be a *stable* equilibrium point.

More specifically, $\bar{x}$ is said to be *locally stable* if $x_t \to \bar{x}$ with increasing t for all x_0 in some neighbourhood of $\bar{x}$. If this assertion holds for all x_0, without restriction, then $\bar{x}$ is said to be *globally stable.*

One classical control problem is that of *regulation* to a set point $\bar{x}$: of finding a control that will make the state variable x seek and hold the value $\bar{x}$. This has classically been regarded as a problem of making $\bar{x}$ a stable equilibrium point of the controlled system. (There may well be secondary design criteria, of course, such as smoothness or rapidity of response and ease of physical realization.)

There is a large literature devoted to the study of these questions and a number of classic criteria for stability. In an optimization approach to the regulation problem one usually does not seek stability as such, but rather relies on the fact that, if stability is possible, then an optimizing solution will automatically achieve it. According to whether stability can or cannot be achieved, minimal costs over an infinite time interval will be bounded or unbounded.

The linear system

$$x_{t+1} = Ax_t \tag{4}$$

has $\bar{x} = 0$ as an equilibrium point; this will be the only one unless $I - A$ is singular. For this finite-dimensional case there is a condition which is necessary and sufficient for both local and global stability of $\bar{x} = 0$ as an equilibrium: that all eigenvalues of A should lie strictly inside the unit circle. That is, that all roots λ_j of

$$|\lambda I - A| = 0 \tag{5}$$

should be strictly less than one in modulus.

In fact, for large t,

$$A^t = O(\mu^t), \tag{6}$$

where μ is the modulus of the eigenvalue largest in modulus, so the necessary and sufficient condition that

$$A^t \to 0 \tag{7}$$

with increasing t is that $\mu < 1$. Under this condition the convergence (7) is exponentially fast. If (7) holds then A is termed a *stability matrix* (in the discrete-time context). By an abuse of nomenclature one sometimes merely says that 'A is stable'. The question of the stability of A reduces then to the algebraic question of whether the polynomial equation (5), which we may write explicitly as

$$\sum_{j=0}^{n} a_j \lambda^{n-j} = 0, \tag{8}$$

has all its roots strictly inside the unit circle. There are a number of classical criteria to test whether this is so for a given A (see Exercises 5 to 8).

Consider again the controlled linear system (1.3), (1.4). Suppose that the current state is fully observable, so that $C = I$. Suppose that one considers a control $u_t = Kx_t$ which is linear in the current state, so that the plant equation reduces to

$$x_{t+1} = (A + BK)x_t = \Gamma x_t,$$

say. The equilibrium $\bar{x} = 0$ will be stable if and only if $\Gamma = A + BK$ is stable. The system $[A, B, I]$ is said to be *stabilizable* if a K can be found such that $A + BK$ is stable for some K.

One might imagine that if stability cannot be achieved by an instantaneous linear control of the form $u_t = Kx_t$, it might perhaps be achievable by some more complicated control, possibly non-linear and involving past x values. In fact, this is not the case (see Exercise 5.5.2).

Exercises and comments

1. Consider the linear system with constant terms inserted in the plant and control equations:

$$x_{t+1} = \alpha + Ax_t + Bu_t,$$
$$u_t = \kappa + Kx_t.$$

Show that the state values $\bar{x}$ which can be made stable operating points are those of the form $\bar{x} = (I - \Gamma)^{-1}(\alpha + B\kappa)$ for some κ where K is chosen so that $\Gamma = A + BK$ is stable.

2. Consider the case $A = \begin{bmatrix} a_{11} & \cdot \\ \cdot & a_{22} \end{bmatrix}$, $B = \begin{bmatrix} 1 \\ \cdot \end{bmatrix}$. Show that $[A, B, I]$ is stabilizable if $|a_{22}| < 1$.

3. Consider the companion matrix A defined in Exercise 1.2. Show that

$$|\lambda I - A| = \sum_0^n a_j \lambda^{n-j}.$$

4. Consider the driven system $x_{t+1} = Ax_t + v_{t+1}$, where $\{v_t\}$ is a known driving sequence. This can be solved in terms of initial conditions as

$$x_t = \sum_{s=0}^{t-1} A^s v_{t-s} + A^t x_0.$$

Consider the system

$$\sum_{j=0}^{n} a_j \xi_{t-j} = v_t.$$

with ξ, v scalar. Show that the analogous solution is

$$\xi_t = \sum_{s=0}^{t-1} b_s v_{t-s} + \sum_{s=t}^{\infty} c_{st} \xi_{t-s}$$

where the coefficients b, c are generated by the long division algorithm

$$\frac{1}{\sum a_j z^j} = \sum_{s=0}^{t-1} b_s z^s + \frac{\sum_{s=t}^{\infty} c_{st} z^s}{\sum a_j z^j},$$

i.e. $\sum b_s z^s$ and $\sum c_{st} z^s$ are quotient and remainder respectively after t steps of division.

5. *The Liapunov condition for stability*. The matrix A is stable if and only if the equation

$$V - AVA' = N \tag{9}$$

with N positive definite admits a positive definite solution for V (more exactly, the condition is necessary for all such N; it is sufficient if it holds for one such N).

Necessity: $V = \sum_0^\infty A^s N(A')^s$ is a solution.
Sufficiency: If A (and so A') has eigenvalues λ_j and the corresponding right eigenvector of A' is α_j, then equation (9) implies that

$$(1 - |\lambda_j|^2)(\alpha_j^\dagger V \alpha_j) = \alpha_j^\dagger N \alpha_j$$

where $\alpha_j^\dagger$ is the transposed complex conjugate of α_j. Hence $|\lambda_j| < 1$.
Interpretation: The criterion is usually derived by the use of so-called Liapunov functions. However, if we consider the randomly driven model

$$x_{t+1} = Ax_t + \varepsilon_{t+1}$$

where $\{\varepsilon_t\}$ is white noise with the covariance matrix N, then the covariance matrix V of x_t will obey equation (9) once a stationary régime has been reached. Such a régime can be reached if and only if A is stable.

6. Show that the Liapunov conditions remain necessary and sufficient even if N is only non-negative definite, provided that $x_t' N x_t$ is not zero for all t, if $x_t = A^t x_0$ and $x_0 \neq 0$.
7. Consider the model

$$\sum_{j=0}^{n} a_j \xi_{t-j} = \varepsilon_t$$

where ξ_t is scalar and ε_t is scalar white noise (of unit variance, say). Suppose the $\{\xi_t\}$ process is stationary (possible only if (8) has all its roots inside the unit circle; see Chapter 21) and let m_{jk} be the jkth element of the *inverse* of the covariance matrix of $(x_t, x_{t-1}, x_{t-2}, \ldots, x_{t-n+1})$. Show by a recursive argument on the distribution of this vector in the case of Gaussian $\{\varepsilon_t\}$ that

$$\min_{\xi_{t-n}} \left[\sum_{j=1}^{n} \sum_{k=1}^{n} m_{jk} \xi_{t-j} \xi_{t-k} + \left(\sum_{0}^{n} a_j \xi_{t-j} \right)^2 \right] = \sum_{j=1}^{n} \sum_{k=1}^{n} m_{jk} \xi_{t-j+1} \xi_{t-k+1}$$

identically in $\{\xi_t\}$. Hence show that

$$\sum_1^n \sum_1^n m_{jk} \lambda^{j-1} \mu^{k-1} = \frac{(\lambda\mu)^n D(\lambda^{-1}) D(\mu^{-1}) - D(\lambda) D(\mu)}{1 - \lambda\mu} \tag{10}$$

where $D(\lambda) = \sum_0^n a_j \lambda^{n-j}$.

8. *The Hermite condition for stability.* Show from Exercises 5 to 7 that A is stable if and only if the matrix $M = (m_{jk})$ determined by (10) is positive definite, where now $D(\lambda) = |\lambda I - A|$.

   A symmetric matrix is positive definite if and only if all its principal minors are positive; by testing M in this way one derives the *Schur–Cohn conditions* for stability of A.
9. For the case $n = 1$, show that $M = 1 - a_1^2$; the condition is the expected one, that $a_1^2 < 1$.
10. For the case $n = 2$, show that

$$M = \begin{bmatrix} 1 - a_2^2 & a_1 - a_1 a_2 \\ a_1 - a_1 a_2 & 1 - a_2^2 \end{bmatrix}$$

and that the conditions for the stability of A are

$$1 - a_2^2 > 0,$$
$$(1 - a_1 + a_2)(1 + a_1 + a_2) > 0.$$

3. CONTROLLABILITY

The properties of the individual matrices A, B and C must very much influence the performance that can be achieved with a given system. The influence can be a question of kind as well as of degree: it may be, for example, that there is *no* choice of control sequence $\{u_t\}$ that will bring all components of x_t on to desired set points. It turns out that there are two fundamental concepts for a system, *controllability* and *observability*. These are of a dual character and are of importance in the consideration of control in general, and optimal control in particular.

We shall discuss only the case of the linear time-homogeneous system $[A, B, C]$ specified by (1.3), (1.4), although these concepts do extend to the time-dependent case and, with much more difficulty, to the non-linear case.

The system is termed *r-controllable* if, given an arbitrary value of x_0, one can choose values $u_0, u_1, \ldots, u_{r-1}$ such that x_r has an arbitrary prescribed value. For example, if $m = n$ and B is non-singular, then the system is 1-controllable, because one can move to any value of x_1 from any value of x_0 by choosing

$$u_0 = B^{-1}(x_1 - Ax_0),$$

As a second example, consider the system

$$\begin{bmatrix} x_{1,t+1} \\ x_{2,t+1} \end{bmatrix} = \begin{bmatrix} a_{11} & \cdot \\ a_{21} & a_{22} \end{bmatrix} \begin{bmatrix} x_{1t} \\ x_{2t} \end{bmatrix} + \begin{bmatrix} u_t \\ \cdot \end{bmatrix}$$

for which $m = 1$, $n = 2$. It is never 1-controllable, because u cannot alter the second component of x in a single step. It is uncontrollable if $a_{21} = 0$, because then u does not affect the second component of x at all. It is 2-controllable if $a_{21} \neq 0$, because

$$x_2 - A^2 x_0 = Bu_1 + ABu_0$$
$$= \begin{bmatrix} 1 & a_{11} \\ \cdot & a_{21} \end{bmatrix} \begin{bmatrix} u_1 \\ u_0 \end{bmatrix},$$

and this has a solution for u_0, u_1 for any x_0, x_2 if $a_{21} \neq 0$.

This argument generalizes.

Theorem 3.1. (*a*) *The n-dimensional system* $[A, B, \cdot]$ *is r-controllable if and only if the matrix* $M_r = [B, AB, A^2B, \ldots, A^{r-1}B]$ *has rank n or*
(*b*), *equivalently, if and only if*

$$\sum_{j=0}^{r-1} A^j BB'(A')^j > 0. \tag{1}$$

(*c*) *If the system is r-controllable, then it is also s-controllable for* $s \geq \min(n, r)$.

We write the system as $[A, B, \cdot]$ rather than as $[A, B, C]$ since the matrix C evidently has no bearing on the question of controllability.

Proof. We must choose $(u_0, u_1, \ldots, u_{r-1})$ to satisfy

$$x_r - A^r x_0 = Bu_{r-1} + ABu_{r-2} + \cdots + A^{r-1}Bu_0. \tag{2}$$

This will be possible for arbitrary values of the n-vector $x_r - A^r x_0$ if and only if the matrix of the equation system (2), M_r, has rank n.

The matrix M_r will have full rank, n, if and only if $M_r M_r'$ does. This is equivalent to requiring that the non-negative definite matrix $M_r M_r'$ be positive definite, which is simply assertion (b).

Since the rank of M_r is non-decreasing in r, then r-controllability certainly implies s-controllability for $s \geq r$. We shall then complete the proof of assertion (c) if we can show that r-controllability, for some $r \geq n$, implies n-controllability. But if

$$|\lambda I - A| = \sum_{j=0}^{n} a_j \lambda^{n-j}$$

then A satisfies

$$\Sigma a_j A^{n-j} = 0$$

(see, for example, Aitken, 1959). Since $a_0 = 1 \neq 0$, this implies that A^r can be written as a linear combination of $(I, A, A^2, \ldots, A^{n-1})$ if $r \geq n$. Thus the last $m(r-n)$ columns of M_r are linearly dependent upon the columns of M_n, and M_n, M_r have the same rank for $r \geq n$. Hence all the assertions of (c) follow. ■

If the system is n-controllable it is simply termed *controllable.* This is a reasonable use of the term, in view of the following immediate deduction from assertion (c) of the theorem:

Corollary 3.1. *If a system of dimension n is r-controllable for some $r \geq n$, then it is controllable, and if it is controllable, then it is r-controllable for all $r \geq n$.*

A way of finding a solution of (2), in the case where one exists, is to choose values of $u_0, u_1, \ldots, u_{r-1}$ minimizing the modulus of the vector difference between the two sides. This leads to the solution

$$u_t = B'(A')^{r-t-1}(M_r M_r')^{-1}(x_r - A^r x_0) \qquad (0 \leq t < r). \tag{3}$$

Appeal to the non-singularity of $M_r M_r'$ is explicit.

One should distinguish between controllability, which implies that one can bring the state vector to a prescribed value in at most n steps, and the weaker property of stabilizability (see section 2), which implies only that the state vector can be made to converge to some relevant equilibrium value ultimately. That controllability implies stabilizability will be proved incidentally in Theorem 5.5.1. That stabilizability does not imply controllability follows from the case where A is a stability matrix and $B = 0$. This is uncontrolled, and so not controllable, but stable, and so stabilizable.

Exercise

1. Let u, u^* be the control vectors for two symmetrically coupled systems with state vectors x, x^* obeying the combined plant equation

$$\begin{bmatrix} x_{t+1} \\ x^*_{t+1} \end{bmatrix} = \begin{bmatrix} A_1 & A_2 \\ A_2 & A_1 \end{bmatrix} \begin{bmatrix} x_t \\ x^*_t \end{bmatrix} + \begin{bmatrix} Bu_t \\ Bu^*_t \end{bmatrix}.$$

Show that the system is not controllable under the constraint $u^*_t = u_t$, but may be controllable under the constraint $u^*_t = 0$.

4. OBSERVABILITY

The time-homogeneous linear system $[A, B, C]$ is said to be *r-observable* if x_0 can be determined from knowledge of $y_1, y_2, \ldots, y_r; u_0, u_1, \ldots, u_{r-2}$ for any initial value x_0. Seeing that control can be based only on the incomplete observations of state y, but is required to be effective in controlling the complete state vector x, some concept such as observability is obviously important in the control context.

One has complete analogues of the controllability results of section 3.

Theorem 4.1. *(a) The n-dimensional system $[A, \cdot, C]$ is r-observable if and only if the matrix $N_r = [C', A'C', \ldots, (A')^{r-1}C']$ has rank n, or* (*b*), *equivalently, if and only if*

$$\sum_{j=0}^{r-1} (A')^j C'CA^j > 0.$$

(*c*) *If the system is r-observable, then it is s-observable for* $s \geq \min(n, r)$.

Assertion (c) will imply the conclusion analogous to Corollary 3.1: if n-observability is termed simply 'observability', then the system is observable if it is r-observable for some $r \geq n$, and is in that case also r-observable for all $r \geq n$.

Proof. We have

$$y_t = C\left[\sum_{j=0}^{t-2} A^{t-j-2}Bu_j + A^{t-1}x_0\right] \qquad (t > 0)$$

so that, if

$$\tilde{y}_t = y_t - C\sum_{j=0}^{t-2} A^{t-j-2}Bu_j, \tag{2}$$

then x_0 is to be determined from the system of equations

$$\tilde{y}_t = CA^{t-1}x_0 \qquad (1 \leq t \leq r). \tag{3}$$

These equations are mutually consistent, by hypothesis; they have a unique solution for the n-vector x_0 under all circumstances if and only if the matrix of the system has rank n. This is assertion (a); the other assertions follow exactly as in the proof of Theorem 3.1. ■

The analogue of (3.3) is the 'least-squares' determination of x_0:

$$x_0 = (N_r N_r')^{-1} \sum_{t=1}^{r} (A')^{t-1} C\tilde{y}_t. \tag{4}$$

Chapter 5 will demonstrate that the concepts of controllability and observability both play a role in the theory of optimal control. They play an even stronger role in what one might term the algebraic or 'finite' theory of systems—the theory of what can be achieved exactly with a deterministic linear system in a finite time. We are not concerned with this, nor with the closely related question of realization, but this is the right place to refer to these important ideas.

One sometimes requires a certain relationship between input u and output y, say

$$y_t = \sum_{j=0}^{\infty} H_j u_{t-j} \tag{5}$$

and asks if there is a system $[A, B, C]$ that will realize this. This is the *synthesis* or *realization* problem. One can also ask for the *minimal realization* for which the system has minimal dimension. One has the important and beautiful theorem, that a system $[A, B, C]$ realizing the relationship (5) is minimal if and only if it is both controllable and observable (see, for example Brockett, 1970, p. 94).

Exercise

1. Give the observability analogues of the controllability examples and exercise of section 3.

5. DETECTABILITY

The property of *detectability* stands in the same relation to observability as does stabilizability to controllability. The system $[A, \cdot, C]$ is detectable if a matrix H can be found such that $A - HC$ is a stability matrix.

The property and its definition arise in the following way. If the value x_t of the state at time t is not fully observed, then one may form an estimate $\hat{x}_t$ of it from the observables y available at time t. It turns out (Chapter 18) that the optimal estimate can be calculated recursively from a relation

$$\hat{x}_{t+1} = A\hat{x}_t + Bu_t + H(y_{t+1} - C\hat{x}_t) \tag{1}$$

This is simply the plant equation, with $\hat{x}_t$ replacing x_t, and with incorporation of a *correction* or *updating* term $H(y_{t+1} - C\hat{x}_t) = HC(x_t - \hat{x}_t)$. This can indeed be seen as a correction term, since $C\hat{x}_t$ is the value of y_{t+1} predicted from the information available at time t; therefore $y_{t+1} - C\hat{x}_t$ is the discrepancy in prediction.

Let $\Delta_t = x_t - \hat{x}_t$ be the error in estimate of the state vector at time t. By subtracting the plant equation from (1) we deduce that

$$\Delta_{t+1} = (A - HC)\Delta_t.$$

Thus $\Delta_t \to 0$ as t increases for arbitrary Δ_0 if and only if $A - HC$ is stable. Detectability thus implies that one can find an updating formula of type (1) which will asymptotically yield a correct estimate of state.

6. CONTINUOUS TIME VERSIONS

In a deterministic continuous time model the state variable, at least, must change continuously with time and take values in a continuous set. We shall suppose for simplicity that $\mathscr{X} = R^n$. The analogue of the plant equation (1.1) will then be a system of first-order differential equations

$$\dot{x} = a(x, u, t) \tag{1}$$

where $x(t)$, $u(t)$ are functions of time, with the t-argument suppressed if this causes no ambiguity, and $\dot{x}(t) = \mathrm{d}x(t)/\mathrm{d}t$. Equation (1) is to be viewed formally as the limit version of the discrete-time recursion

$$x(t+\delta t)=x(t)+a(x(t),u(t),t)\delta t.$$

The quantities a in equations (1.1) and (1) are not in complete analogue, but, for economy of notation, we use the same notation, a, in any case. On the other hand, the relation (1.2) between observables and state we carry over as $y(t)=c(x(t),u(t),t)$—i.e. as an instantaneous rather than as a differential relationship.

For the linear case, relations (1.3) and (1.4) now become

$$\dot{x}=Ax+Bu, \tag{2}$$

$$y=Cx. \tag{3}$$

Since we would regard (2) as the limit version of

$$x(t+\delta t)=x(t)+[Ax(t)+Bu(t)]\delta t \tag{4}$$

we see that it is $I+A\,\delta t$ and $B\delta t$ which are the analogues of the A and the B of relation (1.3).

The results of sections 1 to 4 can now be developed for the continuous case by considering their limit versions for the system with the plant equation (4) as $\delta t\downarrow 0$. The purist may object to this approach, either on the grounds of rigour or of style. However, to argue this way certainly unifies the discrete and continuous time cases, and at the least provides one with one with a means of constructing the assertions one *ought* to be able to prove.

The analogue of the solution $x_t=A^t x_0$ of (2.4) is the statement that

$$x(t)=\mathrm{e}^{At}x(o) \tag{5}$$

solves the time-homogeneous, linear, uncontrolled plant equation

$$\dot{x}=Ax. \tag{6}$$

The matrix exponential in (5) is defined by the convergent series

$$\mathrm{e}^{At}=\sum_0^\infty \frac{t^j A^j}{j!}.$$

The equilibrium at $x=0$ of (6) is stable if and only if all the eigenvalues of A lie in the left complex half-plane, i.e. have a strictly negative real part. A matrix having this property is a *stability matrix* (continuous time context), and if A is a stability matrix of finite order then $\mathrm{e}^{At}\to 0$ exponentially fast as $t\to+\infty$. There are classic criteria for continuous time stability analogous to those deduced in section 2 for discrete time (see Excercises 2 and 3).

Since the r matrices $(I+A\delta t)^j B$ $(0\leq j<r)$ can be written as linear combinations of the r matrices $A^j B$, and vice versa, the rank condition (a) of Theorem 3.1 for controllability can formally be carried over unchanged. However, the notion of steering the state variable in r steps (i.e. within a time $r\,\delta t$) now becomes

unacceptable as $\delta t \downarrow 0$, because the values of u required would in general become infinite. Indeed, the limit control would physically consist of a sequence of impulses delivered in an infinitesimal time. However, the necessary and sufficient condition for controllability over time τ is (analogously to Theorem 3.1 (b)) that the matrix

$$\int_0^\tau e^{At} BB' e^{A't} \, dt$$

be positive definite. This criterion can be met for some (and then for any) $\tau > 0$ if and only if the rank condition of Theorem 3.1 (a) holds for some r (necessarily not greater than n); one then simply terms the system 'controllable'. Note that a control which steers the state variable to the desired value at time τ can be determined as in (3.3); details are left to the reader.

Analogously, the system $[A, \cdot, C]$ is observable if and only if $[C', A'C', \ldots, (A')^{r-1}C']$ is of rank n for some r, or, alternatively, if and only if the matrix

$$\int_0^\tau e^{A't} C'C e^{At} \, dt$$

is positive definite for some $\tau > 0$.

The system will again be stabilizable or detectable according as one can find matrices K, H which make $A + BK$ or $A - HC$ stable (in the continuous time sense) respectively.

Exercises

1. Suppose scalar functions of time $\xi(t)$ and $u(t)$ satisfy the constant coefficient linear differential equation

$$\sum_{j=0}^{n} a_j \frac{d^j \xi}{dt^j} = u. \qquad (*)$$

Recast this relation in the form of (2), defining the state variable by appropriate modification of the argument of Exercise 1.1. What is the 'companion matrix' to the system (*) thus defined?

2. *The Liapunov condition for stability.* Show that A is a stability matrix (continuous time) if and only if the equation for V,

$$AV + VA' + N = 0,$$

possesses a positive definite solution for N positive definite.

3. *The Hermite condition for stability.* Consider the $n \times n$ matrix $M = (m_{jk})$

determined by

$$\sum_1^n \sum_1^n m_{jk}\lambda^{j-1}\mu^{k-1} = \frac{D(\lambda)D(\mu) - D(-\lambda)D(-\mu)}{\lambda + \mu}$$

where $D(\lambda) = |\lambda I - A|$. Show that A is a stability matrix if and only if M is positive definite. By demanding positivity of the principal minors of M one derives the *Routh–Hurwitz conditions* for stability of A.

4. Consider the cases $n = 1, 2$, with

$$D(\lambda) = \sum_0^n a_j \lambda^j$$

and so $a_n = 1$. For the case $n = 1$ one finds $M = 2a_0$ and the condition $a_0 > 0$. For the case $n = 2$ one finds $M = 2\begin{bmatrix} a_0 a_1 & \cdot \\ \cdot & a_1 \end{bmatrix}$ and the conditions $a_1 > 0$, $a_0 > 0$.

7. OPEN- AND CLOSED-LOOP CONTROL

Suppose that for a Markov model one derives the control law

$$u_t = u(x_t, t). \tag{1}$$

This is then a *feedback* or *closed-loop* control in that u_t is made to depend upon the current state; the subsidiary dependence upon time t only reflects any lack of time-invariance in the model. However, one could use the control law (1) and the plant equation (1.1) to solve for x_t and u_t in terms of x_0 and t. The consequent expression for u_t,

$$u_t = u^*(x_0, t), \tag{2}$$

is the *open-loop* expression of control, for which actions are determined by the clock rather than by the current state. The two forms of the control law will indeed be equivalent if a plotted course is always followed exactly. If it is not (the rule, because of unpredictable disturbances, faults in model specification, or cumulative rounding errors in the realization of control) then the two will not be equivalent. The closed-loop form has a potential adaptivity to changing circumstances which the open-loop form lacks.

Consider, for example, a case $x \in Z$, with the plant equation

$$x_{t+1} = x_t + u_t$$

where u_t may take any of the three values $-1, 0, +1$. The cost function over horizon T is $\sum_0^T x_t^2$, so the aim is to keep $|x|$ small. The optimal control is obviously

$$u_t = -\operatorname{sgn}(x_t) \tag{4}$$

in closed-loop form. In open-loop form this could be expressed as

$$u_t = \begin{cases} -\operatorname{sgn}(x_0) & (t \le |x_0|), \\ 0 & (t \ge |x_0|). \end{cases} \tag{5}$$

All would be well if the specification were exactly as stated. Suppose, however, that the plant equation in fact had the disturbed form

$$x_{t+1} = x_t + u_t + \varepsilon_{t+1} \tag{6}$$

where the process noise variables ε_t are independent random variables, taking values $-1, 0, 1$ with probabilities $\alpha/2$, $1-\alpha$, $\alpha/2$ respectively. One finds then that, for large t,

$$E(x_t^2) \sim \alpha$$

if the closed-loop rule (4) is used and

$$E(x_t^2) \sim \alpha t$$

if the open-loop rule (5) is used (see Exercise 2). Comparison of these two expressions makes the point.

Exercises and comments

1. The text indicated how the open-loop law could be derived from the closed-loop law. To carry out the reverse calculation: corresponding to (2) there will be an expression

$$x_t = x^*(x_0, t). \tag{7}$$

Eliminating x_0 from (2) and (7) one derives (1). Alternatively, if

$$u_t = u^*(x_\tau, \tau, t) \qquad (t \ge \tau)$$

is the open-loop control for the initial condition of prescribed x_τ at time τ, then the closed-loop rule is simply

$$u_t = u^*(x_t, t, t).$$

2. Show that, with the plant equation (6) and control law (4), x_t takes the values $-1, 0, 1$ with respective probabilities $\alpha/2$, $1-\alpha$, $\alpha/2$ in equilibrium.

CHAPTER 5

Linear/quadratic models

1. LQ STRUCTURE

The description 'linear/quadratic' implies linear dynamics and quadratic costs, and is abbreviated to LQ. Models with LQ structure, which we shall define more explicitly in a moment, have a very complete optimization theory. If we revert to the general non-Markov description of Section 2.1 with the cost function $C = G(U_{T-1})$ then conditions for LQ structure are simply the following:

(a) The value of horizon T is fixed.

(b) The complete control history $U_{T-1} = (u_0, u_1, \ldots, u_{T-1})$ is a vector free to take any value in some finite-dimensional Euclidean space.

(c) G is a positive definite quadratic function of U_{T-1}.

Suppose that for simplicity we write the composite vector U_{T-1} simply as u, so that we are virtually treating the T-stage problem as a one-stage problem. The assumption is then that C can be written as a quadratic form

$$C = G(u) = u'Qu + 2u'S + R \tag{1}$$

for some scalar R, vector S and positive definite matrix Q. Furthermore, the value $\bar{u}$ of u minimizing C can be chosen without constraint.

Since expression (1) is strictly convex, then $\bar{u}$ is located by the stationarity condition

$$Q\bar{u} + S = 0. \tag{2}$$

The simplifying consequence of LQ structure is thus that the optimality conditions amount to a set of linear equations. Determination of the optimal control $\bar{u}$ amounts to inversion of Q:

$$\bar{u} = -Q^{-1}S. \tag{3}$$

One finds readily that

$$G(u) = G(\bar{u}) + (u - \bar{u})'Q(u - \bar{u}) \tag{4}$$

and

$$G(\bar{u}) = R - S'Q^{-1}S \tag{5}$$

(see Exercise 1).

Suppose we now also postulate state structure. Then the conditions for LQ structure would have to be such that C would reduce to the form (1) if state variables were eliminated (i.e. if x_t were expressed in terms of x_0 and U_{t-1} for $t > 0$). The conventional assumptions are as follows:

(a) The process has Markov structure with additive cost function, as specified by conditions (a) to (d) of Section 2.2.
(b) The value of horizon T is fixed.
(c) $\mathscr{X} = R^n$, $\mathscr{U} = R^m$.
(d) The plant equation is linear in x and u.
(e) Cost components $g(x, u, t)$ and $K(x)$ are non-negative definite quadratic functions of x and u jointly, and g is positive definite quadratic in u alone.

The rest of the chapter consists of a continued demonstration that this model is tractable and that the linear optimization equations (2) can be thrown into a natural recursive form, rich in structural implications. What is more important, the model remains tractable under continued relaxation: addition of process noise (section 12.1), observation noise and incomplete state observation (Chapters 17 and 18), incorporation of risk-sensitivity (Chapter 19) and relaxation of terminal conditions (Chapter 20). As our introduction should have made clear, the issues of state structure and LQ structure are separate; Chapter 21 considers stochastic LQ models without state structure.

The assumptions that cannot be jettisoned without essential loss of LQ character are those of linear dynamics, convex quadratic costs and lack of constraint (other than cost) on control variables. Such is the success of LQ models that these assumptions are sometimes made in situations where they are questionable (see Exercises 6.2 and 6.3).

The tractability of the LQ case is, of course, just a reflection of the naturalness of the theory for linear approximation in the quadratic norm evident in regression theory, prediction theory and so much L^2 theory of mathematics.

Exercises

1. Show that expression (1) is bounded below if and only if Q is non-negative definite and S belongs to the orthogonal complement of the null space N_0 of Q (i.e. $Q\xi = 0$ implies that $S'\xi = 0$). Show that, in this case, (2) has a solution unique to within addition of an element of N_0 and that (4) holds for any such solution.

2. QUADRATIC REGULATION (THE CENTRED CASE)

We shall initially assume a model which is in some respects rather special, in order to isolate some of the essential features. We shall assume a plant equation

$$x_{t+1} = A_t x_t + B_t u_t \tag{1}$$

and a T-horizon cost function

$$C = \sum_{0}^{T-1} (x'Rx + 2u'Sx + u'Qu)_t + x_T' P_T x_T. \tag{2}$$

We have here followed the useful convention of indicating by $(\ )_t$ that all quantities within the brackets are to be given their values at time t; this saves many subscripts. So, it is admitted that the matrices A, B, Q, R and S as well as the variables x and u may be time-dependent; time-homogeneity is not assumed.

Relations (1) and (2) certainly express linear dynamics and quadratic costs respectively. The definiteness conditions expressed in (e) of the last section will require that the matrices Q_t be positive definite and the matrices $\begin{bmatrix} R_t & S_t \\ S_t' & Q_t \end{bmatrix}$ and P_T be non-negative definite.

The cost function penalizes deviations of state from zero (in R^n) and also of control from zero (in R^m). The problem is thus one of regulation to the set point $x = 0$, to be achieved without excessive use of control (i.e. of deviation of u from zero). This is the *quadratic regulation problem.* We shall consider regulation to more general set points, or to moving set points against plant disturbances ('tracking and buffering') in Section 4. The problem expressed by (1), (2) will be termed the 'centred case': centred on the origin, in that C is minimal when x and u are zero and that $x = 0$, $u = 0$ would satisfy the plant equation.

Theorem 2.1. *Optimal quadratic regulation. Suppose that:*

(*a*) *The plant equation has the linear form* (1).

(*b*) *The cost function has the quadratic form* (2)

with $\begin{bmatrix} R_t & S_t \\ S_t' & Q_t \end{bmatrix}$ *and* P_T *non-negative definite and* Q_t *positive definite* $(t < T)$.

(*c*) *The value of* x_t *is observable at time* t.

Then: (i) $F(x_t, t)$ *is a homogeneous quadratic function of state,*

$$F(x_t, t) = x_t' P_t x_t, \tag{3}$$

whose matrix P_t *is non-negative definite, satisfies*

$$P_t = [R + A'P_{t+1}A - (A'P_{t+1}B + S')(Q + B'P_{t+1}B)^{-1}(B'P_{t+1}A + S)]_t \quad (t < T) \tag{4}$$

and equals the P_T *of* (2) *for* $t = T$.

(*ii*) *The optimal choice of control is*

$$u_t = K_t x_t \tag{5}$$

where

$$K_t = -[(Q + B'P_{t+1}B)^{-1}(B'P_{t+1}A + S)]_t. \tag{6}$$

The LQ character thus persists, in that the optimal control (5) is linear in state and the minimal future cost (3) is quadratic in state. The t subscript in (4), (6) indicates that all quantities in the brackets are to receive the subscript t, except P which has the subscript indicated: $t+1$.

Proof. The proof is inductive. Plainly $F(x_r, r) = x_r' P_r x_r$ for $r = T$. Suppose this is true for $t < r \leq T$. Then, by the optimality equation (3.2.4),

$$F(x_t, t) = \min_u [x'Rx + 2u'Sx + u'Qu + (Ax + Bu)'P_{t+1}(Ax + Bu)]_t. \qquad (7)$$

As a quadratic form in x_t and u_t the expression in square brackets in (7) has the matrix

$$\Pi = \begin{bmatrix} \Pi_{xx} & \Pi_{xu} \\ \Pi_{ux} & \Pi_{uu} \end{bmatrix} = \begin{bmatrix} R + A'PA & S + A'PB \\ S' + B'PA & Q + B'PB \end{bmatrix} \qquad (8)$$

where P carries the subscript $t+1$; all other quantities carry the subscript t. This matrix is non-negative definite and the matrix Π_{uu} is positive definite. The minimizing value of u is determined by the stationarity condition

$$\Pi_{ux} x_t + \Pi_{uu} u_t = 0 \qquad (9)$$

which implies (5), (6); the minimal value is the non-negative quadratic form

$$F(x_t, t) = x_t'(\Pi_{xx} - \Pi_{xu}\Pi_{uu}^{-1}\Pi_{ux})x_t, \qquad (10)$$

implying (3), (4). ■

The definiteness assumptions of condition (b) are stronger than is needed for the argument to hold; all that is needed is that the matrix $Q_t + B_t' P_{t+1} B_t$ should be positive definite for all t. One can even weaken the demand to non-negative definiteness if one accepts non-uniqueness of the optimal control. However, these relaxations somewhat complicate the argument.

The attractions of the control rule (5) are that it is a feedback rule (as is always delivered by a dynamic programming argument) and is linear in state (because of the assumptions of state structure and LQ structure).

The matrix recursion (4) is fundamental, as is an understanding of the properties of the sequence of the matrices $\{P_t\}$ it generates. The corresponding differential equation in the continuous time case (20.1.5) is a Riccati equation. For this reason we shall simply refer to (4) (or its continuous time analogue) as 'the Riccati equation'. The term is unambiguous if one is speaking in the LQ context.

The Riccati equation looks singularly unappealing in form (4). One should see it rather as a quasi-linear matrix recursion:

$$P_t = \min_K [R + K'S + S'K + K'QK + (A + BK)'P_{t+1}(A + BK)]_t. \qquad (11)$$

Here K is an arbitrary $m \times n$ matrix. The minimization is in the sense of positive-definiteness: for symmetric matrices M, N of the same order one says that $M \geq N$ if $M - N$ is non-negative definite. The 'minimizing' value is just that given by (6). Relation (11) is quasi-linear in that the form being minimized is linear in P. Most of the properties of $\{P_t\}$ are consequences of this fact.

If the control (5) is used then the plant equation becomes

$$x_{t+1} = \Gamma_t x_t \tag{12}$$

where the matrix

$$\Gamma_t = A_t + B_t K_t \tag{13}$$

is termed the *gain matrix*. The sequence $\{\Gamma_t\}$ turns out to have a fundamental importance, because, if control is successful, then the x_t generated by (12) should converge to zero.

There are a number of normalizing transformations of the problem which somewhat simplify it (Exercises 2 and 3).

In the time-homogeneous version the matrices A, B, Q, R, S will be t-independent for $t < T$. It is natural now to regard P, K as functions $P(s)$, $K(s)$ of time-to-go $s = T - t$, so that

$$P_t = P(T - t),$$
$$K_t = K(T - t).$$

Relations (4), (6) then become

$$P(s) = \phi(P(s-1)), \tag{14}$$
$$K(s) = \psi(P(s-1)), \tag{15}$$

where

$$\phi(P) = R + A'PA - (A'PB + S')(Q + B'PB)^{-1}(B'PA + S),$$
$$\psi(P) = -(Q + B'PB)^{-1}(B'PA + S),$$

with $P(0) = P_T$.

Exercise

1. Show that, under the assumptions of the theorem, P_t is positive definite for $t < T$.
2. Show that if one adopts the transformed control variable $u_t^* = u_t + Q_t^{-1} S_t x_t$ then, in the transformed problem, $S^* = 0$. What are the transformed values of the other matrices?
3. Consider the time-homogeneous case with $S = 0$. Show that by appropriate

separate linear transformations of u and x one can achieve the structure

$$u_t = \begin{pmatrix} u_{1t} \\ u_{2t} \end{pmatrix},$$

$$x_{t+1} = Ax_t + \begin{pmatrix} u_{1t} \\ \cdot \end{pmatrix},$$

$$u_t'Qu_t = u_{1t}'Q_1u_{1t} + u_{2t}'Q_2u_{2t}.$$

That is, one can essentially transform B to the form $\begin{bmatrix} I_r \\ \cdot \end{bmatrix}$, for some r not greater than m or n. The component of control u_{2t} does not appear in the plant equation and does not interact with u_{1t} in the cost function, and so is optimally assigned the value zero.

3. THE SCALAR CASE

The simplest quadratic regulation problem is the scalar case, for which $m = n = 1$, and it is helpful to consider this explicitly. We shall suppose time-homogeneity and that S has been normalized to zero. All matrices are then scalars and, in the time-to-go notation, the Riccati equation becomes

$$P(s+1) = \phi(P(s))$$

where

$$\phi(P) = R + \frac{A^2QP}{Q + B^2P}. \tag{1}$$

The function $\phi(P)$ is monotone increasing and concave (consequences of its quasi-linear character; see (2.11)). We shall make these facts the basis of an informal proof of the following theorem; a formal proof will be given in Section 5 for the vector case.

Theorem 3.1. *Suppose $Q, R > 0$ and $P(0) \geq 0$. Then the sequence $\{P(s)\}$ converges monotonically with increasing s to the limit value P^*, which is the unique positive solution of*

$$P = \phi(P). \tag{2}$$

The gain factor $\Gamma(s)$ correspondingly converges to the numerically smaller root of the quadratic equation

$$Q(1 - A\Gamma)(1 - A\Gamma^{-1}) + B^2R = 0. \tag{3}$$

Proof. In view of the positive, increasing concave character of $\phi(P)$, equation (2) has indeed only one positive root. The sequence $\{P(s)\}$ is generated by the cobweb

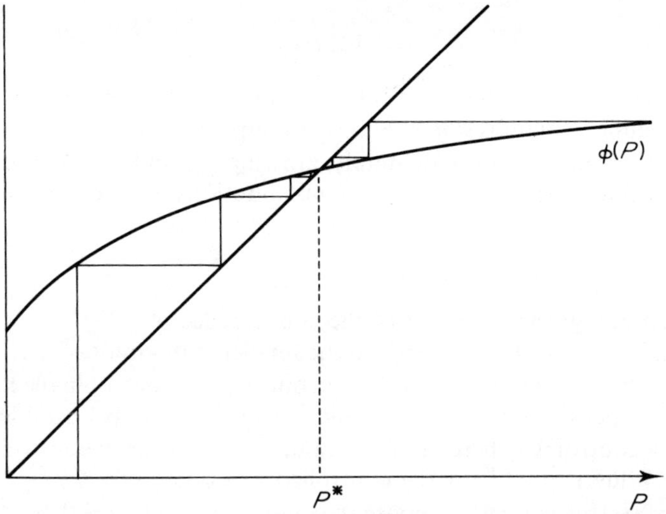

Figure **3.1** Graphical solution of the recursion $P(s+1) = \phi(P(s))$, demonstrating convergence to the limit value P^*

diagram of Figure 3.1. It is graphically apparent from this that $\{P(s)\}$ indeed converges monotonically to P^*, whether it starts above or below. We leave formalization of this argument to the reader.

The value of $\Gamma(s)$ is determined in terms of $K(s)$ and so of $P(s)$, and its limit value Γ^* is correspondingly determined in terms of P^*. We find this determination to be

$$\Gamma^* = A\left(\frac{Q}{Q + B^2 P^*}\right). \tag{4}$$

Eliminating P^* from (2) and (4) we deduce that Γ^* satisfies (3), which has reciprocal real solutions. It also follows from (2) that expression (4) can be re-written as

$$\Gamma^* = \frac{1}{A}\left(\frac{P^* - R}{P^*}\right) \tag{5}$$

and from (2), (4) that $|\Gamma^*| < 1$, whatever the magnitude of A. The asymptotic gain Γ^* can then indeed be identified with the numerically smaller root of (3). ■

That $P(s)$, and consequently $K(s)$, tends to a limit has the important consequence that, as time-to-go tends to infinity, the optimal regulation rule becomes time-homogeneous (at least under the conditions imposed). In this limit situation (which we can imagine being brought about by the recession of the horizon to the infinite future, i.e. $T \to \infty$) the optimally regulated plant equation will be

$$x_{t+1} = \Gamma^* x_t.$$

The fact that $|\Gamma^*| < 1$ then implies that, in this infinite-horizon limit, the optimally regulated plant is *stable*. This is, of course, what one would expect, since an instability would imply indefinitely growing x_t, and so infinite cost. The conclusion is, however, sensitive to cost-assumptions (see Section 5).

Exercises

1. Formalize the graphical proof of the convergence of $\{P(s)\}$.
2. Show that if $Q \to \infty$ then Γ^* tends to the smaller in modulus of A, A^{-1}. That is, if the uncontrolled system is stable then one uses no control in the limit case of infinitely expensive control. If the uncontrolled system is unstable, then one chooses a control which 'reflects' the value $\Gamma = A$ for the uncontrolled process into the value $\Gamma = A^{-1}$ for the controlled process.
3. *Uncosted state deviations.* Suppose that $P(0) > 0$, $Q > 0$, $A \neq 0$, but that $R = 0$, so that state deviations are costed only at termination. Note that recursion (1) can then be written

$$\frac{1}{P(s+1)} = \frac{B^2}{A^2 Q} + \frac{1}{A^2 P(s)}.$$

Solve this. Show that if $|A| < 1$ then $P(s) \to 0$, $\Gamma(s) \to A$. Show that if $|A| > 1$ then $P(s) \to Q(A^2 - 1)$, $\Gamma(s) \to A^{-1}$. Deal with the exceptional cases $P(0) = 0$, $Q = 0$, $A = 0$ or $|A| = 1$.

4. TRACKING AND BUFFERING (UNCENTRED REGULATION)

Let expressions (2.1) and (2.2) for the plant equation and cost be modified to the uncentred forms:

$$x_{t+1} = A_t x_t + B_t u_t + \alpha_{t+1}, \tag{1}$$

$$C = \left[\sum_0^{T-1} (x - x^*)' R (x - x^*) + 2(u - u^*)' S (x - x^*) + (u - u^*)' Q (u - u^*)\right]_t + [(x - x^*)' P (x - x^*)]_T \tag{2}$$

where $\{\alpha_t\}$, $\{x_t^*\}$ and $\{u_t^*\}$ are known sequences. The interpretation is that it is now desired that x_t and u_t should follow the sequences $\{x_t^*\}$ and $\{u_t^*\}$ respectively and that the plant equation is subject to injected disturbances α—hence the title of the section. The sequences $\{x_t^*\}$ and $\{u_t^*\}$ are to be tracked, in that the cost function (2) is to be minimized. This is to be achieved despite the presence of plant disturbances α; i.e. these are to be buffered.

This uncentred case can in fact be seen as a special version of the centred case of

Section 2, if one adopts the augmented state vector $\begin{pmatrix} x \\ 1 \end{pmatrix}$. However, we shall find it simplest to tackle the problem directly. Note first that x_t^* and u_t^* can be normalized to zero. If we define

$$\alpha_{t+1}^* = x_{t+1}^* - A_t x_t^* - B_t u_t^* \tag{3}$$

and

$$\begin{aligned} \tilde{x}_t &= x_t - x_t^*, \\ \tilde{u}_t &= u_t - u_t^*, \\ \tilde{\alpha}_t &= \alpha_t - \alpha_t^*, \end{aligned} \tag{4}$$

then the cost function (2) reduces to 2.2 with $\tilde{x}$, $\tilde{u}$ replacing x, u, and the plant equation (1) reduces to

$$\tilde{x}_{t+1} = A_t \tilde{x}_t + B_t \tilde{u}_t + \tilde{\alpha}_{t+1}. \tag{5}$$

In the new variables $\tilde{x}, \tilde{u}$ the only deviation from centredness is the presence of a disturbing term $\tilde{\alpha}_{t+1}$ in the plant equation; this term combines actual disturbance and those aspects of tracking which are not natural to the dynamics of the plant equation. We shall henceforth drop the tildes and imagine the normalization $x_t^* \equiv 0$, $u_t^* \equiv 0$ achieved.

Theorem 4.1. Optimal uncentred quadratic regulation. *Assume the hypotheses of Theorem* 2.1, *except that the plant equation is generalized to the disturbed form* (1). *Then:*

(*a*) *$F(x_t, t)$ is a quadratic function of the state variable*

$$F(x_t, t) = x_t' P_t x_t + 2\sigma_t' x_t + d_t \tag{6}$$

where P_t is the matrix determined by the Riccati recursion (2.4). *The vector σ_t is determined by the terminal condition and recursion*

$$\begin{aligned} \sigma_T &= 0, \\ \sigma_t &= \Gamma_t'(P_{t+1}\alpha_{t+1} + \sigma_{t+1}) \qquad (t < T) \end{aligned} \tag{7}$$

where Γ_t is the gain matrix defined in (2.13).

(*b*) *The optimal choice of control is*

$$u_t = K_t x_t - D_t^{-1} B_t'(P_{t+1}\alpha_{t+1} + \sigma_{t+1}) \tag{8}$$

where K_t is the matrix previously defined in (2.6) *and*

$$D_t = Q_t + B_t' P_{t+1} B_t \tag{9}$$

Proof. In close analogue to that of Theorem 2.1. Under the inductive hypothesis we have, corresponding to (2.7),

$$F(x_t, t) = \min_u [x'Rx + 2u'Sx + u'Qu + (Ax + Bu + \alpha_{t+1})'P_{t+1}(Ax + Bu + \alpha_{t+1}) + 2\sigma'_{t+1}(Ax + Bu + \alpha_{t+1}) + d_{t+1}]_t. \quad (10)$$

Set $\eta = (P\alpha + \sigma)_{t+1}$. The stationarity condition (2.9) determining u now becomes

$$\Pi_{ux}x_t + \Pi_{uu}u_t + B_t'\eta = 0 \quad (11)$$

whence (8) follows. The minimal value of the bracketed expression in (10) is

$$F(x_t, t) = x_t'\Pi_{xx}x_t + 2\eta' A_t x_t + (\alpha' P\alpha + 2\sigma'\alpha + d)_{t+1} - (x_t'\Pi_{xu} + \eta' B_t)\Pi_{uu}^{-1}(\Pi_{ux}x_t + B_t'\eta).$$

Identification of the quadratic terms in x_t shows that P_t is determined as in Theorem 2.1. Identification of the linear terms shows that

$$\sigma_t = (A_t' - \Pi_{xu}\Pi_{uu}^{-1}B_t')\eta$$

whence (7) follows. ■

Solving the backwards recursion (7) one obtains an explicit form of the optimal control law (8). For explicitness we also revert to the unnormalized formulation of (1), (2).

Corollary 4.1. Combined feedback/feedforward form of the optimal control. *Assume the hypotheses of Theorem 2.1, except that the plant equation and cost function are generalized to the uncentred forms* (1), (2). *Then the optimal control and the optimally controlled plant equation take the respective forms*

$$u_t - u_t^* = K_t(x_t - x_t^*) - D_t^{-1}B_t' \sum_{j=0}^{T-t-1} \Gamma'_{t+1}\Gamma'_{t+2}\cdots\Gamma'_{t+j}P_{t+j+1}(\alpha_{t+j+1} - \alpha^*_{t+j+1}), \quad (12)$$

$$x_{t+1} - x^*_{t+1} = \Gamma_t(x_t - x_t^*) + (\alpha_{t+1} - \alpha^*_{t+1}) - B_t D_t^{-1}B_t' \sum_{j=0}^{T-t-1} \Gamma'_{t+1}\Gamma'_{t+2}\cdots\Gamma'_{t+j}P_{t+j+1}(\alpha_{t+j+1} - \alpha^*_{t+j+1}) \quad (13)$$

where $\{\alpha_t^*\}$ *is defined by* (3).

The term $K_t(x_t - x_t^*)$ in (12) represents feedback, whereas the sum represents a feedforward term: an attempt to anticipate future disturbances and tracking demands intelligently. Since one expects that the gain matrices Γ_t will be stability matrices if the horizon is long enough, then one can argue that the coefficient of α_{t+j+1} in (12) decreases exponentially with increasing j. That is, there is an automatic discounting of future disturbances, allowing for the fact that the further a disturbance lies in the future, the more time there remains to deal with it.

This discounting is at a constant matrix rate in the infinite-horizon limit of the time-homogeneous case (see Section 6).

5. INFINITE-HORIZON LIMITS FOR THE TIME-HOMOGENEOUS CASE

We consider the time-homogeneous centred problem, and are interested in proving, under appropriate conditions, that the matrix $P(s)$ generated by the Riccati equation converges to a finite limit P as $s \to \infty$. Such a result would imply that $K(s)$ converged to a limit K, so establishing optimality of a stationary control rule in the infinite-horizon case. Since the evaluation of $P(s)$ is the same for uncentred as for centred cases, these assertions will carry over to the uncentred case (see the comments at the end of the section).

We shall assume for simplicity that the matrix S has been normalized to zero. If M and N are symmetric matrices of the same order we shall say that $M \geq N$ if $M - N$ is non-negative definite and $M > N$ if $M - N$ is positive definite.

Recall that we have established that in the centred case with $R \geq 0$, $Q > 0$,

$$F_s(x) = x'P(s)x$$

and that the recursion

$$F_s = \mathscr{L} F_{s-1}$$

implies the matrix recursion (2.14).

Lemma 5.1. *If $R \geq 0, Q \geq 0$ and $P(0) = 0$ then $\{P(s)\}$ is monotone non-decreasing.*

Proof. We have $F_1 = x'Rx \geq 0 = F_0$. Thus, by Exercise 2.4.2, $\{F_s\}$ is non-decreasing and so $\{x'P(s)x\}$ is non-decreasing for fixed x. This implies that $\{P(s)\}$ is non-decreasing in the matrix-definiteness sense. ■

Lemma 5.2. *Suppose, in addition to the hypotheses of Lemma 5.1, that the system $[A, B, \cdot]$ is controllable or stabilizable. Then $\{P(s)\}$ is bounded above and has a finite non-negative limit P.*

Proof. To demonstrate boundedness, one must demonstrate that a policy can be found which incurs a finite infinite-horizon cost for any prescribed value x of the initial state x_0. In the controllable case one achieves this by choosing a control u bringing x to zero in a finite number of steps, and by holding u (and so x) zero thereafter. In the stabilizable case one chooses $u_t = Kx_t$ where K is such that $\Gamma = A + BK$ is a stability matrix. Then $x_t = \Gamma^t x$, $u_t = K\Gamma^t x$ and one incurs a bounded total cost

$$C = x'\left[\sum_0^\infty (\Gamma')^t (R + K'QK)\Gamma^t\right]x.$$

The hypotheses thus ensure that $x'P(s)x$ is non-decreasing and bounded above, and so has a finite limit for any prescribed x. Choosing $x = e_j$ (where e_j has zero elements except for a unit in the jth place) one deduces that $p_{jj}(s)$ has a limit p_{jj}. Choosing $x = e_j + e_k$ one deduces that $p_{jk}(s)$ has a limit p_{jk}. The matrix P is thus defined; its construction implies that $x'P(s)x \to x'Px$ for any x and P is finite and non-negative definite. ■

We have now to show that P is indeed the limit of $P(s)$ for any $P(0) \geq 0$. The proof reveals more.

Theorem 5.1. *Suppose that (a) $R > 0$, (b) $Q > 0$, (c) the system $[A, B, \cdot]$ is either controllable or stabilizable. Then: (i) The equilibrium Riccati equation*

$$P = \phi(P) \tag{1}$$

has a unique non-negative definite solution P.
(ii) $P(s) \to P$ for any finite, non-negative definite $P(0)$.
(iii) The limit gain matrix Γ is a stability matrix.

Proof. Define P as $\lim_s \phi^s(0)$; we know by Lemma 5.2 that this exists, is finite, non-negative definite, and solves (1). This last equation can be written

$$P = R + K'QK + \Gamma'P\Gamma \tag{2}$$

where K and Γ are the control and gain matrices corresponding to P (see (2.11)).

Consider the form $V(x) = x'Px$ and a sequence $x_t = \Gamma^t x_0$. Then

$$V(x_{t+1}) - V(x_t) = -x_t'(R + K'QK)x_t \leq 0$$

so that $V(x_t)$ decreases to a limit—finite, because $V(x) \geq 0$. Thus

$$x_t'(R + K'QK)x_t \to 0 \tag{3}$$

which implies that $x_t \to 0$, because $R + K'QK$ is positive definite. Since x_0 is arbitrary, this implies that $\Gamma^t \to 0$, establishing (iii).

We can then deduce from (2) the convergent series expression

$$P = \sum_0^\infty (\Gamma')^j (R + K'QK)\Gamma^j$$

for P.

Note now that, for arbitrary finite non-negative $P(0)$,

$$P(s) = \phi^s(P(0)) \geq \phi^s(0) \to P. \tag{4}$$

Comparing the minimal s-horizon cost with that incurred by using the stationary rule $u_t = Kx_t$ we deduce the reverse inequality

$$P(s) \leq \sum_0^{s-1} (\Gamma')^j (R + K'QK)\Gamma^j + (\Gamma')^s P(0)\Gamma^s \to P. \tag{5}$$

Inequalities (4), (5) establish (ii). Assertion (i) then follows because, if $\hat{P}$ is another finite non-negative definite solution of (1), then

$$\hat{P} = \phi(\hat{P}) = \phi^s(\hat{P}) \to P. \blacksquare$$

The theorem still holds under a weakening of assumption (a). The assumption that $R > 0$ implies that any deviation of state from zero is penalized immediately. All that we really require is some condition assuring that such a deviation is penalized ultimately.

Theorem 5.2. *Theorem* 5.1 *still holds if condition* (*a*) *is weakened to*: (*a'*) $R = L'L$ *where the system* $[A, \cdot, L]$ *is either observable or detectable.*

Proof. Relation (3) now becomes

$$(Lx_t)'(Lx_t) + (Kx_t)'Q'(Kx_t) \to 0$$

which implies that $Kx_t \to 0, Lx_t \to 0$. These convergences, with the relation $x_{t+1} = (A + BK)x_t$, imply that x_t ultimately enters a manifold for which

$$Lx_t = 0, \tag{6}$$

$$x_{t+1} = Ax_t. \tag{7}$$

The observability condition implies that these relations can hold only if $x_t \equiv 0$. The detectability condition implies that we can find an H such that $A - HL$ is a stability matrix, and, since (6), (7) imply that

$$x_{t+1} = (A - HL)x_t,$$

then $x_t \to 0$. So under either assumption $x_t \to 0$. This fact established, the proof continues as in Theorem 5.1. ■

Exercises

1. Note a consequence of (c) and (iii) of Theorem 5.1: controllability implies stabilizability.
2. Note a consequence of the optimality of the linear Markov control $u_t = Kx_t$: if $[A, B, \cdot]$ cannot be stabilized by such a control, then it cannot be stabilized by any other (e.g. non-linear or non-Markov) control.
3. Show that if $P(0)$ satisfies

   $$P(0) - A'P(0)A \geq R$$

   then $P(s)$ is non-increasing.
4. Suppose that $x_1 \neq 0$, $x_1' Rx_1 = 0$, so that x_1 is a state deviation from $x = 0$ carrying no immediate penalty. Suppose it to be uncorrected, so that $u_t = 0$, $x_t = A^{t-1}x_1$ for $t \geq 1$. Show that the condition of Theorem 5.2 implies that $x_t'Rx_t > 0$ for some value of t not greater than $n + 1$. That is, the deviation is ultimately penalized.

6. LONG-TERM TRACKING AND BUFFERING; OFFSET

Revert now to the uncentred formulation of Section 4. Suppose the hypotheses of Theorems 5.1 or 5.2 are satisfied, so that $P(s)$ has an infinite-horizon limit P and the associated limiting gain matrix Γ is stable. One expects then that the optimal control (4.12) will converge in the limit of an infinite horizon to the form

$$u_t - u_t^* = K(x_t - x_t^*) - D^{-1}B' \sum_{j=0}^{\infty} (\Gamma')^j P(\alpha_{t+j+1} - \alpha_{t+j+1}^*), \tag{1}$$

time-invariant in that the dependence of $u - u^*$ upon $x - x^*$ and $\alpha - \alpha^*$ is translation-invariant. The limit form (1) is presumably valid if the sum is absolutely convergent. This will be so if $\alpha_t - \alpha_t^*$ increases not faster than θ^{-t} for increasing t, where θ is greater in modulus than the largest eigenvalue of Γ.

An infinite-horizon formulation implies an interest in the success of regulation or tracking in the long term. One wishes the deviations $x_t - x_t^*$ and $u_t - u_t^*$ to be small. They may not be so initially, because of arbitrariness in the initial conditions, but they should become so ultimately if control is to be regarded as successful. However, it will in general not be possible to bring both these quantities to zero, because a deviation of u_t from u_t^* may be needed to prevent x_t deviating seriously from x_t^*. If $x_t - x_t^*$ and $u_t - u_t^*$ have limit values D_x and D_u as $t \to \infty$, then these values are termed the *offset* in state and control respectively. One would like the offsets to be zero, but, as indicated, this is not in general possible.

Theorem 6.1. *Suppose that $Q, R > 0$ and that $\alpha_t - \alpha_t^*$ has a limit D_α as $t \to \infty$. Then the limit offsets D_x, D_u exist and are given by*

$$D_x = R^{-1}(I - A')\lambda, \tag{2}$$

$$D_u = -Q^{-1}B'\lambda, \tag{3}$$

where

$$\lambda = [(I - A)R^{-1}(I - A') + BQ^{-1}B']^{-1}D_\alpha. \tag{4}$$

Proof. With the control (1) the state equation becomes

$$\begin{aligned} x_{t+1} - x_{t+1}^* &= \Gamma(x_t - x_t^*) + \alpha_{t+1} - \alpha_{t+1}^* \\ &\quad - BD^{-1}B' \sum_{j=0}^{\infty} (\Gamma')^j P(\alpha_{t+j+1} - \alpha_{t+j+1}^*). \end{aligned} \tag{5}$$

If $\alpha_t - \alpha_t^*$ has a limit as $t \to \infty$ then so will $x_t^* - x_t$, generated by (9). So then also will $u_t - u_t^*$, generated by (5).

These limit values will satisfy

$$D_x = AD_x + BD_u + D_\alpha \tag{6}$$

and they will give rise to a cost per unit time of

$$\gamma = D_x{}'RD_x + D_u{}'QD_u. \tag{7}$$

It is this cost which predominates over a long horizon, and D_x, D_u must have the values minimizing (7) subject to (6). Solving this constrained optimization problem by Lagrangian methods, we deduce the assertions of the theorem. ■

The conditions $Q, R > 0$ are imposed so that the inverses occurring in expressions (2) to (4) are proper. The conditions imply that deviations of $x_t - x_t^*$ and $u_t - u_t^*$ from zero are penalized immediately. Expressions (2) to (4) will still be finite if these conditions are relaxed, but Q, R, A and B are such that these deviations are penalized ultimately. However, if there are some directions of deviation that can go unpenalized, then offset in these directions may well be infinite.

One sufficient condition for zero offset is that $D_\alpha = 0$. That is, that the equation

$$x_{t+1}^* = Ax_t^* + Bu_t^* + \alpha_{t+1} \tag{8}$$

hold, at least in the limit of large t. This is the case when the injected disturbance and the ideal paths are consistent with the dynamics of the process, at least asymptotically.

One would also achieve zero offset in state if $A = I$, i.e. if the dynamics were such that the state variable always held its value in the absence of forcing.

The offsets can also be viewed as average values rather than as limiting values. Roughly speaking, if $\alpha_t - \alpha_t^*$ varies about D_α mildly enough, then $x_t - x_t^*$ and $u_t - u_t^*$ vary about D_x and D_u respectively.

Exercises and comments

1. Consider the scalar case, $m = n = 1$. Show that if $A \neq 0$ then relations (1), (5) reduce to

$$u_t - u_t^* = K\left[x_t - x_t^* + A^{-1}\sum_0^\infty \Gamma^j(\alpha_{t+j+1} - \alpha_{t+j+1}^*)\right],$$

$$x_{t+1} - x_{t+1}^* = \Gamma(x_t - x_t^*) + (\alpha_{t+1} - \alpha_{t+1}^*) + (\Gamma A^{-1} - 1)\sum_0^\infty \Gamma^j(\alpha_{t+j+1} - \alpha_{t+j+1}^*),$$

where Γ is the smaller root of (3.3).

2. *The quadratic buffer; dam regulation.* Consider the scalar case with the plant equation

$$x_{t+1} = x_t - u_t + \alpha_{t+1}$$

and cost function $\sum[R(x_t - x^*)^2 + Q(u_t - u_t^*)^2]$. Show that the optimal infinite-horizon control is

$$u_t = (1-\Gamma)\left[x_t - x^* + \sum_{j=0}^{\infty} \Gamma^j \alpha_{t+j+1}\right]$$

where Γ is the smaller root of $Q(1-\Gamma)^2 = R\Gamma$. Why is this rule independent of u^*?

This could be viewed as a simple model of regulation of a dam, in which x represents the amount of water stored, u the amount drawn off and α inflow. One will have a constraint $0 \le x \le C$, where C is the capacity of the dam, which cannot be explicitly included in an LQ model. The cost component $R(x - x^*)^2$ represents an attempt to enforce this constraint approximately, R and x^* being chosen so that excursions of x outside the interval $(0, C)$ are rare. The cost component $Q(u - u^*)^2$ represents a penalty for deviation of draw-off from a target level u^*. All that this can in fact enforce is smoothness in $\{u_t\}$ rather than the absolute level of u_t. The facts that $A = 1$ and water level x must be held finite imply that the long-term average of draw-off u must equal the long-term average of inflow α. Hence the futility of prescribing a target level u^*.

The draw-off u can be regarded as a smoothed version of the inflow α: the role of the dam is precisely to buffer inflow variation.

3. *Production/inventory scheduling*. Dam regulation constitutes a storage problem with prescribed inflow and controlled outflow. For inventory problems one has rather prescribed outflow from stock (demand α) and controlled inflow (order or production u). The plant equation in a simple scalar version is correspondingly

$$x_{t+1} = x_t + u_t - \alpha_{t+1}$$

where x represents the stock held. If the cost function of Exercise 2 is used then $R(x - x^*)^2$ represents costs of excessive positive or negative stock, and $Q(u - u^*)^2$ the cost of varying the order or production rate from a value u^*. Again, the effect of this latter term is to even out production rather than to affect its average level. These cost components are crude quadratic representations of costs and constraints for the more realistic versions of Section 3.2 and Chapter 13.

Deduce the optimal infinite-horizon control rule if the cost function is modified to $\sum[R(x_t - x^*)^2 + Q(u_t - u_{t-1})^2]$. Penalization of variations in the production rate is now explicit.

4. Consider the simultaneous regulation of a number of dams in series; the outflow of one dam is then the inflow of the next.
5. Suppose that $\alpha_t \equiv 0$. Show that the time-invariant relation (5) can be written

$$(I - U^{-1}\Gamma')(B'D^{-1}B)^{-1}(I - U\Gamma)x_t = Rx_t^*$$

where U is the backwards time-translation operator; i.e.

$$Ux_t = x_{t-1}.$$

7. AN ALTERNATIVE FORM OF THE RICCATI RECURSION

The Riccati recursion (2.4) has an alternative form which for a number of purposes is more natural (see Exercises 2 and 3, Section 8 and Chapter 19). The two forms coalesce in continuous time, but transformation from one to the other is non-trivial in discrete time.

We shall assume the formulation of Section 2, but for simplicity shall assume $S_t \equiv 0$. This is achievable by a normalization of the problem (see Exercise 2.2).

Theorem 7.1. *Suppose* $S_t = 0$ *and* $P_t > 0$. *Then the Riccati recursion* (2.4) *can be written*

$$P_t = R_t + A_t'(J_t + P_{t+1}^{-1})^{-1}A_t \tag{1}$$

where

$$J_t = B_t Q_t^{-1} B_t'. \tag{2}$$

The corresponding analogues of formulae (2.6) *and* (2.13) *are*

$$K_t = -Q_t^{-1}B_t'(J_t + P_{t+1}^{-1})^{-1}A_t, \tag{3}$$

$$\Gamma_t = P_{t+1}^{-1}(J_t + P_{t+1}^{-1})^{-1}A_t. \tag{4}$$

Proof. In order to establish (1) we must establish that, if $\phi(P)$ is defined by

$$x'\phi(P)x = \min_u [x'Rx + u'Qu + (Ax + Bu)'P(Ax + Bu)], \tag{5}$$

then

$$\phi(P) = R + A'(J + P^{-1})^{-1}A \tag{6}$$

with $J = BQ^{-1}B'$. We appeal to the fact that, if $P > 0$, then

$$x'Px = \max_y [2y'x - y'P^{-1}y]$$

so that

$$x'\phi(P)x = \min_u \max_y [x'Rx + u'Qu + 2y'(Ax + Bu) - y'P^{-1}y]. \tag{7}$$

The order of the extremizing operations in (7) is immaterial, since the expression in brackets has a saddle-point in u, y. This saddle-point is located by the stationarity conditions

$$Qu + B'y = 0, \tag{8}$$

$$Ax + Bu - P^{-1}y = 0. \tag{9}$$

Solving for u in terms of y from (8) and then y in terms of x from (9) one obtains

$$u = -Q^{-1}B'y = -Q^{-1}B'(J + P^{-1})^{-1}Ax \tag{10}$$

which implies (3). Inserting these solutions for u and y into (7) we verify (6) and so (1). Relation (4) follows from (3) and (2.13). ■

Exercises and comments

1. *The Kailath–Chandrasekhar algorithm.* Show from formulae (1), (4) that, for any $t_1, t_2 \le T$,

$$P_{t_1-1} - P_{t_2-1} = \Gamma'_{t_1}(P_{t_1} - P_{t_2})\Gamma_{t_2} \tag{11}$$

The identity is most used in the case $t_2 = t_1 + 1$, when it supplies a relation

$$P_t - P_{t+1} = \Gamma'_{t+1}(P_{t+1} - P_{t+2})\Gamma_{t+2} \tag{12}$$

which, in conjunction with (4), provides a joint recursion for P_t and Γ_t. The advantage of (12) over the Riccati recursion (1) itself is that it exhibits the definite character of the increment in P_t in the case when $\{P_t\}$ is monotonic. This point is even clearer in the continous time case (see Exercise 38.1.4).

3. Consider the time-homogeneous case and suppose that the limit $P = \lim P(s)$ exists. Show from (11) that

$$P(s) - P = \Gamma(s-1)'\Gamma(s-2)' \cdots \Gamma(0)'(P(0) - P)\Gamma^s$$

which exhibits something like exponential convergence of $P(s)$ to P (see Exercise 4).

4. Under the assumptions of Exercise 3, show that:
(a) If $P(0) \ge P$ then

$$0 \le P(s) - P \le (\Gamma')^s(P(0) - P)\Gamma^s.$$

(b) If $0 \le P(0) \le P$, then there exists a finite integer r and positive scalar β such that $P(r) \ge \beta P$, and hence that

$$0 \le P - P(rs) \le (1 - \beta)^s P$$

(see Exercise 2.4.4).
(c) If $P(0) \ge 0$ then $P(s) \to P$ exponentially fast.

8. THE CASE OF PURELY TERMINAL STATE-COSTING

Suppose that R_t and S_t are zero for $t < T$, implying that it is only at termination that the value of state is material. We know from Section 3.5 that terminal miss-distance for a free trajectory is sufficient for optimal control. Appeal to (7.1) yields further simplification in the LQ case. In fact, if A_t has an inverse then (7.1) implies a linear recursion for P^{-1}:

$$P_t^{-1} = (A_t)^{-1}(J_t + P_{t+1}^{-1})(A_t')^{-1} \tag{1}$$

with solution

$$P_t = \left[\sum_{j=t}^{T-1} \psi_{tj} J_j \psi_{tj}' + \psi_{t,T-1} P_T \psi_{t,T-1}' \right]^{-1} \tag{2}$$

where

$$\psi_{tj} = \begin{cases} I & (j = t-1), \\ (A_j A_{j-1} \cdots A_t)^{-1} & (j \geq t). \end{cases} \tag{3}$$

To assume the matrices A_t non-singular is reasonable if time-increments are small (in relation to the dynamics of the process), so that $I - A_t$ is also small. From (7.1), (7.3) we deduce the optimal control

$$u_t = -Q_t^{-1} B_t (A_t')^{-1} P_t x. \tag{4}$$

As an example, consider the two-dimensional time-homogeneous situation with the plant equation

$$x_{1,t+1} = x_{1t} + x_{2t}, \tag{5}$$

$$x_{2,t+2} = x_{2t} + bu_t, \tag{6}$$

and loss function

$$C = \left[\sum_{1}^{T-1} Qu_t^2 + (c_1 x_{1T}^2 + c_2 x_{2T}^2) \right]. \tag{7}$$

Here x_{1t} is to be regarded as the coordinate of an object on the real line and (5) defines x_{2t} as the rate of change of x_{1t}, i.e. as a discrete-time 'velocity'. Equation (6) gives the rate of change of velocity, i.e. the 'acceleration', as proportional to a control force, u_t. Criterion (7) effectively requires that terminal coordinate and velocity should both be small, but that a cost is also attached to the integrated control effect $\sum_1^{T-1} u_t^2$. The quantities u_t, b and Q are all scalar.

Relations (5), (6) imply that

$$A = \begin{bmatrix} 1 & 1 \\ \cdot & 1 \end{bmatrix}, \qquad B = \begin{bmatrix} \cdot \\ b \end{bmatrix}$$

so that

$$A^j = \begin{bmatrix} 1 & j \\ \cdot & 1 \end{bmatrix}$$

and, by (2),

$$P(s)^{-1} = \frac{b^2}{Q}\sum_1^s \begin{bmatrix} k^2 & -k \\ -k & 1 \end{bmatrix} + \begin{bmatrix} c_1 + c_2 s^2 & -c_2 s \\ -c_2 s & c_2 \end{bmatrix}$$

$$= \frac{b^2}{Q}\begin{bmatrix} \frac{s(s+1)(2s+1)}{6} & -\frac{s(s+1)}{2} \\ -\frac{s(s+1)}{2} & s \end{bmatrix} + \begin{bmatrix} c_1 + c_2 s^2 & -c_2 s \\ -c_2 s & c_2 \end{bmatrix}. \tag{8}$$

Relation (4) with this evaluation yields the optimal control.

Consider the special case $c_2 = 0$, when only the terminal coordinate is costed. Then one can derive the optimal control

$$u_t = -\frac{b(s-1)}{Q(s)}(x_{1t} + sx_{2t})$$

where $s = T - t$ and

$$Q(s) = \frac{Q}{c_1} + \frac{b^2}{6}s(s-1)(2s-1).$$

This follows either by appeal to (4), (8), or by direct appeal to the fact that $F(x_t, t)$ is a quadratic function of the predicted terminal miss-distance $x_{1t} + sx_{2t}$ (see Exercise 1).

For a given value of this predicted miss-distance, u_t first increases with increasing s (as s) and then decreases (as s^{-2}). The reasons why u becomes small for s small and large are, roughly, that if s is small then there is too little time for moderate action to achieve much effect, while if s is large then a sustained small correcting control is optimal.

Exercise

1. Show that for this last example

$$F(x_t, t) = \frac{Q(x_{1t} + sx_{2t})^2}{Q(s)}.$$

9. THE SQUARE-ROOT ALGORITHM

As an algorithm for the calculation of P_t (or $P(s)$) the Riccati recursion (2.4) has the disadvantage that it loses accuracy rather easily, and may even produce a matrix which is not non-negative definite. An alternative algorithm much superior in this respect is the square-root algorithm, first introduced for the updating of estimates of state (see, for example, Morf and Kailath, 1975, or Silverman, 1976).

The algorithm rests on two basic reductions. Firstly, if P is non-negative definite then it can be factorized as $P = M'M$, so that

$$x'Px = |Mx|^2 \tag{1}$$

(see Stewart, 1973). If we write

$$F(x,t) = x'P_t x = |M_t x|^2 \tag{2}$$

then the square-root algorithm gives a recursion for M_t rather than for P_t—hence its name.

The second reduction is the *Householder transformation*: for any matrix N one can find an orthogonal matrix H such that

$$HN = \begin{bmatrix} N_1 \\ \cdot \end{bmatrix}$$

where N_1 is of full row-rank; i.e. its rows are linearly independent (see Stewart, 1973, Bierman, 1977, or Seber, 1977). It is natural that the rows of N can be represented linearly in terms of a number of linearly independent row vectors; the point of the Householder transformation is that there exists such a representation for which the matrix relating the two sets of vectors is orthogonal.

Analogously to the factorization (1), matrices C_t, D_t exist such that

$$(x'Rx + 2u'Sx + u'Qu)_t = |(Cx + Du)_t|^2. \tag{3}$$

The optimal value of u_t is that minimizing

$$V_t = |(Cx + Du)_t|^2 + |M_{t+1}(Ax + Bu)_t|^2, \tag{4}$$

and the minimized value of the form is $|M_t x_t|^2$.

Theorem 9.1. (The Square-root algorithm). *One can find an orthogonal matrix H_t such that*

$$H_t \begin{bmatrix} C_t & D_t \\ M_{t+1}A_t & M_{t+1}B_t \end{bmatrix} = \begin{bmatrix} \cdot & \cdot \\ \phi_t & \cdot \\ \psi_t & \theta_t \end{bmatrix} \tag{5}$$

where the matrix in the right-hand member is characterized by the facts that ϕ_t is a matrix of n columns and full row-rank and θ_t is a matrix of m columns and full row-rank.

One may then choose

$$M_t = \phi_t, \tag{6}$$

and any value of u_t which satisfies

$$\psi_t x_t + \theta_t u_t = 0 \tag{7}$$

is optimal. Equation (7) *always has a solution, and has the unique solution*

$$u_t = -\theta_t^{-1}\psi_t x_t \tag{8}$$

if $Q_t > 0$.

Proof. One can find a Householder transformation H_{1t} such that

$$H_{1t}\begin{bmatrix} D_t \\ M_{t+1}B_t \end{bmatrix} = \begin{bmatrix} \cdot \\ \theta_t \end{bmatrix}. \tag{9}$$

Suppose that this has the effect

$$H_{1t}\begin{bmatrix} C_t \\ M_{t+1}A_t \end{bmatrix} = \begin{bmatrix} \chi_t \\ \psi_t \end{bmatrix} \tag{10}$$

where the row-partitioning is the same in (9) and (10). One can then find a Householder transformation H_{2t} such that

$$H_{2t}\chi_t = \begin{bmatrix} \cdot \\ \phi_t \end{bmatrix}$$

Choosing

$$H_t = \begin{bmatrix} H_{2t} & \cdot \\ \cdot & I \end{bmatrix} H_{1t}$$

one achieves the reduction (5).

Now, expression (4) can be written

$$V_t = \left|\begin{bmatrix} C_t & D_t \\ M_{t+1}A_t & M_{t+1}B_t \end{bmatrix}\begin{bmatrix} x_t \\ u_t \end{bmatrix}\right|^2$$

and, by the orthogonality of the transformation (5), this can be written

$$\begin{aligned} V_t &= \left|\begin{bmatrix} \cdot & \cdot \\ \phi_t & \cdot \\ \psi_t & \theta_t \end{bmatrix}\begin{bmatrix} x_t \\ u_t \end{bmatrix}\right|^2 \\ &= |\phi_t x_t|^2 + |\psi_t x_t + \theta_t u_t|^2 \end{aligned} \tag{11}$$

From (11) it is obvious that u_t satisfying (7) minimizes V_t and also that the minimized value is $|\phi_t x_t|^2$, so that the identification (6) is consistent.

Equation (7) always has a solution, because θ_t is of full row-rank. If $Q_t > 0$ then D_t is of full column-rank, which implies that θ_t is square and non-singular. ■

Of course, (6) is not the unique choice of M_t; one could equally well choose $M_t = H\phi_t$, where H is any matrix (not necessarily square) satisfying $H'H = I$. Indeed, the second stage of transformation after (9), (10) might have been by-passed;

one could have chosen $M_t = \chi_t$. However, the point of this second stage of transformation is to achieve an M_t which has full row-rank and so has at most n rows. Were this step omitted, then the number of rows in M would increase at every stage.

Note that, by an extension of this reduction argument, one could have chosen H_t in (5) so that θ_t (and also ϕ_t, if desired) was lower triangular (more exactly, so that a square matrix $\begin{pmatrix} \cdot \\ \theta_t \end{pmatrix}$ is lower triangular). Solution of the system (7) is very simple in this case.

The square-root algorithm generalizes easily to the uncentred case. Assume the disturbed plant equation (4.1) and an instantaneous cost function

$$g(x_t, u_t, t) = |C_t x_t + D_t u_t + \beta_t|^2.$$

One can then write

$$F(x_t, t) = |M_t x_t + \gamma_t|^2 + \delta_t,$$

and the aim is to find recursions for M, γ and δ and an expression for the optimal control.

Theorem 9.2. The square-root algorithm for the uncentred case. *Suppose that the matrix H_t achieving the reduction* (5) *has the effect*

$$H_t \begin{bmatrix} \beta_t \\ M_{t+1}\alpha_{t+1} + \gamma_{t+1} \end{bmatrix} = \begin{bmatrix} \mu_{1t} \\ \mu_{2t} \\ \mu_{3t} \end{bmatrix} \tag{12}$$

where the row-partitioning is the same in (5) *and* (12). *Then we can make the identifications*

$$M_t = \phi_t,$$

$$\gamma_t = \mu_{2t},$$

$$\delta_t = \delta_{t+1} + |\mu_{1t}|^2,$$

and any u_t satisfying

$$\psi_t x_t + \theta_t u_t + \mu_{3t} = 0$$

is optimal.

Proof. All conclusions follow from the fact that, analogously to (11),

$$V_t = \delta_{t+1} + |\mu_{1t}|^2 + |\phi_t x_t + \mu_{2t}|^2 + |\psi_t x_t + \theta_t u_t + \mu_{3t}|^2.$$

CHAPTER 6

The discrete maximum principle

1. THE DISCRETE MAXIMUM PRINCIPLE

If one is trying to minimize a function $G(x)$ one might in some cases scan $G(x)$ for all values of x. Alternatively, in the case of continuously varying x, one could examine just those values of x which fulfil some necessary condition for local minimality. A necessary condition that G be locally minimal at $\bar{x}$ may be sufficient to ensure that it is in fact globally minimal there if G is, for example, convex.

The method of dynamic programming is like the complete scan, in that it examines all orbits for optimality. Granted, it does this in an economical fashion, by expressing optimal orbits from an arbitrary starting point at time t in terms of those from an arbitrary starting point at time $t+1$. Nevertheless, it is still a complete scan, making no assumption that the criterion function G may vary continuously or differentiably with the orbit.

At least for the case of deterministic processes there is a method akin to the examination of stationary points of a function. This method, the *maximum principle*, provides necessary conditions that an orbit minimize the criterion G locally. These conditions are in fact necessary only if certain structural hypotheses are satisfied, but under stronger hypotheses are also sufficient for the orbit to be globally minimizing. So, suppose that one is operating over a fixed horizon T, so that $Z=(X_T, U_{T-1})$ completely describes the orbit. The criterion function

$$G(Z) = \sum_{t=0}^{T-1} g(x_t, u_t, t) + K(x_T) \tag{1}$$

is to be minimized, subject to the plant equation

$$x_{t+1} = a(x_t, u_t, t) \qquad (0 \le t < T). \tag{2}$$

The minimization may also be subject to other constraints on the orbit, such as prescription of x_0, and any other constraints on x or u, which we shall summarize as $Z \in \mathscr{Z}$.

For the purpose of this chapter we shall suppose that the state variable x takes values in R^n. In the conventional approach optimization is viewed as a matter of choosing u, the values of x being then determined recursively from the plant equation (2). The maximum principle is associated with an approach in which optimization is viewed as a matter of choosing x and u jointly, i.e. of choosing Z in $\mathscr{Z}$, subject to the constraints (2). These constraints are taken into account by the use of Lagrangian multipliers; one minimizes the form

$$G_p = G + \sum_{t=0}^{T-1} p_{t+1}(a(x_t, u_t, t) - x_{t+1})$$

$$= \sum_{t=0}^{T-1} [g(x_t, u_t, t) + p_{t+1}(a(x_t, u_t, t) - x_{t+1})] + K(x_T)$$

freely for Z in $\mathscr{Z}$. Here p_{t+1} is the Lagrangian multiplier associated with constraint (2), a row n-vector. The supposition is that multiplier values can be chosen such that the orbit Z freely minimizing G_p is optimal, and is feasible in that it satisfies (2). The terms $p(a - x)$ added on to G to define G_p must be regarded as penalty terms, which enforce (2) in a free optimization, i.e. which discourage an orbit from being anything but a feasible orbit.

There are two points to be dealt with. The first is the question of whether multipliers p exist such that the orbit freely minimizing G_p is both optimal and feasible. The second concerns the point of the approach: does free minimization of G_p have advantages over constrained minimization of G? We shall give a partial answer to the first point now, reserving a completer treatment until Section 8, and shall then consider the second: application of the maximum principle.

Theorem 1.1. (Validity of the maximum principle). *Define $\mu(b)$ as the minimal value of $G(Z)$ subject to $Z \in \mathscr{Z}$ and the modified plant equation*

$$x_{t+1} = a(x_t, u_t, t) + b_{t+1} \tag{3}$$

where b is the column vector of vectors $b_1, b_2, \ldots, b_T$. Let p be the row vector of vectors $p_1, p_2, \ldots, p_T$.

(a) An orbit freely minimizing G_p which is also feasible exists if and only if $\mu(b) - pb$ is minimal at $b = 0$; such an orbit is optimal.

(b) Suppose that $b = 0$ is an interior point of the set $\mathscr{B}$ of values of b consistent with $Z \in \mathscr{Z}$ and that $\mu(b) - pb$ is minimal and differentiable at this point. Then

$$p = \left[\frac{\partial \mu}{\partial b}\right]_{b=0} \tag{4}$$

Proof. The vector b can be regarded as an artificial forcing term in the plant equation; its magnitude for a given orbit is a measure of the degree and manner of the infeasibility of the orbit. Assertion (a) of the theorem states conditions under which the terms $p(a - x)$ constitute an effective penalty for infeasibility.

Suppose $\bar{Z}$ minimizes G_p and is feasible. Let $\bar{G}$ denote the minimal value of G attained by a feasible orbit. Then $G(\bar{Z}) \geq \bar{G}$. On the other hand, for any feasible Z,

$$G(Z) = G_p(Z) \geq G_p(\bar{Z}) = G(\bar{Z}).$$

Thus $G(\bar{Z}) = \bar{G}$, and we can assert that a value of Z minimizing G_p which is also feasible is optimal. The statement that $\mu(b) - pb$ is minimal at $b = 0$ is certainly equivalent to the statement that G_p is minimal at a feasible orbit, so assertion (a) follows.

Under the hypotheses of (b) we have

$$\begin{aligned}\mu(0) &\leq \mu(b) - pb \\ &= \mu(0) + \left[\left[\frac{\partial \mu}{\partial b}\right]_{b=0} - p\right]b + o(b)\end{aligned}$$

for b in a neighbourhood of the origin. Identification (4) then follows. ■

The condition

$$\mu(b) - \mu(0) \geq pb$$

is often expressed by saying that p is a *sub-gradient* to μ at $b = 0$. We have still to determine conditions which enable us to assert that such a sub-gradient exists. This matter will be discussed more fully in Section 8; let us merely note now that sufficient conditions are that $a(x_t, u_t, t)$ should be linear in x_t and u_t, that $\mathscr{Z}$ should be convex and that G be a convex function of Z. These conditions can be weakened. The conditions of the theorem can also be weakened if we require only that an orbit *locally* minimizing G_p and feasible should be *locally* optimal.

Theorem 1.2. (Consequences of the maximum principle). *Suppose the principle valid, so that for an appropriate p there is an orbit minimizing G_p which is feasible. Then, if $\bar{Z} = \{\bar{x}_t, \bar{u}_t\}$ is such an orbit:*

(a) *$\bar{u}_t$ minimizes $g(\bar{x}_t, u_t, t) + p_{t+1}a(\bar{x}_t, u_t, t)$ freely.*

(b) *If x_t can be varied within a neighbourhood of $\bar{x}_t$ consistent with $Z \in \mathscr{Z}$ then*

$$p_t = \frac{\partial g(x_t, u_t, t)}{\partial x_t} + p_{t+1}\frac{\partial a(x_t, u_t, t)}{\partial x_t} \qquad (1 \leq t < T), \tag{5}$$

$$p_T = \frac{\partial K(x_T)}{\partial x_T}, \tag{6}$$

where these differentials (supposed to exist) are evaluated on the orbit $\bar{Z}$.

Proof. These are immediate consequences of the minimality of G_p at $\bar{Z}$, assertion (a) following from a global variation of u_t and assertion (b) from a local variation of x_t. ■

The term 'maximum principle' derives from assertion (a) for the case of a maximizing (rather than a minimizing) problem. Assertion (a) exhibits the optimization of u at time t as apparently decoupled from the optimization of u at other times, although the coupling is in fact provided by the values of the variables x_t, p_t.

Recursions (2), (5) and assertion (a) provide determining relations for (x_t, p_t, u_t) along an optimal orbit. These are supplemented by an initial condition, prescription of x_0, and the terminal condition (6). If the orbit $\bar{Z}$ meets the boundary of $\mathscr{X}$ in such a way that x_t cannot be varied within a neighbourhood of $\bar{x}_t$, then relations (5), (6) must be modified. If (5) is valid then the determining recursions can be given compact expression.

Theorem 1.3. (Hamiltonian formulation). *Define the Hamiltonian function*

$$H_t = H(x_t, p_{t+1}, u_t, t) = g(x_t, u_t, t) + p_{t+1} a(x_t, u_t, t). \tag{7}$$

The determining relations for x_t, p_t and u_t in the case of a free optimal orbit $\bar{Z}$ can be written

$$x_{t+1} = \frac{\partial H_t}{\partial p_{t+1}}, \tag{8}$$

$$p_t = \frac{\partial H_t}{\partial x_t}, \tag{9}$$

$$H_t = \min_{u_t} H(\bar{x}_t, \bar{p}_{t+1}, u_t, t), \tag{10}$$

where H_t and its derivatives are evaluated on the orbit $\{\bar{x}, \bar{p}, \bar{u}\}$.

Here by a 'free' orbit we mean one in which x_t is not constrained by encounter of the orbit with the boundary of $\mathscr{X}$, for $t > 0$.

2. THE DISCRETE MAXIMUM PRINCIPLE AND DYNAMIC PROGRAMMING

A local form of the maximum principle can be derived immediately, if unrigorously, from the dynamic programming equation. A rigorous and complete treatment requires considerably more work. Nevertheless, this derivation is an excellent one, in that it immediately suggests motivation, interpretation, and the appropriate treatment of problems presented by boundary constraints, etc. It also suggests the appropriate mode of generalization to, for example, time-invariant problems (necessarily with infinite horizon) or stochastic problems (see Sections 16.3 and 16.4). For reasons which will transpire in Chapters 7 and 8 a dynamic programming derivation is much more powerful in continuous time, when also a maximum principle is valid under much weaker conditions.

Theorem 2.1. *Let $\{\bar{x}_t, \bar{u}_t\}$ be an optimal trajectory. Suppose the trajectory free, in that $\bar{x}_t$ belongs to the interior of the permitted set of values of x_t for $0 < t \leq T$ and that differentiability may be assumed whenever required.*

(a) If a conjugate variable is defined by

$$p_t = \left[\frac{\partial F(x_t, t)}{\partial x_t}\right]_{x=\bar{x}} \tag{1}$$

then $\{p_t\}$ satisfies equations (1.5), (1.6).

(b) $H(\bar{x}_t, p_{t+1}, u_t, t)$, as defined in (1.7), is non-decreasing for local variations of u_t from $\bar{u}_t$ and

(c) is globally minimal in u_t at $\bar{u}_t$ if $g(x_t, u_t, t)$ is convex in u_t, $a(x_t, u_t, t)$ linear in u_t and $\mathscr{U}(x_t, t)$ convex.

Proof. Define the quantity

$$J(x_t, u_t, t) = g(x_t, u_t, t) + F(a(x_t, u_t, t), t+1).$$

The optimality equation can then be written

$$\begin{aligned} F(x_t, t) &= \min_{u_t} J(x_t, u_t, t) \\ &= J(x_t, \bar{u}_t, t). \end{aligned} \tag{2}$$

This implies the relations

$$\frac{\partial J}{\partial u_t}\Delta u_t \geq 0, \tag{3}$$

$$\frac{\partial J}{\partial u_t} = \frac{\partial H}{\partial u_t} \tag{4}$$

with p_{t+1} determined by definition (1). Here derivatives are evaluated at $u_t = \bar{u}_t$, and Δu_t is a permitted direction of variation of u_t from $\bar{u}_t$. From (3), (4) we deduce

$$\frac{\partial H}{\partial u_t}\Delta u_t \geq 0$$

and so assertion (b). Under the assumptions of assertion (c) H is convex in u_t, so that assertion (b) can be strengthened to (c).

Differentiating relation (2) we have

$$\frac{\partial F(x_t, t)}{\partial x_t} = \frac{\partial J}{\partial x_t} + \frac{\partial J}{\partial u_t}\frac{\partial \bar{u}_t}{\partial x_t} \tag{5}$$

where the derivatives are again evaluated at $u_t = \bar{u}_t$. But the final term in (5) is zero, because, as x_t varies locally, then $\bar{u}_t$ will vary only in the direction in which J

is stationary. Hence

$$\frac{\partial F(x_t, t)}{\partial x_t} = \frac{\partial J(x_t, \bar{u}_t, t)}{\partial x_t}$$

which indeed implies (1.5). The definition (1) itself implies (1.6). ■

The conclusions of the theorem are somewhat cautious (i.e. local, except for assertion (c)) because it cannot be asserted that $J(\bar{x}_t, u_t, t)$ and $H(\bar{x}_t, p_{t+1}, u_t, t)$ are actually minimal at the same value $\bar{u}_t$ of u_t. In continuous time J and H turn out to be identical, which makes the maximum principle and its relation to the dynamic programming principle more natural.

Under the conditions of (c) it can be stated that the value $\bar{u}_t$ minimizes $g(x_t, u_t, t) + c_t u_t$, where c_t is independent of u_t—actually is equal to $(\partial/\partial u_t)$ $F(a(x_t, u_t, t), t+1)$ evaluated on the trajectory. Under these conditions the quantity c_t is sometimes termed a *stimulating price*. For short-term optimization one would choose u to minimize g; one can induce long-term optimization by modifying this to minimization of $g + cu$. That is, by imposing a unit cost of c on the value of u chosen one can induce a short-term optimizer to in fact take long-term optimal decisions.

3. SOME SPECIAL CASES; CONSUMPTION/INVESTMENT

One special class of cases particularly amenable to the use of the maximum principle specifies the plant equation as linear in x_t and u_t, the instantaneous cost g as independent of x_t and the constraint $Z \in \mathscr{Z}$ as wholly enforceable by the cost functions themselves. The consumption/investment problem of Section 3.1 is an example. To rework this example using the maximum principle is somewhat repetitive, but instructive.

Theorem 3.1. *Suppose the plant equation to be linear*

$$x_{t+1} = A_t x_t + B_t u_t, \tag{1}$$

the instantaneous cost to be equal to $g(u_t, t)$, independent of x_t, and that the only constraint is the specification of x_0. Then the optimal u_t is that minimizing $g(u_t, t) + p_{t+1} B_t u_t$, where

$$p_t = p_T A_{T-1} A_{T-2} \cdots A_t.$$

Proof. The assertions follow from the maximum principle (which is in fact valid, if $K(x_t)$ is convex and $g(u_t, t)$ convex in $u_t (t < T)$) and from the fact that (1.5) reduces in this case to $p_t = p_{t+1} A_t$. ■

For such a case the optimal controls are thus determined completely in terms of the final multiplier p_T. Solving (1) with this control and x_0 prescribed, one then determines p_T from the terminal condition (1.6).

Consider, for example, the consumption/investment problem of Section 3.1. This is a maximization problem with the linear plant equation (3.1.1) and concave reward functions $g(u,t) = \beta^t u^{1-v}$, $K(x) = A_0 \beta^T x^{1-v}$. For this time-homogeneous problem it is simpler to take time-to-go $s = T-t$ as the time argument; let us write $x_t = x[T-t]$, $p_t = p[T-t]$. The reward functions can then be taken as $A_0 x^{1-v}$ for terminal reward and $\beta^{-s} u^{1-v}$ for instantaneous reward with time s to go.

We have then

$$p[s] = \alpha^s p[0]$$

and $u[s]$ has the value maximizing $\beta^{-s} u^{1-v} - \alpha^s p[0] u$, so that

$$u[s] = \left[\frac{1-v}{(\alpha\beta)^s p[0]} \right]^{1/v} = b\phi^s, \tag{2}$$

say. Substituting (2) in the plant equation (3.1.1) we deduce that

$$x_T = x_{T-s}\alpha^s + \frac{\alpha b \phi}{\alpha\phi - 1}[1 - (\alpha\phi)^s], \tag{3}$$

and, from the terminal condition (1.6), that

$$b = \frac{x_T}{A_0^{1/v}}. \tag{4}$$

Eliminating b from (3), (4) we determine x_T in terms of a prescribed x_t at an arbitrary time $t = T-s$ as

$$x_T = \frac{x_t \alpha^s}{1 + (1/A_0^{1/v})[(\gamma^{-s} - 1)/(1-\gamma)]} \tag{5}$$

where

$$\gamma = (\alpha\phi)^{-1}$$

is the quantity defined in (3.1.6). Inserting the evaluation of b from (4), (5) into (2) we deduce the optimal control in its feedback form (3.1.4), (3.1.7).

Two points of this treatment are to be noted as significant generally. Firstly, a *backwards* calculation of the optimal orbit is natural; one tries to choose the terminal value x_T so that the orbit passes through the prescribed initial value x_0 (or x_t). Secondly, at various stages, such as relation (2), one deduces an expression for the optimal control which is simple, but not in the preferred feedback form. To take this particular example, it is not until the unknown b has been determined in terms of initial values (x_t, t) that one has the optimal control in feedback form.

Exercise

1. Carry through the same analysis for the choice of reward functions

$$g(u,t) = \beta^t \log(u - \bar{u}), \qquad K(x) = \beta^T A \log(x - \bar{x})$$

where $\bar{u}, \bar{x}$ are prescribed constants.

4. THE REGULATION OF A DAM WITH A DETERMINISTIC INFLOW

Let x_t be the amount of water in a dam of capacity C at time t, so that

$$0 \leq x_t \leq C. \tag{1}$$

This obeys the recursion

$$x_{t+1} = x_t + v_t - u_t \tag{2}$$

where v_t is the inflow, presumed known, and u_t the draw-off, which is to be chosen to maximize $\sum_0^{T-1} g(u_t)$. The reward function $g(u)$ represents the utility of a draw-off (for power generation, agriculture or urban use); we shall suppose it concave and non-decreasing. We have set $K = 0$, supposing water remaining at time T to have zero utility, although constraints (1) must be maintained.

This problem differs from that of the previous section in the occurrence of the explicit constraints (1) on the state variable. To see the effect of these it is best to return to the original Lagrangian form of the maximum principle. Appeal to the maximum principle will be justified in Section 8. We maximize then the Lagrangian form

$$G_p = \sum_{t=0}^{T-1} [g(u_t) + p_{t+1}(x_t + v_t - u_t - x_{t+1})]$$

with respect to x, u. The variable x_t enters linearly, with coefficient $p_{t+1} - p_t$, so that then

$$\begin{Bmatrix} x_t = 0 \\ 0 \leq x_t \leq C \\ x_t = C \end{Bmatrix} \quad \text{according as } p_{t+1} \begin{Bmatrix} < \\ = \\ > \end{Bmatrix} p_t$$

or

$$p_{t+1} \begin{Bmatrix} \leq \\ = \\ \geq \end{Bmatrix} p_t \text{ according as } \begin{Bmatrix} x_t = 0 \\ 0 < x_t < C \\ x_t = C \end{Bmatrix} \qquad (t = 1, 2, \ldots, T-1). \tag{3}$$

The three cases in (3) will be referred to as the dam's being in an *empty*, *intermediate* or *full* phase respectively. Relation (3) holds also for $t = T$ if we formally set $p_{T+1} = 0$. In the case of indefinite continuation, with a period of length T, $p_T = p_0$ would be the appropriate boundary condition.

Since G_p is also maximal with respect to u_t, then u_t must maximize $g(u) - p_{t+1}u$. Collect now the assertions that can be made:

(a) *The outflow u_t maximises $g(u) - p_{t+1}u$.* Note that if the u maximizing $g(u) - pu$ is denoted by $u(p)$, then the function $u(p)$ is decreasing (see Figure 4.1).

(b) if $0 < x_t < C$ then $p_{t+1} = p_t$, so that $u_t = u_{t-1}$. Thus, *the outflow u_t is constant in an intermediate phase* (although in general having different values in distinct intermediate phases).

(c) If $x_t = 0$ then $p_{t+1} \leq p_t$ so that $u_t \geq u_{t-1}$. Since $u_t \leq v_t$ and $u_{t-1} \geq v_{t-1}$ if neither x_{t+1} nor x_{t-1} is to be negative, it also follows that $v_t \geq v_{t-1}$. That is, *inflow is increasing in an empty phase.*

(d) Correspondingly, *inflow is decreasing in a full phase.* Explicitly, if $x_t = C$, then $p_{t+1} \geq p_t$ and $u_t \leq u_{t-1}, v_t \leq v_{t-1}$.

(e) If *$g(u)$ is strictly increasing, then $p_t > 0$ $(t = 1, 2, \ldots, T)$ and $x_T = 0$.* It follows from (a) that if $p_{t+1} \leq 0$, then $u_t = +\infty$, which cannot happen. Thus $p_T > 0$, but since $p_{T+1} = 0$ it follows from (3) that $x_T = 0$. Essentially, the fact that marginal utility $g'(u)$ is always positive implies that the 'shadow price' for water is always positive and that all water will be used by the end of the period.

The natural question now is whether the above conditions are sufficient to

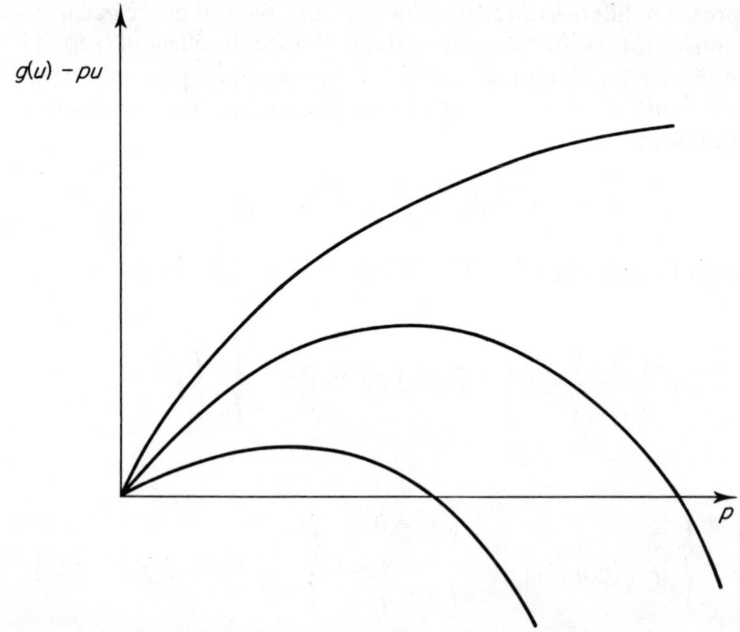

Figure 4.1 Graphs of $g(u) - pu$ for varying p. The curve drops as p increases

determine the optimal policy for the release of water. In fact, they are not. At least one further condition is required, which is simpler to specify in the case of continuous than of discrete time. For this reason, the continuous time version of the problem will now be considered, with $x(t)$ denoting the amount of water in the dam at time t and $u(t)$ and $v(t)$ the outflow and inflow *rates*. We have then, corresponding to (1) and (2),

$$0 \geq x(t) \leq C, \tag{4}$$

$$\dot{x} = v - u, \tag{5}$$

where $\dot{x}$ is the time rate-of-change of x. It is supposed that the object is now to choose $u(t)$ so as to maximize $\int_0^T g(u(t))\,dt$.

Theorem 4.1 *Suppose that $g(u)$ is strictly concave, finite for finite positive u, that $v(t)$ is continuous and that x, v are not both initially zero. Then:*

(*a*) *u is constant in an intermediate phase.*
(*b*) *If $x = 0$ then $u = v$, and v is increasing.*
(*c*) *If $x = C$ then $u = v$, and v is decreasing.*
(*d*) *u is continuous at a change of phase, except possibly at $t = T$.*
(*e*) *If $g(u)$ is strictly increasing, then $x(T) = 0$.*

Proof. With one exception these assertions are direct analogues of results already established in the discrete case, and it will be assumed that they carry over from that case. (To prove the point requires a treatment of constrained maximization in function space.) The exception is the vital assertion (d), which gives the transition rule at a change of phase. Assertion (b) of the discrete case states that $u_t = u_{t-1}$ if the dam is in an intermediate phase at time t, which implies, in the continuous case, that $u(t)$ will be continuous when the dam *enters* an intermediate phase from a full or empty phase. It is now desired to establish continuity at *exit* from an intermediate phase. This continuity condition at exit seems to have no simple analogue in the discrete time case and requires special proof.

Consider an intermediate/empty transition (see Figure 4.2). Suppose that the dam contains amount x at time t_0 and that it becomes empty at time τ. If between t_0 and τ the water is being released at rate u (necessarily constant) then we must have

$$\int_{t_0}^{\tau} (u - v(t))\,dt = x. \tag{6}$$

If u were changed, then so would τ be, and it would follow from (6) that

$$\tau - t_0 + (u - v(\tau))\frac{\partial \tau}{\partial u} = 0.$$

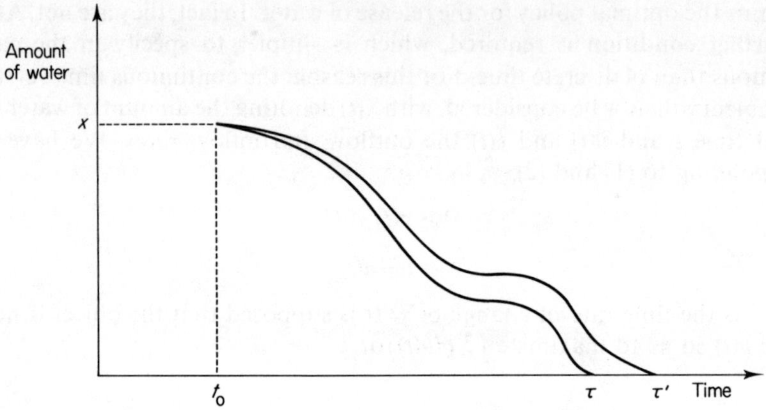

***Figure* 4.2** The variation of passage to an empty phase with variation in release rate

Thus, if u is changed to $u' = u + \delta$, then τ is changed to

$$\tau' = \tau - \Delta = \tau - \frac{(\tau - t_0)\delta}{u - v(\tau)},$$

to first order in δ. The change in utility is

$$\begin{aligned}\int_{t_0}^{\tau'} g(u')\mathrm{d}t - \int_{t_0}^{\tau} g(u)\mathrm{d}t + \int_{\tau'}^{\tau} g(v)\mathrm{d}t &\sim g'(u)(\tau - t_0)\delta + (g(v(\tau)) - g(u))\Delta \\ &\sim (\tau - t_0)\delta\left[g'(u) - \frac{g(u) - g(v(\tau))}{u - v(\tau)}\right].\end{aligned} \quad (7)$$

However, the maximal utility must be stationary with respect to variations in u, so that the expression in square brackets in (7) must be zero. Since g is strictly concave, the equality

$$g'(u) = \frac{g(u) - g(v(\tau))}{u - v(\tau)}$$

is possible only if $u = v(\tau)$. Since $v(\tau)$ is certainly the value adopted by u immediately *after* entry into the empty phase, one sees that u is indeed continuous at the transition. ■

The reason why this argument has no simple analogue in the discrete-time case is that the moment of transition cannot then be varied continuously, as was possible here. The point $t = T$ is excepted in (d), of Theorem 4.1, for the same reason.

These conditions will in most cases determine the policy completely, and in any case narrow it down to a small number of alternatives. Starting from (x, t) a release rate u is chosen which ensures that either (a) the dam becomes empty at time T, without previous change of phase, or (b) $u = v(\tau)$ at the moment τ of the first change of phase. In the event that there are several possible τ, the latest will presumably be chosen, but this is a matter for individual test. The same rules guide the decision of when to leave a full or empty phase (u being continuous on exit), with the additional provisos that rules (b) and (c) of Theorem 4.1 must be observed.

Some particular cases are drawn in Figures 4.3 and 4.4 on the assumption of a

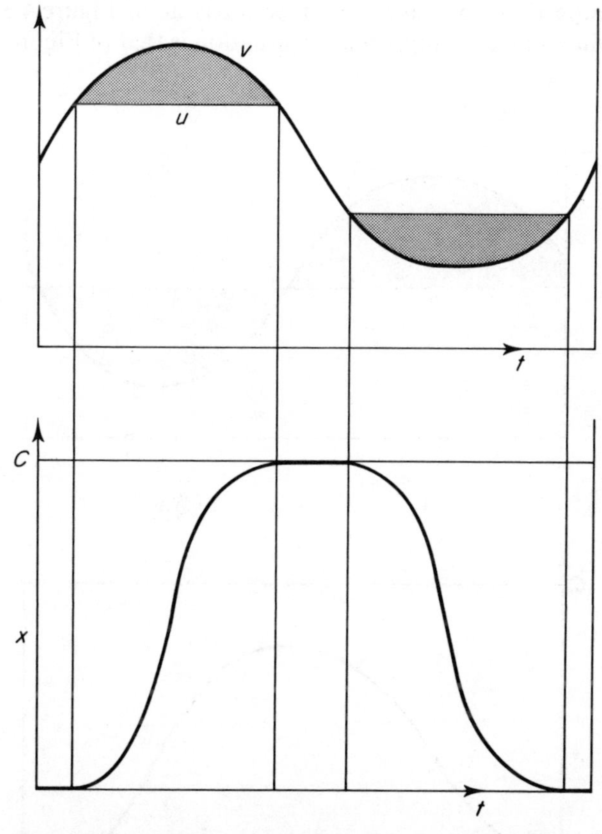

***Figure* 4.3** A diagram of inflow rate (v), release rate (u) and water level (x) for a situation periodic in time. The dam is brought from the empty state to the full state, at constant release rate, during an interval of maximum inflow rate. The reverse occurs during a period of minimum inflow rate. The effect is that the graph of u is derived from that of v by removal of compensating peaks and troughs, each area removed being equal to the dam capacity C

periodic v (i.e. a seasonal pattern) and the assumption that an equilibrium policy has been reached in which u and x are also periodic. The condition of periodicity thus replaces the terminal condition $x(T) = 0$ in this case.

In the case of Figure 4.3 there are one empty, one full and two intermediate phases in each cycle. The transition points are fully determined by continuity of u and by the requirement that x change by $\pm C$ over each intermediate phase (i.e. that the shaded regions in Figure 4.3 each have area C). It is seen that the effect of the dam is to slice peaks and troughs off the graph of $v(t)$, removing compensating areas equal to C.

If C is so large that one cannot remove areas as in Figure 4.3 without the consequent values of u crossing, then the situation is that of Figure 4.4. Here the

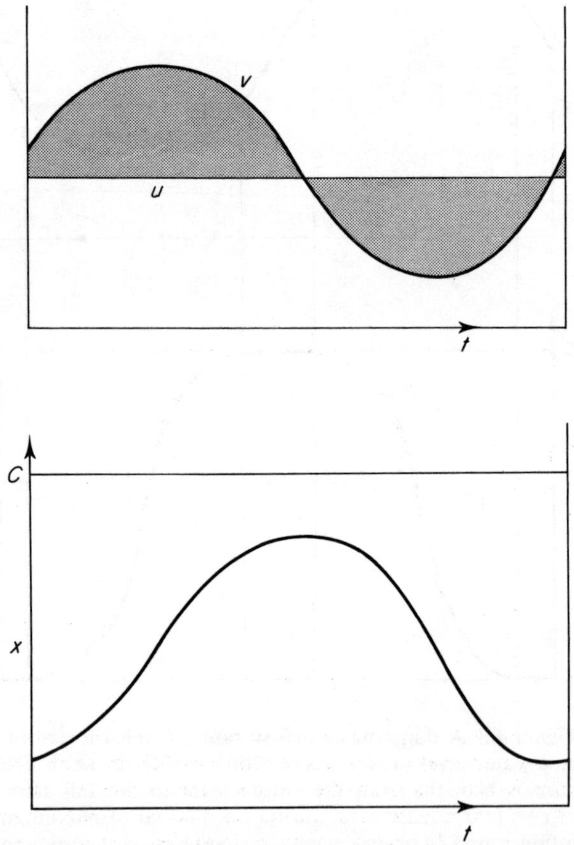

Figure **4.4** As in Figure 4.3, except that the dam is sufficiently large relative to inflow variations that it becomes neither empty nor full throughout the period and a constant release rate can be maintained

dam is so large that it can buffer all v variations. The dam can be held in an intermediate phase, with u constant throughout the season. In this case the shaded areas above and below the u line are equal, but less than C in value.

Suppose that v shows the minor variations of Figure 4.5. Because of rule (c) of Theorem 4.1, it is not optimal to hold $x = C$ and $u = v$ throughout the final local maximum and minimum on the v curve: at some stage the water level is allowed to fall and then to rise again. The moment of transition is determined by

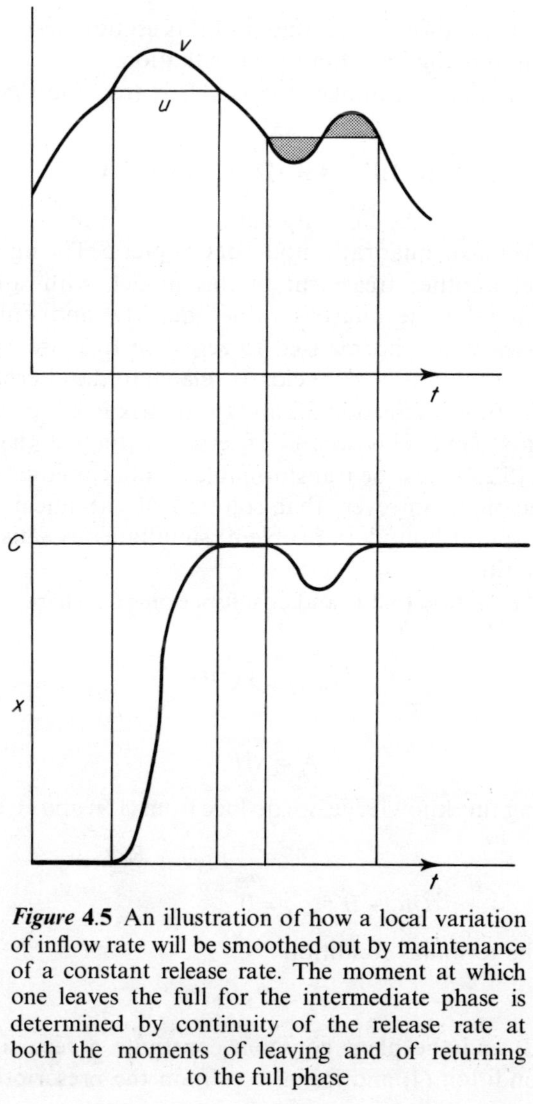

Figure 4.5 An illustration of how a local variation of inflow rate will be smoothed out by maintenance of a constant release rate. The moment at which one leaves the full for the intermediate phase is determined by continuity of the release rate at both the moments of leaving and of returning to the full phase

continuity of u and equality of the two shaded areas. The effect of the manoeuvre is that these minor variations are smoothed out on a constant u value.

Exercises

1. Suppose the dam is subject to evaporation and leakage, so that equation (2) must be modified to

$$x_{t+1} = \rho x_t + v_t - u_t$$

where $0 < \rho < 1$. Modify the treatment of this section accordingly. Consider also the continuous time version of this situation.

2. Note that continuity of u implies that $\dot{x} = 0$ at transition points.

5. THE LINEAR/QUADRATIC CASE

A model which obeys all the linearity/convexity conditions of Theorem 2.1 or Theorem 8.1 is the linear/quadratic model of Chapter 5. The maximum principle thus provides yet another treatment of this model, with some independent interest. We shall take the analysis rather quickly, and only for the time-homogeneous case, with S normalized to zero (see Exercise 1). Two points of interest emerge: the analysis is more closely related to the alternative algorithms of Section 5.7 than those of Section 5.2 and the matrix P_t is seen as determined by a linear equation system. This second observation implies that the non-linear Riccati equation (5.2.4) must be transformable to a linear equation. That is true, with the complication, however, that solution of the linear system requires solution of a two-point boundary problem; simultaneous satisfaction of initial and terminal conditions.

With the plant equation (5.2.1) and cost function $\frac{1}{2}C$, where C is prescribed by (5.2.2), we know in fact that

$$F(x, t) = \tfrac{1}{2}x'P_t x,$$

so that, by (2.1),

$$p_t = x_t' P_t. \tag{1}$$

However, ignoring this knowledge, we deduce from (1.9) and (1.10) the equations

$$p_t' = A'p_{t+1}' + Rx_t, \tag{2}$$

$$Qu_t + B'p_{t+1}' = 0, \tag{3}$$

for $t < T$, with the terminal condition

$$p_T' = P_T x_T. \tag{4}$$

Equations (2), (3) and the plant equation constitute a recursion for $\{x_t, p_t, u_t\}$ with terminal condition (4) and initial condition the prescription of the initial value of x. If we solve (3) for u (assuming $Q > 0$) and substitute into (2) and the plant equation we deduce an equation system

$$(x_t' \, p_t)\begin{pmatrix} A' & -R \\ \cdot & I \end{pmatrix} = (x_{t+1}' \, p_{t+1})\begin{pmatrix} I & \cdot \\ J & A \end{pmatrix} \tag{5}$$

where

$$J = BQ^{-1}B'.$$

If A is non-singular then this can be rewritten as

$$\begin{aligned}(x_{t+1}' \, p_{t+1}) &= (x_t' \, p_t)\begin{pmatrix} A' + RA^{-1}J & -RA^{-1} \\ -A^{-1}J & A^{-1} \end{pmatrix} \\ &= (x_t' \, p_t)\Phi,\end{aligned}$$

say. We thus have

$$(x_T' \, p_T) = (x_T' \, x_T' P_T) = (x_t' \, p_t)\Phi^s = (x_t' \, p_t)\begin{pmatrix} \Phi_{11}^{(s)} & \Phi_{12}^{(s)} \\ \Phi_{21}^{(s)} & \Phi_{22}^{(s)} \end{pmatrix}, \tag{6}$$

say, where $s = T - t$.

Relation (6) can be regarded as expressing the unknowns p_t and x_T in terms of x_t, t. Solving for p_t we deduce that

$$p_t = x_t'(\Phi_{12}^{(s)} - \Phi_{12}^{(s)} P_T)(\Phi_{22}^{(s)} - \Phi_{21}^{(s)} P_T)^{-1}.$$

By comparison with (1) we then deduce that

$$P_{T-s} = (\Phi_{12}^{(s)} - \Phi_{11}^{(s)} P_T)(\Phi_{22}^{(s)} - \Phi_{21}^{(s)} P_T)^{-1}$$

so that P_t is calculable in terms of the elements of Φ^{T-t}.

Exercises

1. Show that if $S \neq 0$ then the calculations of this section indeed hold with the substitutions

$$A \to A - BQ^{-1}s, \qquad R \to R - S'Q^{-1}S.$$

   Consider the problem without time-homogeneity.
2. Deduce from (1), (5) the dual form (5.7.1) of the Riccati recursion in the time-homogeneous case.

6. ECONOMIC DEVELOPMENT PROGRAMMES

Consider a problem to be treated in other versions in Chapter 36: the optimization of a dynamic economy. The treatment (due to Gale, 1968) is interesting, in that it successfully applies Lagrangian methods while allowing non-linearity in the relation between the variables of consecutive stages. The device employed may seem to sidestep the issue of non-linearity, but it nevertheless leads to fairly specific conclusions.

It will be assumed that x_t represents the vector of amounts of resources available at time t. The variable u_t, which represents the element of choice available in the passage from x_t to x_{t+1}, then enters in two ways. Firstly, part of x_t can be diverted for immediate consumption; secondly, the remaining part can be allocated over the processes of production in various ways. It is desired to optimize both of these compound decisions.

Let the vector amounts consumed and reinvested at time t be denoted v_t and w_t so that

$$v_t + w_t \leq x_t. \tag{1}$$

The relationship between w_t and x_{t+1}, the amounts reinvested and then produced, can take various forms according as the allocation of reinvestment is varied. However, it is not necessary to assume a linear relationship, nor, indeed, to put the process in the form of a functional relationship (a 'plant equation') at all. Rather, it is better to assume that w_t, x_{t+1} can jointly take any value-pair in a convex set $\mathscr{T}(t)$, known as the *technology* at time t. The term 'technology' is appropriate, in that, by specifying $\mathscr{T}(t)$, we specify what range of x_{t+1} could actually be produced from a given w_t at time t.

The approach is now to regard $\mathscr{T}(0) \times \mathscr{T}(1) \times \cdots \times \mathscr{T}(T-1)$ as specifying the basic region of variation $\mathscr{X}$ and to take account of restrictions (1) by Lagrangian multipliers. That is, the non-linear part of the $x_t \rightarrow x_{t+1}$ relationship has been sunk back into the specification of $\mathscr{X}$, leaving only the relatively trivial linear constraints (1) to be accounted for by Lagrangian methods.

This seems like an evasion of the problem; to some extent it is. However, even for this weakened version of the problem, Lagrangian methods furnish strong conclusions, and the notion of representing the recursion by a set rather than by a family of functional relationships seems a valuable one.

Suppose then that we wish to maximize the utility function

$$G = \sum_{0}^{T-1} g(v_t, t) + K(x_T) \tag{2}$$

within the set $\mathscr{X}$ (x_0 being specified), subject to constraints (1). Here $g(v_t, t)$ represents the utility derived from consumption v_t at time t and $K(x_T)$ the utility represented by an accumulation of resources x_T at the end of the programme. If it is assumed that these functions are concave, then G is concave, the set is convex and the constraints (1) linear, so that one can appeal to the theorems of Section 8 and assert the existence of non-negative vectors p_t such that (x, v, w) maximizes

$$G_p = \sum_{t=0}^{T-1} [g(v_t, t) + p_t[x_t - v_t - w_t]] + K(x_T)$$

in $\mathscr{X}$. A number of conclusions then follow immediately.

Theorem 6.1. *Under the given assumptions there exist non-negative row vectors p_t such that :*

(a) (x_{t+1}, w_t) maximizes $p_{t+1}x_{t+1} - p_t w_t$ in $\mathscr{T}(t)(0 \leq t \leq T-2)$ and (x_T, w_{T-1}) maximizes $K(x_T) - p_{T-1}w_{T-1}$ in $\mathscr{T}(T-1)$.

(b) v_t maximizes $g(v_t, t) - p_t v_t$ $(0 \leq t < T)$.

(c) p_t is interpretable as an effective marginal price for resources at stage x, in that, if (1) is modified to

$$v_t + w_t \leq x_t + b_t, \tag{3}$$

then p_t is the derivative of maximal return with respect to b_t at $b = 0$ if this derivative exists; more generally, $p = (p_1, p_2, \cdots, p_T)$ is a super-gradient to this function at $b = 0$.

All these assertions are formally immediate; only interpretation is necessary. Assertion (c) states the familiar shadow-price interpretation of the multiplier p; that p_t is the rate of change of maximal return with resource availability at time t (or an appropriate generalization if derivatives do not exist). Assertion (a) then states that production is to be operated at the tth stage so as to maximize immediate profit in terms of these prices. Assertion (b) states that consumption at stage t is to be chosen at a level which maximizes net immediate utility, if consumption must be paid for at a price (utility rate) of p_t.

It should be noted that the p_t of this section are not quite the multipliers of other sections, because they correspond to the constraint (1) rather than to the recursion between x_{t+1} and x_t. Nor does the theorem provide a means of determining them. Nevertheless, the problem is interesting as an application of multiplier methods to a dynamic problem in which the 'plant equation' is highly non-linear. Note that constraint (1) is an inequality rather than an equality. Lagrangian methods are still applicable, and it is because (1) is an inequality that one can assert that the multipliers p_t are non-negative (see, for example, assertion (vi) on p. 60 of Whittle, 1971).

Some of the consequences of the theorem are given in the exercises. They reflect the general tendency, already noted in Section 3.1, to hold consumption down until the end of the planning period approaches.

Exercises

1. (Gale). Consider the case of a single good and suppose that the technology takes the constant form of the set of points (x, w) such that $w \geq 0$, $x \leq G(w)$ where G is concave. Suppose further that the model is *productive*, in the sense that

$$G(w + \delta) - G(w) > \delta$$

for all $w \geq 0$, $\delta > 0$. Suppose also that $g(v, t)$ is independent of t and that both it and $K(x)$ are increasing. Show then that v_t is a non-decreasing function of t in the optimal scheme.

2. (Gale). Show, under the conditions of Exercise 1 and for the optimal scheme, that if w_t once begins to decrease then it continues to do so. Thus, inputs increase up to some point in time and decrease thereafter.

7. AN EXAMPLE OF INAPPLICABILITY OF THE MAXIMUM PRINCIPLE

It is perhaps wise to give an example which corrects any impression that the maximum principle is universally valid and illustrates invalidating features. Consider the production scheduling example of Section 3.2. If the maximum principle were valid then, for some p_{t+1}, the optimal u_t would minimize the expression

$$ai(u_t) + bu_t + p_{t+1}u_t \tag{1}$$

over non-negative values. Here

$$i(u) = \begin{cases} 0 & (u = 0), \\ 1 & (u > 0). \end{cases}$$

We know that the optimal policy is such that $u_t = 0$ if current stocks are sufficiently large, while if stocks are low then u_t is chosen so that stocks are brought up to a level adequate to meet demand for an integral (but finite) number of periods. However, in minimizing expression (1) one would choose $u_t = 0$ if $b + p_{t+1} > 0$, $u_t = +\infty$ if $b + p_{t+1} < 0$, which is unrealistic. The trouble is the non-convexity of the cost function, caused by the discontinuous factor $i(u)$.

Note, nevertheless, that the policy recommended by maximization of (1) could be acceptable if u_t were a *rate* of stock replenishment in continuous time, satisfying a constraint $u_t \leq \bar{u}$ for physical reasons. The recommended policy would then be to choose $u_t = 0$ if $b + p_{t+1} > -a/\bar{u}$ and $u_t = \bar{u}$ if $b + p_{t+1} < -a/\bar{u}$. That is, one replenishes stock at either zero rate or the maximal feasible rate, save in the usually exceptional case $b + p_{t+1} = -a/\bar{u}$. Such a policy would be acceptable, and could even be optimal.

8. CONDITIONS OF A GLOBAL MAXIMUM PRINCIPLE

The maximum principle was characterized in Section 1 by embedding the problem with the plant equation (1.2) in the class of problems with the plant equation (1.3), and demanding that the minimal cost function $\mu(b)$ then defined possessed a sub-gradient at $b = 0$. Theorem 1.1 asserts the equivalence of this latter condition to validity of the maximum principle. A rather natural and complete extension of the condition is to demand that $\mu(b)$ possesses a sub-gradient for all relevant b, so that a maximum principle is valid for all problems in this extended class. To do this is to demand more than required, but the extension is indeed a natural one.

Let $\mathscr{B}$ denote the set of all b such that a trajectory consistent with (1.3) and $Z \in \mathscr{Z}$ exists; this is the domain on which $\mu(b)$ is naturally defined.

Theorem 8.1. *Suppose $\mathscr{B}$ convex, $\mu(b)$ convex in $\mathscr{B}$ and finite in the interior of $\mathscr{B}$. Then the maximum principle is valid for all b in the interior of $\mathscr{B}$.*

Proof. Consider the set of points above the graph of $\mu(b)$, the *hypograph* $\mathscr{H}$ of $\mu(b)$. That is, if c is a scalar then we are considering the set of points (b, c) such that $b \in \mathscr{B}$, $c \geq \mu(b)$. This set is convex, and so, by the supporting hyperplane theorem (see, for example, Whittle, 1971, p. 49), there exists $(p, q) \neq 0$ (p an n-dimensional row vector, q a scalar) such that

$$qc - pb \geq qc_0 - pb_0 \tag{1}$$

for all (b, c) in $\mathscr{H}$ and (b_0, c_0) an assigned boundary point of $\mathscr{H}$. Taking the boundary point $(b_0, \mu(b_0))$, we then have

$$q(\mu(b) + d) - pb \geq q\mu(b_0) - pb_0 \tag{2}$$

for $b \in \mathscr{B}$ and d any non-negative scalar. Since d may be arbitrarily large, we deduce that $q \geq 0$. Setting $d = 0$ in (2) we see that we have deduced that, for any b_0 of $\mathscr{B}$, there exists a $(p, q) \neq 0$ with $q \geq 0$ such that $q\mu(b) - pb$ is minimal at b_0. If we could only assert that $q = 1$ then this statement would imply that $\mu(b)$ had sub-gradient p at b_0, and so validity of the maximum principle at b_0. Seeing that the scale of (p, q) is irrelevant we could make this assertion if we knew only that $q \neq 0$.

However, if q were zero, then we would deduce from (1) that

$$p(b - b_0) \geq 0. \tag{3}$$

If b_0 is an interior point of $\mathscr{B}$ then (3) must hold for b in a neighbourhood of b_0, which implies $p = 0$ and so $(p, q) = 0$, which we know not to be the case. Thus q is non-zero, and so strictly positive, and so can be normalized to unity. There thus exists a p such that $\mu(b) - pb$ is minimal in $\mathscr{B}$ at b_0. ■

The proof relies on the fact that existence of a sub-gradient to $\mu(b)$ at b_0 and existence of a 'non-vertical' supporting hyperplane to the hypograph of $\mu(b)$ at $(b_0, \mu(b_0))$ are equivalent. By 'non-vertical' we mean that $q \neq 0$ in (1). The hypotheses that b_0 be interior to $\mathscr{B}$ and $\mu(b)$ finite in its neighbourhood, in addition to being convex, are sufficient to assure that $\mu(b)$ has a finite directional derivative in all directions from b_0 and that the supporting hyperplane cannot be vertical.

A near-converse to Theorem 8.1 is trivially valid (see Exercise 1).

We now need conditions on the prescribed quantities a, C and $\mathscr{Z}$ which ensure convexity of $\mathscr{B}$ and $\mu(b)$.

Theorem 8.2. *The set $\mathscr{B}$ and the function $\mu(b)$ are convex if $\mathscr{Z}$ is convex, $a(x_t,u_t,t)$ linear in x_t,u_t and C convex in $Z=\{x_t,u_t\}$.*

Proof. Consider two values $b^{(i)}$ in $\mathscr{B}$ of b ($i=1,2$) and let $Z^{(i)}=\{x_T^{(i)},u_T^{(i)}\}$ denote the corresponding values of an optimal trajectory. Consider an averaged value

$$b^*=\alpha b^{(1)}+\beta b^{(2)}$$

with $\alpha,\beta>0$; $\alpha+\beta=1$. Then b^* belongs to $\mathscr{B}$ since, if we define $Z^*=\alpha Z^{(1)}+\beta Z^{(2)}$, then Z^* belongs to $\mathscr{Z}$, by convexity of $\mathscr{Z}$, and satisfies the plant equation with $b=b^*$, by linearity of a. Thus $\mathscr{B}$ is convex. Further

$$\begin{aligned}\mu(b^*)\le C(Z^*)&\le \alpha C(Z^{(1)})+\beta C(Z^{(2)})\\&=\alpha\mu(b^{(1)})+\beta\mu(b^{(2)}),\end{aligned}$$

which proves $\mu(b)$ convex. ■

These conditions are far from necessary for convexity of $\mu(b)$, even less for existence of a sub-gradient (and so validity of the maximum principle) at $b=0$ alone. We indicate some relaxations in Exercises 2 and 3.

Exercises

1. Suppose the maximum principle valid for all b in $\mathscr{B}$. Show then that the greatest convex minorant of $\mu(b)$ in the convex hull of $\mathscr{B}$ equals $\mu(b)$ in $\mathscr{B}$.
2. Assume the problem normalized to the purely terminal form, as in Section 2.4. Suppose that $\mathscr{Z}$ has the specification: x_t must lie in a convex set $\mathscr{X}_t$ and u_t in a set $\mathscr{U}_t (t=0,1,2,\ldots)$. Suppose C is convex and that for any x_t', x_t'' in $\mathscr{X}_t$ and u_t', u_t'' in $\mathscr{U}_t$ and any α,β (conventions above) there exists a u_t in $\mathscr{U}_t$ such that

$$a(\alpha x_t'+\beta x_t'',u_t,t)=\alpha a(x_t',u_t',t)+\beta a(x_t'',u_t'',t). \tag{4}$$

Show that $\mathscr{B}$, $\mu(b)$ are convex.
3. The convexity-type assumption (4) can be relaxed at the expense of stronger conditions in other directions. Consider the following modification of the conditions of the second-last sentence of Exercise 2. Suppose there exists a cone $\mathscr{C}$ such that, if x_t belongs to $\mathscr{X}_t$ and y to $\mathscr{C}$ then x_t+y also belongs to $\mathscr{X}_t$. Suppose that a is $\mathscr{C}$-increasing in the sense that

$$a(x_t+y,u_t,t)-a(x_t,u_t,t)$$

belongs to $\mathscr{C}$ for y in $\mathscr{C}$ and that K is convex and such that $K(x_T+y)\le K(x_T)$ for y in $\mathscr{C}$. Suppose finally that the condition associated with (4) is replaced by the following: there exists a u_t in $\mathscr{U}_t$ such that the left-hand member of (4) exceeds the right-hand member by an element of $\mathscr{C}$. Show that $\mathscr{B}$, $\mu(b)$ are convex.

CHAPTER 7

The continuous maximum principle; concepts and examples

The continuous maximum principle provides a criterion of optimality in the case when time t is continuous and the state variable x takes values in a Euclidean space R^n. One considers variations of trajectory $\{x(t), u(t)\}$ which cause infinitesimal variations in $\{x(t)\}$. By demanding stationarity of cost with respect to such variations one deduces conditions which are necessary for optimality, if certain differentiability assumptions can be justified.

We begin with a heuristic derivation of the principle in Section 1. This follows, together with terminal conditions and a number of associated statements, rather easily from the dynamic programming principle, and in a stronger form than was possible in the discrete-time case. One is then led to hope that the principle might have a more general validity in continuous than in discrete time. For example, validity might be less dependent on assumptions of convexity, which seem to play no role in the dynamic programming proof. The closer analysis of Chapter 8 shows this optimism to be justified. The continuous time case turns out to have two special features. Firstly, because the x-orbit is determined by a differential equation rather than a difference equation, one can test radical variations in u (for short times—the so-called 'strong variation') consistent with producing only local variations in x. Secondly, even if u does not take values in a convex set, it can be varied so rapidly in time that there is a time-averaging effect, with an effect on the orbit much as if the permissible u-set were indeed convex (the convex hull of the actually permitted set). The point is pursued in Chapter 8.

1. A HEURISTIC DERIVATION OF THE PRINCIPLE, BY DYNAMIC PROGRAMMING ARGUMENTS

The argument which follows is heuristic only in that it assumes differentiability whenever such an assumption is convenient, although this is an assumption which certainly requires some backing. It can well be claimed as the natural approach to the maximum principle. The best justification of this assertion is that, in cases of doubt as to the formulation or application of the principle, resort to dynamic programming arguments will provide a guide, even if other arguments are needed to confirm its correctness.

We assume that the state variable x takes values in R^n, that the plant equation takes the differential form

$$\dot{x} = a(x, u, t) \tag{1}$$

(where $\dot{x} = \mathrm{d}x/\mathrm{d}t$) and the cost function the additive form

$$C = \int_0 g(x, u, t)\,\mathrm{d}t + \text{terminal costs.} \tag{2}$$

For the moment we leave the formulation of the stopping rule and termination costs open. For many purposes it is convenient to combine x and t into a joint state variable $\xi = (x, t)$; for others it is better to keep x and t as distinct arguments. For the moment we take the second approach, but shall, when convenient, adopt the first.

We shall use the subscript notation for partial derivatives, so that $F_t = \partial F/\partial t$, and $F_x = \partial F/\partial x$ is the row vector of partial derivatives $\partial F/\partial x_j$ $(j = 1, 2, \ldots, n)$. The total time-derivative of $F(x, t)$ would thus be

$$\dot{F} = \frac{\mathrm{d}F}{\mathrm{d}t} = F_t + F_x \dot{x}.$$

For a given orbit, $\{\bar{x}(t), \bar{u}(t)\}$, define p, p_0 as the values of F_x and F_t evaluated on the orbit. More explicitly

$$p(t) = \left[\frac{\partial F(x, t)}{\partial x}\right]_{x = \bar{x}(t)}, \tag{3}$$

$$p_0(t) = \left[\frac{\partial F(x, t)}{\partial t}\right]_{x = \bar{x}(t)}. \tag{4}$$

Finally, we define the *Hamiltonian*

$$H(x, u, p, p_0) = g(x, u, t) + pa(x, u, t) + p_0. \tag{5}$$

We shall tend to suppress arguments where there is no ambiguity, and write (5) simply as

$$H = g + pa + p_0.$$

Theorem 1.1. (The maximum principle on a free orbit). *If differentiation is permissible where employed, then the following assertions hold on an optimal orbit*:

(*a*) $H = 0$ *and* u *adopts a value which minimizes* H.

(*b*) *The dual variables* p *and* p_0 *obey the equations*

$$\dot{p} = -(g_x + pa_x), \tag{6}$$

$$\dot{p}_0 = -(g_t + pa_t). \tag{7}$$

(c) *The plant equation* (1) *and the equations* (6), (7) *can be written*

$$\begin{aligned} \dot{x} &= H_p, \\ \dot{p} &= -H_x, \\ \dot{p} &= -H_t. \end{aligned} \tag{8}$$

Proof. The optimality equation is, in infinitesimal form,

$$F(x, t) = \min_u [g(x, u, t) + F(x + a(x, u, t)\delta t, t + \delta t) + o(\delta t)]. \tag{9}$$

In the limit this becomes

$$\min_u [g + F_x a + F_t] = 0, \tag{10}$$

where a minimizing u value is known to be optimal. Relation (10) implies both assertions of (a) if we make the identifications (3), (4) of the conjugate variables. Differentiating relation (9) with respect to x, t and taking the limit $\delta t \downarrow 0$ one derives relations (6), (7). In doing this one must observe that either the minimizing value of u does not change (as x, t change) or that the square-bracketed expression in (9) is stationary with respect to any change (cf. the proof of Theorem 6.2.1). Assertion (c) is immediate. ■

As for the discrete-time case of Chapter 6, we appeal to equations (8) and the H-minimizing property of u to determine x, u, p and p_0 along an optimal orbit. The equation $H = 0$ supplies one end-condition; initial and terminal conditions (to be determined in the next section) supply others. The conditions thus implied for an orbit have the character of being necessary, if regularity conditions are satisfied. That is, if these regularity properties do not obtain, then the calculations are of nil or doubtful meaning. If they do, then the calculations may only partially determine the optimal orbit.

In the case of a time-homogeneous problem, when a and g are independent of t, then $\dot{p}_0 = 0$ and

$$p_0 = \text{constant}$$

along an optimal orbit.

In the case of a time-invariant problem (for which $F(x, t)$ depends only upon x) then $F_t = 0$ and

$$p_0 = 0.$$

As noted at the beginning of the section it is often convenient to achieve a formally time-homogeneous formulation by taking the column vector $\xi = \begin{pmatrix} x \\ t \end{pmatrix} = \begin{pmatrix} x \\ x_0 \end{pmatrix}$, say, as the state variable, with the conjugate variable $\eta = (p, p_0)$. We then have

$$H = g(\xi, u) + \eta\begin{pmatrix} a(\xi, u) \\ 1 \end{pmatrix}$$

and the two final relations of (8) become simply

$$\dot{\eta} = -H_{\xi}.$$

The equation

$$\dot{\xi} = H_{\eta}$$

implies the plant equation (1) and its supplement

$$\dot{x}_0 = 1.$$

2. TERMINAL CONDITIONS

If the stopping rule is $t = T$ with the terminal loss $K(x)$ prescribed, then $F(x, T) = K(x)$, and it follows from the identification (1.3) that the terminal condition is

$$p = K_x \qquad (t = T). \tag{1}$$

The general case is best formulated in terms of the combined state variable $\xi = (x, t)$. Consider first the case of obligatory stopping, when the process stops at *first entry* to a set $\xi \in D$ on which a terminal cost $K(\xi)$ is defined. Then

$$F(\xi) = K(\xi) \qquad (D) \tag{2}$$

We suppose D to be a closed set with boundary ∂D. Seeing that the path is continuous and stops on first entry to D, it necessarily stops on ∂D, and prescription of K in $D - \partial D$ is immaterial.

It does not necessarily follow from (2) that $\lim F(\xi) = K(\xi)$ as ξ approaches a point $\bar{\xi}$ of ∂D by an arbitrary path outside D. If $K(\bar{\xi})$ is sufficiently large and the process sufficiently controllable then the point $\bar{\xi}$ will not be the terminal point of an optimal trajectory and there will be a discontinuity in F between $\bar{\xi}$ and points close to it outside D (see Exercises 1 and 2 and Chapters 20 and 37).

Therefore, let $F_{\lim}$ denote the limiting value of F as one approaches D along an optimal trajectory. The stronger version of (2)

$$F_{\lim} = K \tag{3}$$

will hold only for points in the set we might denote D_{opt}—those points of D (in fact of ∂D) which are the terminal points of some optimal trajectory. From relation (3) we deduce the generalized form of (1).

Theorem 2.1. (Terminal conditions with obligatory stopping). *If the required* *derivatives* K_ξ *exist then the relation*

$$(\eta - K_\xi)\sigma = 0 \tag{4}$$

holds for points $\bar{\xi}$ *of* D_{opt} *such that* $\pm\sigma$ *are directions from* $\bar{\xi}$ *into* D_{opt}

By saying that σ is a direction from $\bar{\xi}$ into D_{opt} we mean that there is a continuous path $\xi(\theta)$ in D_{opt} parametrized by the scalar $\theta \geq 0$ such that

$$\xi(\theta) = \bar{\xi} + \theta\sigma + o(\theta)$$

for small positive θ.

Suppose that the surface of D is specified by the vector constraint

$$\psi(\xi) = 0$$

where this amounts to restriction of ξ to a locally linear sub-space in the neighbourhood of the point of ∂D in question. Then the assertion of Theorem 2.1 is sometimes alternatively expressed.

Corollary 2.1. *Under the regularity conditions indicated, relation* (4) *can be alternatively expressed as* (*a*)

$$\eta - K_\xi \perp \partial D \tag{5}$$

at points in the interior of $(\partial D)_{\text{opt}}$ *or* (*b*)

$$\eta = [K(\xi) + \lambda\psi(\xi)]_\xi \tag{6}$$

at any point $\bar{\xi}$ *in the interior of* $(\partial D)_{\text{opt}}$, *where* λ *is a suitable multiplier matrix specific to* $\bar{\xi}$.

If stopping is to some extent optional then a further degree of optimization is possible, and the assertion of Theorem 2.1 can be strengthened. Let the points on an optimal orbit be parametrized by a scalar τ, so that as τ varies then $\xi(\tau)$ traces out the orbit. Suppose termination takes place at $\tau = \bar{\tau}$. Then

$$F(\xi(\tau)) = \int_\tau^{\bar{\tau}} c(\zeta)\,\mathrm{d}\zeta + K(\xi(\bar{\tau})) \tag{7}$$

where $\int_{\tau_1}^{\tau_2} c(\zeta)\,\mathrm{d}\zeta$ is the minimal cost of transfer from $\xi(\tau_1)$ to $\xi(\tau_2)$. Let us write (7) as

$$f(\tau) = \int_\tau^{\bar{\tau}} c(\zeta)\,\mathrm{d}\zeta + k(\bar{\tau}). \tag{8}$$

Suppose that one may terminate (at cost $k(\tau) = K(\xi(\tau))$ at any point τ of the orbit. The optimal $\bar{\tau}$ will then minimize expression (8), subject to $\bar{\tau} \geq \tau$. If we assume that the starting point τ is such that it is not optimal to stop immediately, so that the orbit is indeed begun, then $\bar{\tau} \geq \tau$ is no restriction. If c is bounded and k

differentiable, and the minimizing value of $\bar{\tau}$ is interior to the permitted interval, then it will be at a stationary point of expression (8):

$$c(\bar{\tau}) + k'(\bar{\tau}) = 0,$$

Since

$$f'(\tau) = -c'(\tau) \qquad (\tau \leq \bar{\tau})$$

it then follows that

$$f'(\bar{\tau}) = k'(\bar{\tau}). \tag{9}$$

That is, relation (4) *holds also for directions* σ *along an optimal orbit.*

In general, the stopping rule may have both obligatory and optimal elements, but we can collect conclusions to make a simple general assertion.

Let $\tilde{D}$ be the set of ξ at which one *may* terminate, so that

$$F(\xi) \leq K(\xi) \qquad (\tilde{\mathrm{D}}).$$

Let $D \subset \tilde{D}$ be the set of ξ at which it is either obligatory or optimal to terminate, so that

$$F(\xi) = K(\xi). \qquad (\mathrm{D}) \tag{10}$$

Then

$$(F - K)_\xi \sigma = 0 \tag{11}$$

at a point $\bar{\xi}$ of D for directions σ from $\bar{\xi}$ into D, by (10), and for directions σ out along an optimal orbit (if one choose termination at $\bar{\xi}$), by (9). But F_ξ can be identified with terminal η only at points which are terminal points of an optimal orbit. We are thus led to:

Theorem 2.2. *Let D_{opt} denote the set of ξ which are terminal points of optimal orbits. Let $\tilde{D}_{\text{opt}}$ denote the set of points on optimal orbits at which the termination option is available. Then*

$$(\eta - K_\xi)\sigma = 0$$

at points $\bar{\xi}$ of D_{opt} such that $\pm\sigma$ are directions from $\bar{\xi}$ into $\tilde{D}_{\text{opt}}$.

Of course, $D_{\text{opt}} \subset \tilde{D}_{\text{opt}}$, with equality in the case of obligatory stopping.

Exercises

1. Consider the problem $x \in R$ with $\dot{x} = u, u = \pm 1$ and $g = 1$. The termination set is $\{0, 1\}$ with $K(0) = 0, K(1) = M$. Then for $0 < x < 1$, evidently

$$F(x) = \min(x, 1 - x + M).$$

So, if $M > 1$ then $x = 1$ cannot be the termination point of an optimal orbit from the left, and there is a discontinuity there: $F(1) = M, F(1-) = 1$.

2. Consider the problem $x \in R^2$ with $\dot{x} = u$, $|u| \leq 1, g = 1$ with the obligatory termination set D the axis $x_2 = 0$ and $K(x) = x_1^2$. Show that only the points $|x_1| \leq 1$ of D are termination points of optimal orbits, that $F(x_1, 0) \neq F(x_1, 0+)$ for $|x_1| > 1$, but that both $F(x_1, 0+)$ and $\partial F(x_1, 0+)/\partial x_1$ are continuous in x_1.

3. EIKONAL PROBLEMS

Consider the problem discussed in Section 3.4, of cost-minimizing paths in R^d, with the additional feature that a terminal cost $K(x)$ is prescribed in the obligatory stopping region $x \in D$. The problem is time-invariant, with

$$H = c(x) + pu. \tag{1}$$

The minimizing value of u subject to $|u| \leq 1$ is

$$u = -\frac{p'}{|p|} \tag{2}$$

and, along an optimal free trajectory, p obeys the equation

$$\dot{p} = -c_x. \tag{3}$$

If in a neighbourhood of the termination point the stopping boundary ∂D is locally planar and belongs to ∂D_{opt}, and K is differentiable, then the condition

$$p - K_x \perp \partial D \tag{4}$$

holds at the terminal point.

If c is constant then (3) implies that $\dot{p} = 0$, so that u is constant along a free trajectory and the optimal free trajectories are straight lines in the direction of u. If, in addition, $K(x)$ is constant in D, then (4) reduces simply to $u \perp \partial D$. That is, the optimal trajectories meet the boundary orthogonally if the regularity conditions associated with (4) are met. The case $c = 1$ is a relatively simple one to visualize because of the linear character of the optimal orbits. This makes it useful for examination of the less regular cases. For example, in the case $c = 1, K = 0$ one obviously has

$$F(x) = \inf_{\bar{x} \in D} |x - \bar{x}|. \tag{5}$$

The minimizing value $\bar{x}$ in (5) designates the optimal target point in D from x, towards which one moves on a constant heading. When one moves on this heading, $\bar{x}$ will be the first point of D one encounters.

If some boundary point of D is such that there is no hyperplane tangent to the surface of D at the point (as at points P, Q of Figure 3.1), then several things can happen. An extroverted point, such as P, has a multitude of locally supporting

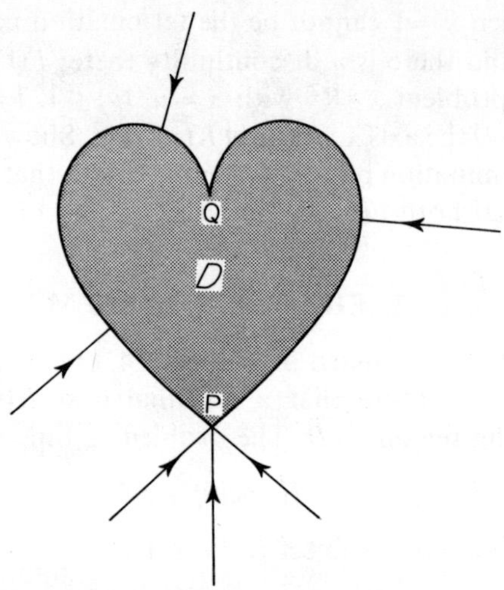

Figure 3.1 Points of D as possible terminal points of optimal orbits. The extroverted point P is the terminal point of infinitely many such orbits; the introverted point Q is the terminal point of none

hyperplanes and is the 'nearest point of D' for points in many directions. It consequently is the terminal point of many optimal trajectories. An introverted point, such as Q, has no locally supporting hyperplanes, is not a 'nearest point of D' for any point strictly outside D and is the terminal point of no optimal trajectory.

If one allows a variable termination cost $K(x)$ then there can be two new features. Firstly, the transversality condition (4) may fail because there are points of ∂D which are not termination points for any optimal orbit. Secondly, the optimal termination point may not be reachable by a straight line path ('free orbit').

To illustrate the first point, consider the case $d = 2, x = (x_1, x_2)$, with D prescribed by $x_2 = 0$ and terminal cost

$$K(x) = \begin{cases} 1 & (x_1 < 0) \\ 0 & (x_1 \geq 0). \end{cases}$$

Then

$$F(x) = \inf_{\bar{x}_1} [\sqrt{[(x_1 - \bar{x}_1)^2 + x_2^2]} + K(\bar{x}_1, 0)],$$

where the minimizing value gives the x_1 coordinate of the ideal termination point

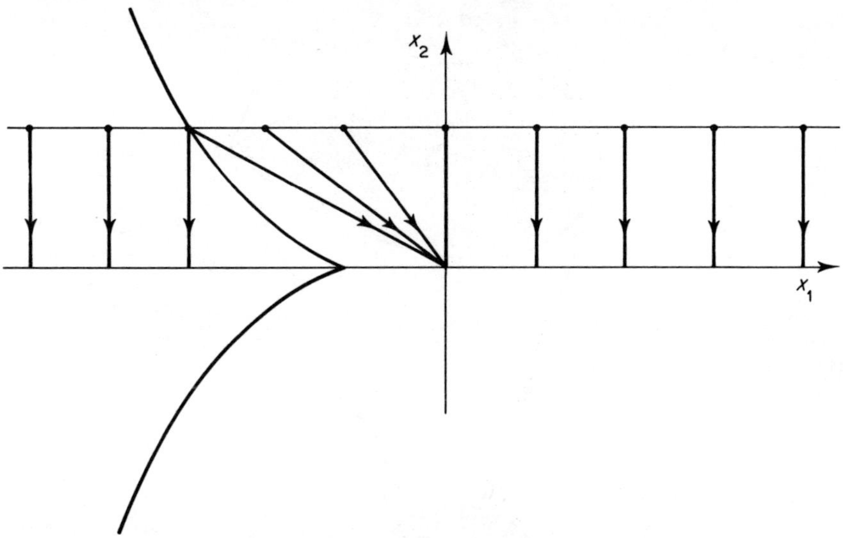

Figure 3.2 Optimal trajectories for the plane problem of the text

from x. We then have (see Figure 3.2)

$$F(x)=\begin{cases} |x_2| & (x_1 \geq 0), & (\mathscr{X}_1) \\ \sqrt{(x_1^2+x_2)} & (-\sqrt{(1+2|x_2|)} \leq x_1 \leq 0), & (\mathscr{X}_2) \\ |x_2|+1 & (x_1 \leq -\sqrt{(1+2|x_2|)}). & (\mathscr{X}_3) \end{cases}$$

In the regions marked $\mathscr{X}_1$ and $\mathscr{X}_3$ one has $\bar{x}_1 = x_1$; in $\mathscr{X}_2$, $\bar{x}_1 = 0$. Only for initial points in $\mathscr{X}_1$ and $\mathscr{X}_3$ do optimal trajectories meet D normally. The segment $-1 < x_1 < 0$ of D is never entered by the optimal trajectory.

Note that F changes analytical form but is continuous at the $\mathscr{X}_1/\mathscr{X}_2$ and $\mathscr{X}_2/\mathscr{X}_3$ boundaries. For the first case both F_x and the action rule are also continuous; for the second they are not.

The example of Exercise 2.2 also shows the features of the $\mathscr{X}_1/\mathscr{X}_2$ transition, although $K(x)$ is continuous and differentiable in that case.

For an example of the second type of behaviour, consider the case illustrated in Figure 3.3, where D splits into the two sets D_1 and D_2. We suppose that $K = 0$ on D_2, while K is so large on D_1 that entry to D_1 is never optimal. Effectively, then, D_1 is a forbidden part of state space.

Paths such as *abcd* and *efg* are fairly clearly optimal. They both skirt D_1 but do not enter it. The path *efg* consists of two linear segments with a discontinuity in direction (and so in p) at f. The path *abcd* consists of two linear segments plus a traverse of part of the boundary of D_1 (without entering D_1). Because, as drawn, the points at which this traverse is begun and finished can be varied infinitesimally, the direction of the orbit (and so p) is continuous at both points. This illustrates what can happen when optimal orbits meet the boundary of $\mathscr{X}$.

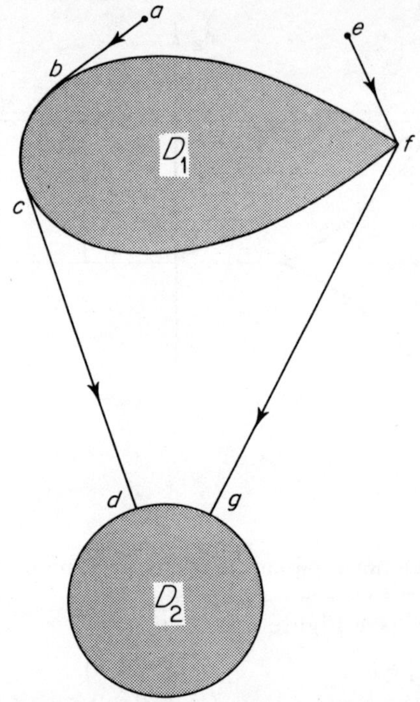

Figure 3.3 If terminal cost varies, then optimal orbits can break up into segments of different characters. In this case the terminal cost is zero in D_2, but infinite in D_1, which is thus effectively forbidden. The optimal orbits abcd and efg have different types of grazing encounter with D_1 before terminating in D_2

Exercises

1. Consider the case $d = 2, c(x) = 1$ with D consisting of the axis $x_1 = 0$ (where $K = 0$) and the line segment $x_1 = 1, -1 \leq x_2 \leq 1$ (where $K = a$). Calculate $F(x)$ for general a.
2. Suppose that $u \in R^m$ and $g(x, u) = g(x) + \frac{1}{2}u'Qu$, and that the problem is time-invariant. Use the conditions $H = 0$ and $H_u = 0$ to determine the optimal control on a free orbit in the scalar case $m = n = 1$.

4. THE BUSH PROBLEM

This is one of the celebrated early problems of optimal control: to bring a massive particle to rest at a given position in minimal time, using a force of bounded magnitude. One example would be that of bringing the rollers of a rolling mill to rest in a standard position in minimal time: the angular momentum of the rollers corresponding to the linear momentum of the particle. In the one-dimensional case the solution turns out to be simple: to apply maximal force first in one direction, then in the other, the order and duration being such that the particle will come to rest and to the desired position simultaneously. However, to prove the optimality of this manoeuvre, extreme and discontinuous in character, was beyond classical methods of variational calculus.

Consider the one-dimensional case; let x denote the coordinate and v the velocity. Note that (x, v) is then the state variable, with the plant equation

$$\begin{aligned} \dot{x} &= v, \\ \dot{v} &= u, \end{aligned} \tag{1}$$

where u can be interpreted as the force applied per unit mass of the particle. We suppose that

$$|u| \le M, \tag{2}$$

that $g = 1$ and that D consists of the single point $(x, v) = (0, 0)$. That is, the particle is to be brought to rest at the origin in minimal time.

If p, q are taken as the variables conjugate to x, v respectively then

$$H = 1 + pv + qu, \tag{3}$$

$$\left.\begin{aligned} \dot{p} &= 0 \\ \dot{q} &= -p \end{aligned}\right\}, \tag{4}$$

$$u = -M \operatorname{sgn}(q), \tag{5}$$

If the terminal values of p and q are denoted α and β, then (4) implies that

$$\begin{aligned} p &= \alpha, \\ q &= \beta + \alpha s. \end{aligned} \tag{6}$$

Since $u = -M \operatorname{sgn} \beta$ and $v = 0$ at termination, the condition $H = 0$ implies that

$$|\beta| = \frac{1}{M}. \tag{7}$$

Consider first the positive option, $\beta = M^{-1}$. If $\alpha \ge 0$, then $q > 0$ always, by (6), so that $u = -M$, and, by backwards integration along the orbit,

$$\begin{aligned} v &= Ms, \\ x &= -\tfrac{1}{2}Ms^2. \end{aligned} \tag{8}$$

Thus x, v lie on the parabolic locus

$$x + \frac{v^2}{2M} = 0 \qquad (x \le 0; v \ge 0). \tag{9}$$

If, on the other hand, α is negative and equal to $-(Ms_0)^{-1}$, say, then $u = -M$ for $s \leq s_0$ and (8) holds in this range. However, by (5) to (7), $u = M$ for $s > s_0$, and in this range the solution (8) of (1) continues as

$$\begin{aligned} v &= M(2s_0 - s), \\ x &= M(\tfrac{1}{2}s^2 - 2ss_0) + Ms_0^2. \end{aligned} \tag{10}$$

Thus, for $s > s_0$ the point (x, v) lies on the shifted parabolic locus

$$x + Ms_0^2 - \frac{v^2}{2M} = 0. \tag{11}$$

For $\beta < 0$ the loci corresponding to (9) and (11) are

$$x - \frac{v^2}{2M} = 0 \quad (x \geq 0;\, v \leq 0), \tag{9'}$$

$$x - Ms_0^2 + \frac{v^2}{2M} = 0, \tag{11'}$$

on which $u = M$ and $u = -M$ respectively.

The two half-parabolae described by (9) and (9′) constitute the so-called *switching locus*, drawn heavily in Figure 4.1. Once (x, v) is on this, the application

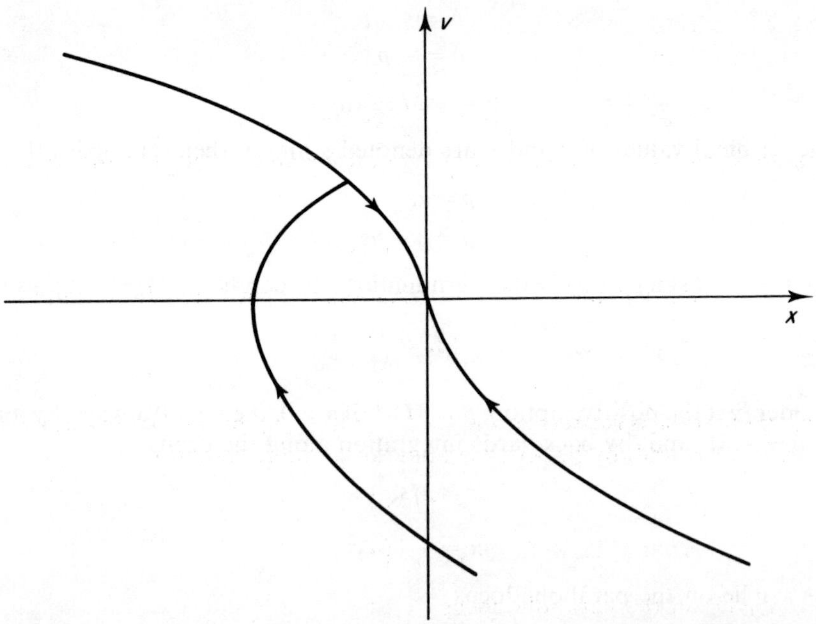

***Figure* 4.1** The Bush problem. Optimal loci in state space and the switching locus

of a constant maximal force ($-M$ on the left branch, $+M$ on the right) brings (x, v) to the target value $(0, 0)$. If (x, v) is not on the locus then one applies a maximal force of $+M$ or $-M$, depending upon whether the initial (x, v) is below or above the switching locus, and so moves on a parabola (11) or (11′) respectively. The value of s_0 corresponding to the parabola passing through the initial point represents time-to-go at the moment when (x, v) hits the switching locus; at this moment one reverses the direction of applied thrust u.

Figure 4.1 shows locus (11) for the case of an initial (x, v) below the switching locus. The arrows denote the direction of movement of the representative point in real time t.

The optimal control rule is thus completely determined. However, $F(x, v)$ can be calculated; its value for (x, v) below the switching locus will be obtained by eliminating s, s_0 from relations (10) and

$$F(x, v) = s,$$

The elimination yields

$$F = \sqrt{\left(\frac{2v^2}{M^2} - \frac{4x}{M}\right)} - \frac{v}{M} \tag{12}$$

below the locus, and

$$F = \sqrt{\left(\frac{2v^2}{M^2} + \frac{4x}{M}\right)} + \frac{v}{M} \tag{12′}$$

above it. We verify from (12) that

$$\frac{\partial F}{\partial x} = -\frac{2}{M\Delta} = -\frac{1}{Ms_0} = p,$$

$$\frac{\partial F}{\partial v} = \frac{2v}{M^2\Delta} - \frac{1}{M} = \frac{v}{M^2 s_0} - \frac{1}{M} = \frac{s_0 - s}{Ms_0} = q,$$

where Δ denotes the square-root term in (12). We thus confirm the identity of F-derivatives and conjugate variables on the orbit.

However, we meet in a more acute form than in the last section the fact that the F-derivatives may not exist, although the conjugate variables seem well defined. On the right branch of the switching locus, when Δ is zero, the F-derivatives are infinite, or undefined. This is again because a small change in x implies a discontinuous change in action. Starting from the point A of Figure 4.2 we will follow the path ABO. Starting from the not very different point A′ we will follow the very different path A′B′O. The values of p, q on the switching locus itself are finite, but not uniquely defined, being dependent upon the parameter s_0, which labels the point at which the orbit meets the switching locus. This non-uniqueness is matched by the fact that F-derivatives on the switching locus are direction-dependent. The matter has been investigated by Morton (1968) who replaces the

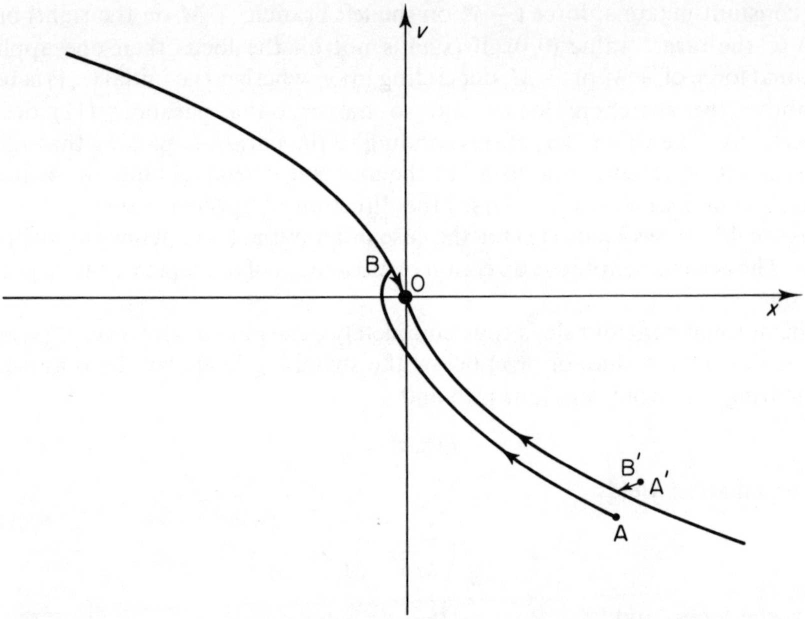

Figure 4.2 The Bush problem. The optimal policy and path changes radically if the starting point is perturbed to either side of the switching locus

hard constraint $|u| \leq M$ by incorporation of a control cost $|u/M|^{\gamma}$; all differentials then exist. He explains how the maximum principle continues to give a correct solution in the limit $\gamma \to +\infty$. However, our treatment of the next chapter demonstrates the continuing validity of the maximum principle in such cases.

5. ROCKET THRUST PROGRAMME OPTIMIZATION; THE OPTIMIZATION OF A SOUNDING ROCKET

The solution of the Bush problem of Section 4 is very much what intuition might suggest, but is nevertheless remarkable for its extreme and discontinuous ('bang-bang') character. It is just for this reason that the problem was so resistant to classical optimization methods, such as the calculus of variations. The maximum principle has been significant in that it provides a technique for just such cases. A source of significant problems of this type over the past few decades has been the incentive to determine a rocket thrust programme which is optimal in that it minimizes the amount of fuel or the amount of time (for example) needed to achieve some given transfer of the rocket.

We shall assume the rocket to be effectively a point mass, so that its state is specified by its coordinates in physical space (x), its velocity (v) and its mass (m). That is, we neglect rotation, yaw, vibration or any aspect of the rocket other than translation of its centre-of-mass through space and wasting of its mass through fuel consumption. Note that x is in this case *not* the state vector, but simply that part of the state vector (x, v, m) which describes the Euclidean coordinates of the rocket in physical space.

Suppose that the rocket jet has a vector velocity k and that the rocket is subject to external forces of magnitude $h = h(x, v, m, t)$. Then the condition of momentum conservation over a time interval of length δt gives the equation

$$(m - \delta m)(v + \delta v) + (v - k)\delta m = mv + h\,\delta t \tag{1}$$

or

$$m\dot{v} = k\dot{m} + h \tag{2}$$

the so-called *rocket equation.* By definition

$$\dot{x} = v. \tag{3}$$

We shall assume that the direction of the jet vector k can be freely controlled and that the rate $\dot{m}$ at which mass can be expelled in the rocket jet can also be controlled within limits. If we set

$$k\dot{m} = u, \tag{4}$$

$$|k|^{-1} = c, \tag{5}$$

then we can regard the thrust vector u as the control vector, to be chosen freely subject to

$$|u| \le M, \tag{6}$$

say. In terms of this control variable, the equations describing the rate of change of velocity and mass become

$$m\dot{v} = u + h, \tag{7}$$

$$\dot{m} = -c|u|, \tag{8}$$

and (3), (7), (8) constitute the state equations for the rocket.

If conjugate variables p, q, r are associated with x, v, m respectively, then equations (1.6) become for this case

$$\begin{aligned} \dot{p} &= \frac{-qh_x}{m}, \\ \dot{q} &= -p - \frac{qh_v}{m}, \\ \dot{r} &= \frac{q(u + h - mh_m)}{m^2} \end{aligned} \tag{9}$$

and u minimizes the expression

$$\frac{qu}{m} - cr|u|. \tag{10}$$

For a given value of $|u|$ the thrust vector u will thus be chosen in the opposite direction to q', i.e.

$$u = -q'\frac{|u|}{|q|}, \tag{11}$$

and $|u|$ will then be chosen to minimize

$$-|u|\left(\frac{|q|}{m} + cr\right). \tag{12}$$

That is,

$$|u| = \begin{Bmatrix} 0 \\ \text{indeterminate} \\ M \end{Bmatrix} \text{according as } \frac{|q|}{m} + cr \begin{Bmatrix} < \\ = \\ > \end{Bmatrix} 0. \tag{13}$$

The vector q is often termed the *primer*; its direction determines the optimal thrust direction and the magnitudes of q and r determine the optimal thrust magnitude. The paths on which $|u|$ is respectively 0, M or intermediate in value are often called *maximal thrust*, *null thrust* and *intermediate thrust* arcs respectively, or simply MT, NT and IT arcs.

Equations (9) simplify somewhat if the rocket is moving in a purely gravitational field, when

$$h = m\gamma, \tag{14}$$

where $\gamma = \gamma(x)$ is the gravitational vector. One is thus neglecting aerodynamic effects such as drag. The vector γ will have the form

$$\gamma(x) = \left(\frac{\partial V(x)}{\partial x}\right)' \tag{15}$$

if the field is a conservative one, V being then the potential energy associated with the field.

In the case (14) equations (9) reduce to

$$\dot{p} = -q\gamma_x, \tag{16}$$

$$\dot{q} = -p, \tag{17}$$

$$\dot{r} = \frac{qu}{m^2}, \tag{18}$$

and the primer obeys the equation system

$$\ddot{q} = q\gamma_x = qV_{xx} \tag{19}$$

if (15) also holds. Here V_{xx} is the matrix of second differentials of $V(x)$. Equations (19) become particularly simple if V may be assumed quadratic, when V_{xx} is independent of x.

As a very special case, consider the problem of maximizing the height reached by a sounding rocket, assuming constant gravity, and neglecting (implausibly!) effects such as aerodynamic drag. The problem can be taken as a one-dimensional one, in which x represents height measured upwards from the starting point (the ground). The quantities v, p, q and r will all be scalar, and $\gamma = -g$, where g is the gravitational constant. The various equations are, in full,

$$\begin{aligned} \dot{x} &= v, \\ \dot{v} &= \frac{u}{m} - g, \\ \dot{m} &= -c|u|, \end{aligned} \tag{20}$$

$$\begin{aligned} \dot{p} &= 0, \\ \dot{q} &= -p, \\ \dot{r} &= \frac{qu}{m^2}, \end{aligned} \tag{21}$$

$$H = pv + q\left(\frac{u}{m} - g\right) - cr|u| = 0. \tag{22}$$

Since the problem is one of maximization, then

$$u = \begin{Bmatrix} 0 \\ \text{intermediate in the direction of } q \\ \text{maximal in the direction of } q \end{Bmatrix} \text{according as } \kappa \begin{Bmatrix} < \\ = \\ > \end{Bmatrix} 0 \tag{23}$$

where

$$\kappa = \frac{|q|}{m} - cr.$$

The problem is one of maximizing x at termination. Let m_0 denote the mass of the rocket structure, so that $w = m - m_0$ is the mass of fuel remaining. The terminal conditions holding are then

$$\begin{aligned} p &= 1, \\ q &= 0, \\ r &\leq 0 \quad \text{if} \quad w > 0, \\ r &\geq 0 \quad \text{if} \quad w = 0. \end{aligned} \tag{24}$$

It follows from (21) and (24) that

$$p = 1, \qquad q = s. \tag{25}$$

Thus

$$\kappa = \frac{s}{m} - cr$$

and so, by (20) and (21),

$$\frac{\mathrm{d}\kappa}{\mathrm{d}s} = \frac{1}{m} + \frac{cs(u - |u|)}{m^2}.$$

But u, if non-zero, is in the direction of $q = s$, by (23), and so positive (upward-directed). Thus

$$\frac{\partial \kappa}{\partial s} = \frac{1}{m} > 0$$

so that κ decreases strictly through time. The thrust programme then must take the form of a period of maximal thrust followed by a period of null thrust, either or both of these periods possibly being zero.

Denote the terminal value of r by r_0. If $r_0 > 0$ (so that $w = 0$ and $m = m_0$ at termination, by (24)) then thrust is zero for $s < cm_0 r_0$ and maximal for larger values of s, and it follows from (22) that $v = 0$ at termination. This is the usual case, for which an MT arc is followed by an NT arc in which the rocket coasts to a maximal height, when its velocity is zero.

If $r_0 \leq 0$ then $\kappa > 0$ at termination. There is thus no NT arc; maximal thrust is applied throughout. If $r_0 = 0$ then it follows again from (22) that $v = 0$ at termination. If $r_0 < 0$ then $v < 0$ at termination. These are the cases in which thrust is insufficient to lift the rocket. If initially the rocket happens to be already rising, maximal thrust is applied until velocity is on the point of reversing, which is taken as the terminal instant. If the rocket happens to be already falling, termination is immediate.

6. TRANSFER OF A ROCKET BETWEEN COPLANAR CIRCULAR ORBITS IN A CENTRAL INVERSE-SQUARE FIELD

The whole subject of rocket guidance in the field of a planetary system is a complicated one, now, of course, brought to a high degree of sophistication. In this section we shall merely give an indication of the simplest result.

Suppose one has a single spherical planet, which then has an inverse-square force field. It is known that the free motion of a point mass in such a field must be a conic section (ellipse, parabola or hyperbola) in a plane including the centre of the planet, and with the centre as focus. The null thrust arcs of a 'point' rocket must then be of this character. The closed orbits must be ellipses, with circles as special cases.

We shall consider the case of a single planet, with rocket motion restricted to a particular plane through the centre of the planet, and also adopt other simplifying assumptions later in the analysis. If Cartesian coordinates x_1, x_2 are adopted in this plane then the equations of motion of the rocket become

$$\begin{aligned} m\ddot{x}_1 &= f_1, \\ m\ddot{x}_2 &= f_2, \end{aligned} \tag{1}$$

where m is the rocket mass and f_1, f_2 the components of applied force. Transform now to polar coordinates r, θ by the relations

$$\begin{aligned} x_1 &= r\cos\theta, \\ x_2 &= r\sin\theta. \end{aligned} \tag{2}$$

Denoting radial velocity $\dot{r}$ and angular velocity $\dot{\theta}$ by ρ, ϕ respectively, we then rewrite equations (1) as a system of state equations in the new coordinates:

$$\dot{r} = \rho \tag{3}$$

$$\dot{\theta} = \phi, \tag{4}$$

$$\dot{\rho} = r\phi^2 - \frac{\gamma}{r^2} + \frac{u^{(r)}}{m}, \tag{5}$$

$$\dot{\phi} = -\frac{2\rho\phi}{r} + \frac{u^{(\theta)}}{mr}, \tag{6}$$

$$\dot{m} = -c|u|. \tag{7}$$

The term $-\gamma/r^2$ in (5) represents the central inverse-square gravitational field, and $u^{(r)}, u^{(\theta)}$ represent the components of rocket thrust u in the mutually orthogonal directions of increasing r and increasing θ respectively. The final equation (7) completes the set by relating mass to fuel consumption.

The variables p_1, p_2, p_3, p_4, p_5, respectively conjugate to r, θ, ρ, ϕ, m, obey the equations

$$\dot{p}_1 = -\left(\phi^2 + \frac{2\gamma}{r^3}\right)p_3 + \left(\frac{u^{(\theta)}}{m} - 2\rho\phi\right)\frac{p_4}{r^2}, \tag{8}$$

$$\dot{p}_2 = 0, \tag{9}$$

$$\dot{p}_3 = -p_1 + \frac{2\phi p_4}{r}, \tag{10}$$

$$\dot{p}_4 = -p_2 - 2r\phi p_3 + \frac{2\rho p_4}{r}, \tag{11}$$

$$\dot{p}_5 = \frac{1}{m^2}\left(u^{(r)}p_3 + \frac{u^{(\theta)}p_4}{r}\right). \tag{12}$$

By the maximum principle u will maximize

$$\frac{p_3 u^{(r)}}{m} + \frac{p_4 u^{(\theta)}}{mr} - cp_5|u| \tag{13}$$

subject to

$$|u| \leq M \tag{14}$$

if the function to be optimized is wholly terminal.

We shall consider what would seem to be the simplest manoeuvre: transfer from one circular orbit to another (coplanar) with minimal fuel consumption. Suppose we wish to transfer from $r = R_1$ to $r = R_2$, with the terminal value of θ regarded as irrelevant and the terminal value of m to be maximal. On a ballistic orbit $r = R$ one will, of course, have

$$\begin{aligned} \rho &= 0, \\ \phi &= \sqrt{(\gamma R^{-3})}. \end{aligned} \tag{15}$$

The terminal conditions imply that

$$p_2 = 0, \tag{16}$$

$$p_5 = 1 \tag{17}$$

at termination, and it follows from (9) that (16) will hold throughout all time.

On the ballistic orbit $r = R$ one sees then from equations (8) to (12) that p_2 is zero, that p_5 is constant and that p_1, p_3, p_4 obey a linear system. This linear system yields, by elimination,

$$\ddot{p}_1 + \phi^2 p_i = 0 \qquad (i = 1, 3, 4), \tag{18}$$

so that these three quantities vary sinusoidally with the period $2\pi/\phi$. One then has

$$p_3 \propto \sin(\phi t - \psi)$$

for some ψ, and finds that

$$p_3^2 + \frac{p_4^2}{R^2} \propto 1 + 3\cos^2(\phi t - \psi). \tag{19}$$

The significance of expression (19) is that it is proportional to the squared magnitude of the primer, as we see from (13). In fact, full thrust should be applied at these moments when the expression

$$\frac{\sqrt{(p_3^2 + p_4^2/r^2)}}{m} - cp_5$$

is positive.

Suppose, however, that $M = \infty$ in (14), so that there is no physical limit on

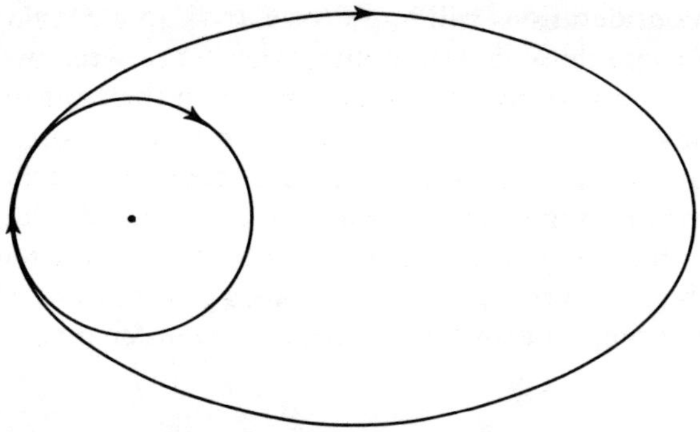

Figure **6.1** The effect of a tangential impulse applied to a craft in a circular ballistic orbit

thrust. This corresponds to the case of an infinitely powerful rocket motor. Since neither m nor p_5 change except when thrust is applied, then the optimal thrust will in this case take the form of impulses, applied when expression (19) is maximal.

Since $p_3 = 0$ on a circular orbit at the moment when the primer is maximal, it follows from the form of expression (13) that also $u^{(r)} = 0$, i.e. that the impulse, when it is applied, is wholly tangential to the circle. The effect of this impulse will be to perturb the trajectory from a circular one to an elliptic one, tangential to the circle at one end of the major axis of the ellipse (see Figure 6.1).

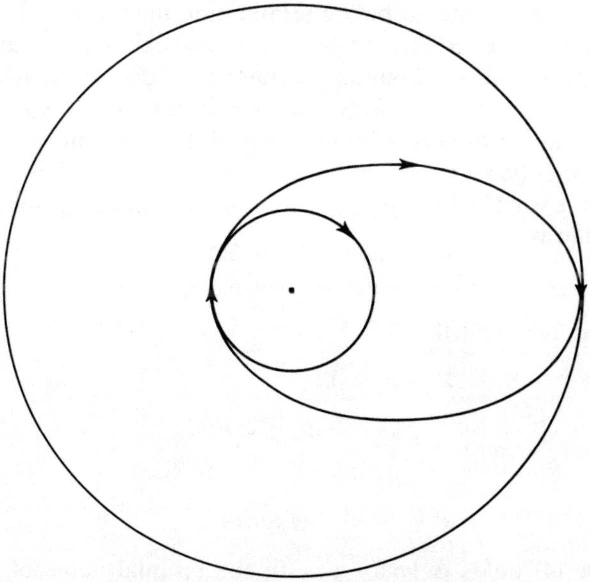

Figure **6.2** Optimal transfer between coplanar circular ballistic orbits in a central force field

As the same considerations will apply for entry into a circular orbit from an elliptic one, one might think that the optimal transfer between two circular orbits would be via an elliptic orbit, as in Figure 6.2, with the appropriate tangential impulses at points A and B. This is, in fact, the optimal transfer mode, if the two circular orbits are to have the same direction of rotation and if the ratio R_1/R_2 falls within a certain range (see Lawden, 1963). For other values of R_1/R_2 the optimal transfer may be multi-stage with an intermediate circular orbit. This intermediate orbit may even have infinite radius, a possibility ruled out however if the optimization criterion also takes account of transfer time.

7. CONTROL OF TEMPERATURE PEAKING

This is the problem set out in Section 3.9, of minimum-time passage to a given coordinate consistent with a constraint $\theta(t) \leq \bar{\theta}$ on the temperature reached. This is then a situation in which the orbit may not be wholly free, in that it may reach the boundary of permitted state space.

Because only finite accelerations are feasible, velocity v can only change continuously, and hence $\dot{\theta} = \alpha v^2 - \theta$ can only change continuously. This implies that if ever $\theta = \bar{\theta}$ then $\dot{\theta} = 0$, except possibly at termination, where one demands only that $\dot{\theta} \geq 0$. The reason is that $\theta = \bar{\theta}$ and $\dot{\theta} \neq 0$ at a given time would imply $\theta > \bar{\theta}$ either just before or just after that time, which is forbidden (except that overheating may be accepted after termination). Otherwise expressed, if the boundary of $\mathscr{X}$ is encountered before termination then it must be both entered from $\mathscr{X}$ and left for $\mathscr{X}$. Furthermore, both entry and exit must be tangential, since the rate of approach to the boundary cannot vary discontinuously.

Such a sojourn on the boundary will be referred to as a *boundary orbit*, in distinction to a free orbit. On a boundary orbit the relations $\theta = \bar{\theta}, |v| = \bar{v}$ then hold, where $\bar{v} = \sqrt{(\bar{\theta}/\alpha)}$.

Let us take the variables (p_1, p_2, p_3) as conjugate to (x, v, θ) respectively, so that the Hamiltonian is

$$H = 1 + p_1 v + p_2 u + p_3(\alpha v^2 - \theta).$$

On an optimal free orbit then

$$\dot{p}_1 = 0, \tag{1}$$

$$\dot{p}_2 = -p_1 - 2\alpha p_3 v, \tag{2}$$

$$\dot{p}_3 = p_3, \tag{3}$$

$$u = -M \operatorname{sgn}(p_2), \tag{4}$$

where the rule (4) holds only for $p_2 \neq 0$; the optimal value of u if $p_2 = 0$ is undetermined.

We see from (1) that $p_1 = F_x$ is constant. We shall suppose we are in the region $x < 0$, so that this constant value must be negative.

A free orbit can terminate in two ways: either by true termination ($x = 0$) or by entry into a boundary orbit.

In the case of true termination one must have $\theta \leq \bar{\theta}$ and $v > 0$ terminally; also, by (2.5), $p_2 = p_3 = 0$. If we express quantities as a function of time-to-go s, then it follows from (3) that $p_3[s] = 0$, from (2) that $p_2[s] = p_1 s$ and from (4) that $u[s] = M$ for $s > 0$.

Therefore, if true termination occurs at the end of a free orbit then the history is one of constant and maximal acceleration towards the target point. The only way such an orbit could emerge from a boundary orbit would be if one went back in time sufficiently far that v had so large a *negative* value that $\theta = \bar{\theta}$.

The only boundary orbit which could immediately precede termination is that for which $v = \bar{v}$. At a time t_0 when a free orbit enters such a boundary orbit one must have $p_3(t_0) \geq 0$, because an increase in temperature could not but imply an earlier constraint on orbit.

The following argument shows also that $p_2(t_0) \geq 0$. Since θ rises to $\bar{\theta}$ at t_0, then it must be true that $v > \bar{v}$ immediately before t_0. If $p_2(t_0)$ were negative then one would have $u = M, \dot{v} > 0$ at t_0, and so $v > \bar{v}$ at and after t_0. This would imply that θ rises above $\bar{\theta}$ after t_0, which is forbidden except at termination. Consequently, $p_2(t_0)$ must be non-negative.

Assume $p_2(t_0) > 0$. Then $u = -M$ at t_0, and deceleration is maximal. This will also be the case on the free orbit immediately prior. As t decreases from t_0 then

$$v(t) = \bar{v} + M(t_0 - t),$$

$$p_3(t) = p_3(t_0)e^{t - t_0}.$$

It follows then from these two relations and (2) that $\dot{p}_2$ ultimately becomes positive as t is decreased. As one continues back in time, p_2 will thus ultimately change sign, and one enters a phase of maximal acceleration, $u = M$. The quantity $\dot{p}_3$ cannot surpass its value at this point as t decreases further, so that $\dot{p}_2$ and p_2 remain positive and negative respectively, and there is no further reversal of thrust in free orbit. The situation continues thus until one again goes so far back in time that v is sufficiently large and negative that $\theta = \bar{\theta}$.

If $p_2(t_0) = 0$ and $\dot{p}_2(t_0) > 0$ then $p_2(t_0 -) > 0$, and we can follow the argument just set out. The case $p_2(t_0) = 0$ and $\dot{p}_2(t_0) < 0$ can be excluded by the same arguments as those which excluded $p_2(t_0) < 0$.

We thus see that termination can be preceded by just two types of behaviour: (a) maximal acceleration without temperature constraint or (b) a phase of maximal acceleration to a velocity v exceeding $\bar{v}$, followed at an appropriately chosen instant by a phase of maximal deceleration, so that v falls to $\bar{v}$ and θ rises to $\bar{\theta}$ simultaneously; the orbit is then completed at constant velocity and temperature under zero thrust. An orbit of one of these two types passes through any initial (x, v, θ) for which ultimate overheating is not inevitable.

Exercise

1. The drag which raises temperature presumably also brakes motion. Treat the case in which the equation $\dot{v} = u$ is modified to

$$\dot{v} = u - \beta v$$

with β positive.

CHAPTER 8

The continuous maximum principle; derivation

A completely rigorous proof of the maximum principle requires a similarly rigorous proof of certain properties of solutions of perturbed differential equations, formulated as assertions A1 and A2 of Sections 2 and 3. If these assertions are taken for granted (and their validity in regular cases is clear) then the treatment of this chapter is sufficient to demonstrate the validity of the maximum principle in continuous time and to show why this is independent of convexity assumptions.

1. CONTINUOUS TIME AND CONVEXITY

In establishing conditions for the validity of the discrete maximum principle in Section 6.8 we modified the plant equation by introduction of a forcing term b_t as in (6.1.3), and regarded the minimized loss as a function $\mu(b)$ of the forcing terms. A sufficient condition for validity of the discrete principle is that $\mu(b)$ be convex, this condition also being necessary if the principle is to hold for all relevant b. It follows as a special case that, if $\mu(b)$ is convex in b, then $F(x, t)$ is convex in x.

For maximization problems one of course demands concavity of a maximal criterion function.

It is useful to consider a simple problem, non-convex in prescription, for which the maximum principle is invalid in discrete time. This illustrates how non-convexities are automatically remedied in continuous time.

Consider the plant equation

$$x_{t+1} = x_t + u_t + b_{t+1} \tag{1}$$

in scalar x and u, and suppose that u_t can adopt just one of the two values ± 1 for any t. Suppose that $\sum_1^T |x_t|$ is to be minimized, for $x_0 = 0$. If b_t itself takes values ± 1 then u_t can be chosen to cancel it, and $\mu(b) = 0$. If, however, b_t takes the value 0 for all t then the best one can do is to alternate u_t between $+1$ and -1, so that $|x_t|$ alternates between 0 and $+1$, and

$$\mu \sim \frac{T}{2}.$$

In this case, then,

$$\mu(0,0,\ldots,0) > \tfrac{1}{2}[\mu(-1,-1,\ldots,-1) + \mu(1,1,\ldots,1)]$$

so that μ is not convex, and the maximum principle is invalid. The failure could be ascribed to the fact that u takes values in a non-convex set: $\{-1,+1\}$.

Consider now the continuous time analogue of (1):

$$\dot{x} = u + b \tag{2}$$

with $u = \pm 1$ again. It is intuitively clear that, if $|b(t)| \leq b < 1$ and $b(t)$ is continuous, then u can be chosen to annul any b sequence, in the sense that $\int_0^T |x(t)|\,dt$ can be made arbitrarily small for $x(0) = 0$ and any finite T. Indeed, consider the rule under which u adopts the value -1 as $x(t)$ enters $x \geq \varepsilon$, the value $+1$ as $x(t)$ enters $x \leq -\varepsilon$, and does not change otherwise. Then, for fixed positive ε, $u(t)$ changes value a bounded number of times, and $\int_0^T |x|\,dt < T\varepsilon$. In the ideal (if practically unrealizable) limit $\varepsilon \downarrow 0$ the value of x is held exactly at zero, perhaps by an infinitely rapid alternation of the values ± 1 permitted to u.

In fact, by varying u sufficiently rapidly one can effectively achieve an arbitrary distribution $\rho_t(u)$ of values of u over $\mathscr{U}$ in the course of an infinitesimal time interval $(t, t+\delta t)$. If $a(x,u,t)$ varies sufficiently continuously with x and t the plant equation will then effectively become

$$\dot{x}_t = \int \rho_t(u)a(x,u,t)\nu(du)$$

for an appropriate u-measure ν, and the instantaneous cost rate at t correspondingly becomes $\int \rho_t(u)g(x,u,t)\nu(du)$. The formal proof of these statements under appropriate regularity conditions is often referred to as 'the chattering lemma'. The choice of u from a possibly non-convex set $\mathscr{U}$ can thus be conceived as being replaced by the choice of a distribution $\rho(u)$, drawn from the convex set of such distributions over $\mathscr{U}$. In terms of the 'randomized control' $\rho(\cdot)$ the problem then has all the linearity and convexity properties required to establish a formal continuous time equivalent of Theorem 6.8.2.

Exercise

1. Show that a formal application of the maximum principle to the discrete-time problem considered in this section yields the equations

$$p_t - p_{t-1} + \operatorname{sgn}(x_t) = 0,$$
$$u_t = -\operatorname{sgn}(p_t).$$

Show that the corresponding continuous time equations have solutions corresponding to an optimizing control.

2. PROOF OF THE MAXIMUM PRINCIPLE UNDER CONVEXITY ASSUMPTIONS

We shall prove the maximum principle under convexity assumptions for the case of the stopping rule $t = T$ and purely terminal costs. The terminal-cost formulation is purely a matter of normalization (see Section 2.4) and the proof of this section adapts in an obvious way to more general stopping rules.

For simplicity we shall suppose the set of admissible controls to be the same for all x, t. This assumption then excludes the case when the orbit may be constrained by encounter with the boundary of $\mathscr{X}$. The maximum principle can cover this case, at the expense of some complication.

Under these specializations the maximum principle will amount to the following assertion: if $\{\bar{x}(t), \bar{u}(t)\}$ is an optimal orbit then there exists a row vector $p(t)$ such that

$$\dot{p}(t) = -p(t)a_x(\bar{x}(t), \bar{u}(t), t) \qquad (0 \leq t < T), \tag{1}$$

$$p(T) = K_x(\bar{x}(T)) \tag{2}$$

and

$$p(t)a(\bar{x}(t), u(t), t) \geq p(t)a(\bar{x}(t), \bar{u}(t), t) \qquad (0 \leq t < T) \tag{3}$$

for any alternative control $u(t)$. (Note: inequality (3) is in the direction appropriate for a *minimization* problem.)

The treatment following, unlike that of Section 7.1, does not invoke the idea of the cost function $F(x, t)$ or presume its differentiability. A convexity assumption is made provisionally, to be discarded in the next section.

The following lemma for linear equation systems is well known.

Lemma 2.1. *The linear equations system in* $x \in R^n$

$$\dot{x}(t) = a(t)x(t) \tag{4}$$

has the solution

$$x(t) = \phi(t, \tau)x(\tau) \qquad (t \geq \tau) \tag{5}$$

where the square matrix ϕ *satisfies*

$$\phi(t, t) = I, \tag{6}$$

$$\frac{\partial}{\partial t}\phi(t, \tau) = a(t)\phi(t, \tau), \tag{7}$$

$$\frac{\partial}{\partial \tau}\phi(t, \tau) = -\phi(t, \tau)a(\tau), \tag{8}$$

for $t \geq \tau$. *The non-homogeneous equation*

$$\dot{x}(t) = a(t)x(t) + y(t) \tag{9}$$

has the solution

$$x(t) = \phi(t, \tau)x(\tau) + \int_{\tau}^{t} \phi(t, \sigma)y(\sigma)\,\mathrm{d}\sigma \qquad (t \geq \tau). \tag{10}$$

Proof. There must be a linear relationship between $x(t)$ and $x(\tau)$; let it be (5). Then (6) is a necessary consequence, and (7), (8) follow by differentiation of (5) with respect to t and τ respectively. The matrix $\phi(t, \tau)$ is the Green's function of the system (4); solution (10) is easily verified. ■

Consider now the function $a(x, u, t)$ appearing in the plant equation (7.1.1), and let $\mathscr{A}$ be the set of functions $A(x, t)$ of x and t generated by $a(x, u, t)$ as different admissible choices are made of $u(t)$. This is well defined, since $\mathscr{U}$ is assumed independent of x and t. In this section we shall make the provisional assumption that $\mathscr{A}$ is *convex*, in that, if $\bar{A}$ and A belong to $\mathscr{A}$, then so does $(1 - \varepsilon)\bar{A} + \varepsilon A$ for any ε in $[0, 1]$. That is, for any $\bar{u}, u$ in $\mathscr{U}$ there exists a u_ε in $\mathscr{U}$ such that

$$(1 - \varepsilon)a(x, \bar{u}, t) + \varepsilon a(x, u, t) = a(x, u_\varepsilon, t) \tag{11}$$

for all x, t.

Let $\bar{u}(t)$ be a prescribed control for $t \geq 0$ and let $\bar{x}(t)$ be the corresponding state path for prescribed $x(0) = \bar{x}(0)$. Then $\bar{x}(t)$ is the solution of the plant equation

$$\dot{x}(t) = a(x(t), \bar{u}(t), t). \tag{12}$$

Let $u(t)$ be an alternative control and let $x_\varepsilon(t)$ be the solution, with the same initial condition $x(0) = \bar{x}(0)$, of the perturbed plant equation

$$\dot{x}(t) = (1 - \varepsilon)a(x(t), \bar{u}(t), t) + \varepsilon a(x(t), u(t), t). \tag{13}$$

By the convexity assumption (11), relation (13) is the actual plant equation for some control $u_\varepsilon(t)$.

Consider the following assumption:

Assumption A1. *If $\bar{x}(t)$ and $x_\varepsilon(t)$ are the state paths just defined, then, for small ε,*

$$x_\varepsilon(t) = \bar{x}(t) + \varepsilon\zeta(t) + o(\varepsilon) \tag{14}$$

where $\zeta(t)$ satisfies the differential equation

$$\dot{\zeta} = a_x(\bar{x}, \bar{u}, t)\zeta + a(\bar{x}, u, t) - a(\bar{x}, \bar{u}, t) \tag{15}$$

for arbitrary prescribed $\bar{u}(t), u(t)$ and $x(0)$.

If the perturbation relation (14) is assumed then (15) is just what one would obtain by substituting (14) into (13) and considering the terms which are formally of order ε. The assumption (A1) that this formal calculation is valid amounts to a regularity condition on the plant equation.

Theorem 2.1. *Suppose that the set of functions $\{a(x,u,t): u \in \mathscr{U}\}$ is convex, that the regularity condition (A1) on the plant equation holds and that the terminal loss $K(x)$ is differentiable in a neighbourhood of the terminal point. Then the maximum principle holds.*

Proof. If $\{\bar{x}, \bar{u}\}$ is optimal then the perturbation $\bar{u} \to u_\varepsilon$, inducing the perturbation $\bar{x} \to \bar{x} + \varepsilon\zeta + o(\varepsilon)$, cannot decrease the value of K at termination, so that

$$p(T)\zeta(T) \geq 0 \tag{16}$$

with $p(T)$ defined by (2). Since $x(0)$ is prescribed then $\zeta(0) = 0$, and formula (10) applied to the system (15), linear in ζ, yields

$$\zeta(T) = \int_0^T \phi(T,t)(a(\bar{x},u,t) - a(\bar{x},\bar{u},t))\,\mathrm{d}t. \tag{17}$$

Here ϕ is the Green's function of the linear system

$$\dot{\zeta} = a_x(\bar{x},\bar{u},t)\zeta. \tag{18}$$

If we define $p(t)$ by

$$p(t) = p(T)\phi(T,t) \tag{19}$$

then this indeed satisfies (1), by (8), (18). From (16), (17), (19) we deduce that

$$\int_0^T p(t)(a(\bar{x},u,t) - a(\bar{x},\bar{u},t))\,\mathrm{d}t \geq 0. \tag{20}$$

Since $u(t)$ can be chosen arbitrarily in $\mathscr{U}$ for each t, then (20) implies (3). ■

3. PROOF OF THE MAXIMUM PRINCIPLE WITHOUT CONVEXITY

The convexity property demanded of the plant equation in the last section can effectively be achieved if the chattering lemma is valid, so that the choice of u can be conceived as being replaced by the choice of a distribution, $\rho(u)$. However, rather than prove and invoke some form of the chattering lemma we shall appeal directly to the following strengthened form of the regularity condition A1.

Assumption A2. *Consider two controls $\bar{u}(t), u(t)$ and the control $u_\varepsilon(t)$ under which $\bar{u}(x)$ and $u(t)$ are applied for fractions $1-\varepsilon$ and ε of the time respectively, in any time interval, however small. Let the orbits under $\bar{u}$ and u_ε be denoted $\bar{x}$ and x_ε. Then the assertions (2.14), (2.15) of assumption A1 hold for arbitrary prescribed $\bar{u}(t)$, $u(t)$ and $x(0)$.*

The assumption is that the 'mixed' control $u_\varepsilon(t)$ will yield a plant equation (2.13) and that the assertion of A1 will then again apply. This is what one would expect on formal grounds, so that A2 is again a regularity condition on the plant equation. Theorem 3.1 below follows by the same proof as Theorem 2.1, but without invocation of convexity, because A2 implies that the perturbed plant equation (2.13) is the actual plant equation if the mixed control $u_\varepsilon(t)$ is used.

Theorem 3.1. *Suppose that the regularity condition (A2) on the plant equation holds, and that the terminal loss $K(x)$ is differentiable in a neighbourhood of the terminal point. Then the maximum principle holds.*

4. INSUFFICIENCY OF THE MAXIMUM PRINCIPLE

The continuous maximum principle has a mixed local/global character. It optimizes under global variations of u at a given time, but only under local variations of the x-orbit itself. For this reason orbits derived by the principle always have some kind of extremal significance, but not necessarily the desired one.

Consider the following example, due to Bather (unpublished). Let x, u be scalars satisfying

$$\dot{x} = u \tag{1}$$

subject to

$$|u| < 1 \tag{2}$$

and suppose that the only cost is the terminal cost

$$K(x) = \cos x \tag{3}$$

incurred at $t = T$. The Hamiltonian is then

$$H = pu + p_0$$

so that application of the maximum principle yields

$$u = -\operatorname{sgn} p,$$

$$\dot{p} = 0$$

and p has a constant value on an orbit

$$p = -\sin x(T).$$

If $p \neq 0$ one thus deduces the orbit

$$x(t) = x(T) + (t - T)\operatorname{sgn}(\sin x(T)). \tag{4}$$

That is, $x(t)$ will change at a constant rate, the rate being $+1$ or -1 according to whether $K(x)$ is decreasing or increasing at $x(T)$. The line ABC in Figure 4.1 represents one such path. However, in such a case one would do better to follow the path ABD, which gives a terminal cost of -1.

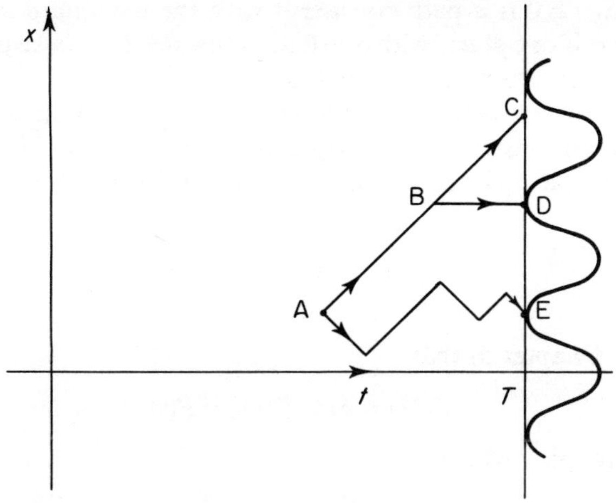

Figure **4.1** An example of the insufficiency of the maximum principle. The termination point C is optimal for starting points on ABC which are near enough to it. However, if the starting point is sufficiently far removed then a trajectory such as ABD will give the global extreme. (The cosine curve on the right is intended to indicate the magnitude of the terminal-cost function, as a function of x, and should properly be drawn in a plane perpendicular to the paper)

The extremal significance of the path ABC is that it is optimal with respect to local variations of the terminal point C. It is even globally optimal if one starts on this path close enough to C. However, if the starting point is sufficiently remote from C then one does better to aim for another termination point. The path ABD is, of course, consistent with the maximum principle in that, on this path, $p = 0$ and all values of u minimize the Hamiltonian.

This extreme lack of determination of the optimal policy corresponds to the fact that the optimal policy is indeed far from unique. Any $\{u(t)\}$ which brings $x(T)$ to a value minimizing $K(x)$ is optimal; consider the path AE.

The exercises illustrate the same phenomenon for some cases where the costs are not purely terminal.

Exercises

1. Consider a problem with the plant equation (1) and constraint (2), zero termination cost at $t = T$ and instantaneous cost rate $g(x, u) = h(x)|u|$, where $h(x)$ is a differentiable approximation to the specification

$$h(x) = \begin{cases} 1 & (|x| \leq 1), \\ -1 & (|x| > 1). \end{cases}$$

Show that $x(t) \equiv 0$ is a path consistent with the maximum principle. Show, however, that if one starts with $x = 0$ at t, and $s = T - t$ is large enough, then there is an alternative path of smaller cost.

2. Consider a problem with the plant equation (1), unconstrained u, terminal cost kx^2 at $t = T$ and instantaneous cost rate $g(x, u) = u^2 - c^2x^2$.

Show that the orbits through $x(T) = 0$ determined by the maximum principle are

$$x(T - s) = \alpha \sin(cs) \tag{5}$$

for some α.

Show (cf. Chapter 5) that

$$F_s(x) = F(x, T - s) = P(s)x^2$$

for small enough s where

$$P(s) = c \tan\left[\tan^{-1}\left(\frac{k}{c}\right) - cs\right].$$

Show that if $k = +\infty$ (so that $x(T) = 0$ is demanded) then

$$F_s(x) = \begin{cases} cx^2 \cot(cs) & (0 \le s \le \pi/c), \\ -\infty & (s \ge \pi/c), \end{cases}$$

and that the path (5) is optimal for $s < \pi/c$, but not for $s > \pi/c$.

Suppose, in particular, that $x(0) = 0$, and consider a control of the form

$$u(t) = \begin{cases} a & (0 \le t < b), \\ 0 & (b \le t < T - b), \\ -a & (T - b \le t < T). \end{cases}$$

Show, by an appropriate choice of a, b, that one can achieve a path of cost $-\infty$ if $T > \sqrt{(32/3)}/c$. The value of $\sqrt{(32/3)}$ is about 3.26—only slightly larger than the true threshold value, π.

CHAPTER 9

The minimax criterion

The formulation of the optimization problem must be supplemented if, as is usual, knowledge of the values of relevant variables is incomplete at the time a decision is called for. Two approaches to such problems of incomplete information are the *minimax* and *stochastic* formulations. This chapter is devoted to the former.

1. DECISION PROCESSES WITH INCOMPLETE INFORMATION

The first part of this book is devoted to deterministic problems, the remainder to stochastic problems. However, the correct contrast is not that between deterministic or stochastic problems, but that between problems with complete or incomplete information.

We have hitherto assumed complete information, in that it has been assumed that all relevant quantities are known when needed. The motorist of Section 1.4 knew exactly where he was and knew that a route-plan could be followed exactly as decided; the production manager of Section 3.2 knew the course of future demand; the warehouseman of Section 3.3 knew the course of future prices. In general, it was assumed that the optimizer could choose actions secure in the knowledge that these would have a unique and predictable outcome.

However, in general one does not have this complete information. Variables generated outside the system (*exogenous variables*), such as demand in the production problem or prices in the warehousing problem, may well have a course which the optimizer can know or predict only imperfectly. Special cases of such exogenous variables are the various disturbances to which the system may be subject. The motorist's intended route-plan may be upset by various mishaps ('process noise'), or he may not be able to determine exactly where he is ('observation noise'). There may even be aspects of the system itself which are unknown; the motorist may, for example, be driving a strange car of unknown performance.

It can be said that temporal optimization problems reveal their full structure first when one is compelled to take account of incomplete information. On the other hand, the optimization problem is then ill-posed, unless the criterion is also modified to take this into account.

2. OBSERVABLES AND POLICIES

An interesting aspect of a problem with incomplete information is that information is generally acquired in the course of time, as observations become available. The fact that knowledge thus increases with time is one of the characteristic features of a temporal optimization problem.

The *observational history* of the process at time t will be denoted W_t. This is a catalogue of all values and facts relevant to the problem which are known at time t. Any quantity which is known at time t must then be determined by knowledge of W_t, i.e. it must be a function of W_t. We shall say that such a quantity is *observable* at time t, or *W_t-measurable*. W_t is itself the *maximal observable* at time t.

If information once acquired is never lost (e.g. forgotten or discarded) then W_{t-1} must be W_t-measurable. This fact expresses the monotonic character of the observables. We regard W_t as specifying one's state of knowledge immediately before action u_t is taken. We shall assume this includes knowledge of t and of past actions, so that t and u_{t-1} are W_t-measurable. That is, one knows the time and remembers what one has done.

Actions at time t can depend only upon known quantities at time t, so u_t is restricted to being W_t-measurable. The choice of an action rule would thus seem to be equivalent to the choice of the functional dependence $u_t(W_t)$ of u_t upon W_t. This concept of an action rule turns out to be adequate, although it can be broadened, as we shall see in the next section.

The specification of an action rule for all relevant t constitutes a *policy*. If the rule specifies a functional dependence $u_t(W_t)$ for all relevant t then the policy is *determinate*.

It is useful to define a variable W whose value prescribes the values of all variables relevant to the process. To know W would be to know everything: W_t is W-measurable for all t.

Exercises and comments

1. A problem of complete information on $t \geq 0$ could be characterized as one for which there is no information unavailable at $t = 0$ whose provision at $t = 0$ could decrease the minimal cost attainable.

   Show that this condition is fulfilled if the cost function C is a function of W_0 and U_∞ at most.

3. STOCHASTIC AND MINIMAX FORMULATIONS

In all cases we consider an optimization over $t \geq 0$, i.e. a minimization of the cost function C with respect to $U_\infty = (u_0, u_1, u_2, \ldots)$. However, if C is a function of quantities unobservable at time $t = 0$ as well as of U_∞, then the value of C for a given U_∞ is not calculable and the exhortation to 'minimize C' is meaningless.

There are at least two ways of taking account of the dependence of the cost C upon unobservables. In the *stochastic formulation* one takes $E_\pi(C|W_0)$ as the criterion—the expectation of C conditional on initial observables W_0 and calculated on the basis that a prescribed policy π is to be used. That is, one eliminates unobservables from the criterion by calculating the average cost over variation of those unobservables. For this calculation to be possible, one must be able to supply a *stochastic model* of the process—a prescription of the distribution of unobservables as random variables for given π.

For such a criterion the optimal policy π, if one exists, will be that which minimizes $E_\pi(C|W_0)$. The remainder of the book, from Chapter 10 onwards, will be concerned with optimization under this stochastic criterion.

A second approach is the *minimax formulation*, in which one takes $M_\pi(C|W_0)$ as the criterion—the maximum of C with respect to unobservables at time $t = 0$, calculated on the basis that a prescribed policy π is to be used (actually, a supremum value of C, but we shall loosely use the term 'maximum'). This could be characterized as a maximum conditional on W_0; one substitutes for u_t in C from the prescribed policy rule $u_t = u_t(W_t)$ and then maximizes C over all arguments except the known quantities W_0. For this calculation to be possible one must be able to supply a 'fuzzy model'—a prescription of the set of values over which variables can jointly range.

A minimax approach corresponds to the idea of a 'worst-case analysis'. For a given policy π one calculates the worst that could happen, so evaluating the worst-case cost $M_\pi(C|W_0)$. The policy π is then chosen to minimize this, yielding a policy optimal in the minimax sense.

The minimax approach is more pessimistic than the stochastic one, but requires less in the way of additional specification of the model. Which approach is appropriate must depend upon circumstances; certainly the stochastic formulation is far commoner. Mathematically, the two approaches develop in almost complete analogue, as will be evident from a comparison of this chapter with the next. We take advantage of this similarity to give a relatively cursory treatment of the minimax case and a more careful treatment of the equivalent stochastic material in Chapters 10 and 11. However, this is also consistent with the greater practical importance attached to the stochastic formulation. The extensive development of the stochastic case which occupies the rest of the book would almost certainly have a faithful minimax analogue, awaiting the interested investigator.

The two approaches each allow a generalization of the concept of policy. For the stochastic case the idea of a functional dependence $u_t(W_t)$ could be generalized to that of a conditional distribution, $P(u_t|W_t)$. That is, u_t is determined at time t as the sample value of a random variable with this distribution. The expectation in $E_\pi(C|W_0)$ also covers this additional random variation. Specification of $P(u_t|W_t)$ for all relevant t and W_t constitutes a *randomized policy*. A randomized policy will reduce to a determinate policy (generally termed a *deterministic policy* in this context) if each distribution is a one-point distribution, concentrated as a single value of u_t for given W_t.

For the minimax case the idea of a functional dependence $u_t(W_t)$ could be relaxed to the statement that u_t shall take some value in a W_t-dependent set $\mathscr{V}(W_t)$. Specification of $\mathscr{V}(W_t)$ for all relevant t and W_t constitutes a *fuzzy policy*, reducing to a determinate policy in the case when every $\mathscr{V}(W_t)$ has only a single member. The maximization in $M_\pi(C|W_0)$ also covers this additional indeterminacy.

In fact, for the single-optimizer cases with which we are concerned, a determinate policy can be found which is at least as good as any given indeterminate policy. Attention can thus be restricted to determinate policies. This is in general no longer true for multi-person problems, however, for which there may be an advantage if one's actions are to some extent unpredictable.

4. RECURSIONS AND TERMINAL CONDITIONS

For any two variables x, y whose joint set of variation is prescribed, the quantity $M_\pi[\phi(x)|y]$ is defined. If this is independent of π then we shall write it simply as $M[\phi(x)|y]$. This is the 'max' analogue of a condition expectation. Because of the monotonicity of $\{W_t\}$ one has

$$M_\pi[\phi|W_{t_1}] = M_\pi[M_\pi[\phi|W_{t_2}]|W_{t_1}] \tag{1}$$

for $t_1 \leq t_2$.

Define now

$$V(\pi, W_t) = M_\pi(C|W_t). \tag{2}$$

This is the maximal loss which could be incurred by operating policy π from time t with an observational history of W_t. It is fairly clearly independent of policy *before* time t, a point confirmed for the stochastic case in Theorem 10.3.1, when we formally invoke causality. The dependence of V upon t is indicated sufficiently by its dependence upon W_t.

It follows then from (1) and (2) that V obeys the recursion

$$V(\pi, W_t) = M_\pi[V(\pi, W_{t+1})|W_t] \qquad (t \geq 0). \tag{3}$$

The optimal policy is that which minimizes $V(\pi, W_0)$.

Let us say that the problem has horizon T if $V(\pi, W_T)$ is independent of the specification of π from time T onwards. This is to say that it is independent of π completely, so that we can write

$$V(\pi, W_T) = V(W_T). \tag{4}$$

Recursion (3) then holds over the range $0 \leq t < T$ with the terminal condition (4). In fact, $V(W_T)$ can be taken as the effective-cost function, whence we derive the following assertion.

Theorem 4.1. *Suppose the problem has horizon T. Then it can be so normalized that the cost function is a function $C(W_T)$ of W_T alone, defined as $M[\text{Cost}|W_T]$ if necessary. The quantities $V(\pi, W_t)$ then obey recursion (3) for $0 \le t < T$ with the terminal condition $V(\pi, W_T) = C(W_T)$.*

Exercise

1. Solve the minimax analogue of Exercise 10.6.1. It is probably best to read as far as Section 10.6 before doing so.

5. THE OPTIMALITY EQUATION

Define

$$G(W_t) = \inf_\pi V(\pi, W_t), \tag{1}$$

the minimax cost conditional on an observational history W_t at t.

Theorem 5.1 (The minimax optimality equation). *Suppose the problem has finite horizon T, so that we may suppose $C = C(W_T)$. Then the functions $G(W_t)$ obey the recursion (the optimality equation)*

$$G(W_t) = \inf_{u_t} M[G(W_{t+1})|W_t, u_t] \qquad (0 \le t < T) \tag{2}$$

with the terminal condition

$$G(W_T) = C(W_T), \tag{3}$$

and the minimizing value of u_t in (2) is minimax-optimal.

The minimizing value is a function of W_t alone, and so admissible as defining a policy. The last statement of the theorem is to be regarded as shorthand for the statement that the infimal cost can be approached if u_t is chosen to approach the infimum in (2), for all relevant t.

Proof. Suppose, for simplicity, that the infimum is attained in (2) for all relevant t, W_t by $u_t = u_t^*(W_t)$, say. The argument has an obvious ε-modification otherwise. Then this action rule defines a policy π^*, and it follows from Theorem 4.1 that the functions satisfying (2), (3) can be identified as $V(\pi^*, W_t)$. We shall show that

$$V(\pi^*, W_t) \le V(\pi, W_t) \tag{4}$$

for all t, W_t and π; this implies optimality of π^* and identity of $V(\pi^*, W_t)$ with $G(W_t)$.

Relation (4) holds at T, when both members equal $C(W_T)$. Suppose it holds at $t+1$. Then

$$\begin{aligned} V(\pi, W_t) &= M_\pi(V(\pi, W_{t+1})| W_t) \\ &\geq M_\pi(V(\pi^*, W_{t+1})| W_t) \\ &\geq M(V(\pi^*, W_{t+1})| W_t, u_t^*) \\ &= V(\pi^*, W_t) \end{aligned}$$

which completes the induction. ■

6. RECURSIVELY GENERATED COST FUNCTIONS

One is now interested in whether the optimality equation (5.2) can be reduced to a simple recursion by assumption of an appropriate version of state structure. Such a reduction is best accomplished in two stages: first, to assume that a 'future cost' can be defined, which is generated recursively; second, to assume a recursive structure for the dynamics.

Let us assume that a 'cost from time t', $C^{(t)}$, can be defined and that this obeys a backwards recursion

$$C^{(t)} = \psi(W_{t+1}, C^{(t+1)}). \tag{1}$$

For the problem with horizon T we may assume $C^{(T)}$ to be a function $C^{(T)}(W_T)$ of W_T. Define the minimax future cost

$$F(W_t) = \inf_\pi M_\pi(C^{(t)}| W_t). \tag{2}$$

Theorem 6.1. *Suppose that $C = C^{(0)}$, where the sequence $\{C^{(t)}\}$ is generated by recursion* (1), *with ψ strictly increasing in $C^{(t+1)}$ for all t, W. Suppose also that the problem has horizon $T < \infty$. Then the optimality equation takes the form*

$$F(W_t) = \inf_{u_t} M[\psi(W_{t+1}, F(W_{t+1}))| W_t, u_t] \qquad (0 \leq t \leq T) \tag{3}$$

with the terminal condition

$$F(W_T) = C^{(T)}(W_T). \tag{4}$$

The minimising u in (3) *defines the minimax value of u_t.*

Proof. Iteration of (1) for $t = 0, 1, \ldots, j-1$ yields

$$C = \psi_j(W_j, C^{(j)}), \tag{5}$$

say, where $\psi_j(\cdot, \cdot)$ is strictly increasing in its second argument. It then follows that

$$G(W_t) = \psi_t(W_t, F(W_t)). \tag{6}$$

Now let $\mathcal{M}_t$ denote the operator $\inf_{u_t} M[\cdot|W_t, u_t]$. Inserting expression (6) for $G(W_t)$ in the optimality equation (5.2) we deduce that

$$\begin{aligned}\psi_t(W_t, F(W_t)) &= \mathcal{M}_t\psi_{t+1}(W_{t+1}, F(W_{t+1}))\\ &= \mathcal{M}_t\psi_t(W_t, \psi(W_{t+1}, F(W_{t+1})))\\ &= \psi_t(W_t, \mathcal{M}_t\psi(W_{t+1}, F(W_{t+1}))).\end{aligned} \tag{7}$$

This last commutation of $\mathcal{M}_t$ and ψ_t is valid because W_t is a constant as far as $\mathcal{M}_t$ is concerned, and ψ_t is increasing in its second argument. Since it is *strictly* increasing, relation (7) can be inverted to (3). ■

7. STATE STRUCTURE

Assume that there exists a sequence of variables $\{x_t\}$ such that:

(a) x_t is W_t-measurable.

(b) For any t, W and ϕ the quantity $M[\phi(x_{t+1})|W_t, u_t]$ is a function of x_t, u_t and t alone, and is consequently identifiable with $M[\phi(x_{t+1})|x_t, u_t]$.

(c) The set of values u_t may adopt depends on x_t and t at most; $\mathscr{U}(x_t, t)$, say.

(d) $C = C^{(0)}$, where $\{C^{(t)}\}$ is generated by the backward recursion

$$C^{(t)} = \psi(x_t, x_{t+1}, u_t, t, C^{(t+1)}) \qquad (0 \le t < T), \tag{1}$$

for which ψ is strictly increasing in its last argument and $C^{(T)}$ is specified as a function $K(x_T)$ of x_T.

These conditions are the minimax analogues of conditions (a) to (d′) of Sections 2.2 and 2.3, and will similarly imply that x is a state variable. Condition (a) requires that x_t be observable at time t. Condition (b) is the 'fuzzy' analogue of the assumption (2.2.1) of a simply recursive plant equation. In the case of incomplete information, x_{t+1} is not in general determined by W_t, u_t and even less by x_t, u_t, t. However, condition (b) implies that its maximal excursions conditional on these two sets of variables are the same.

Theorem 7.1 (The optimality equation with minimax state structure). *Under conditions (a) to (d) above $F(W_t)$ is a function of x_t, t alone, $F(x_t, t)$, say, and the optimality equation reduces to*

$$F(x_t, t) = \inf_{u_t} M[\psi(x_t, x_{t+1}, u_t, t, F(x_{t+1}, t+1))|x_t, u_t] \qquad (0 \le t < T) \tag{2}$$

with the terminal condition

$$F(x_T, T) = K(x_T). \tag{3}$$

The minimizing value in (2) determines a minimax value of u_t.

Proof. This follows the now-familiar inductive course, establishing that the optimality equation (6.3) reduces to (2) under the given assumptions. ■

We conclude with four examples, three of which have state structure.

8. MINIMAX LOCATION OF THE ZERO OF A FUNCTION

Suppose that all that is known of a function $f(\xi)$ of a scalar variable ξ is that it is strictly monotonic increasing and has a root in $[a, b]$. Thus $f(\xi)$ has a single but unknown zero, ξ_0, in $[a, b]$. One is allowed T function evaluations and wishes to choose these so as to locate ξ_0 optimally in the minimax sense, so that the maximal length of the interval within which the root can be guaranteed to lie should be minimal. From evaluation of f at ξ one can of course determine whether $\xi_0 > \xi$ or $\xi_0 < \xi$ (or, exceptionally, that $\xi_0 = \xi$).

One's initial information, W_0, is specification of the interval $[a, b]$. Suppose that one is now allowed T function evaluations; the policy π is a rule for choosing the values of ξ at which evaluations are made. These evaluations yield information; the cost is the length of the interval within the zero is definitely located by this information.

Suppose all T function evaluations must be made simultaneously. The maximal cost incurred is then

$$V(\pi) = \max_j (d_j)$$

whose d_j is the length of an interval between consecutive evaluation points under policy π. Clearly the minimax policy is to choose the T evaluation points and the end-points equispaced in $[a, b]$, and

$$\inf_\pi V(\pi) = \frac{b-a}{T+1}. \tag{1}$$

Suppose, however, that evaluations may be made sequentially. Then, fairly obviously, one chooses the next evaluation in the centre of the sub-interval within which the root has currently been located, and achieves

$$\inf_\pi V(\pi) = 2^{-T}(b-a). \tag{2}$$

We shall treat the problem formally, however, partly to exhibit it as indeed a sequential minimax problem with state structure and partly to prepare for the more difficult problem of the next section.

The information W_t after t evaluations is the set of evaluation points ξ_j and evaluations $f(\xi_j)$, for $1 \le j \le t$. The state variable might be taken as the coordinates of the ends of the sub-interval within which the root has currently been located. However, as far as costs are concerned, one may as well take a new origin and take the state variable x_t as being the length of this sub-interval, broken into yet smaller intervals of lengths u_t and $x_t - u_t$ by the next function evaluation. Then all that one can predict of x_{t+1} is that it will be either u_t or $x_t - u_t$. The cost function is purely terminal.

$$C^{(t)} = C^{(t+1)} = C^{(T)} = x_T.$$

Let us write $F(x, t)$ rather as $F_{T-t}(x)$, where $T - t$ is the number of further evaluations permitted. Then

$$F_0(x) = x \tag{3}$$

and the optimality equation (7.2) becomes

$$F_s(x) = \min_{0 \le u \le x} \max \left[F_{s-1}(u), F_{s-1}(x-u)\right]. \tag{4}$$

From (3), (4) one readily establishes recursively that

$$F_s(x) = 2^{-s}x,$$

consistent with (2), and that the optimal value of u is $x/2$ for all s.

9. MINIMAX LOCATION OF THE MAXIMUM OF A FUNCTION

Suppose that $f(\xi)$ is a scalar function of a scalar variable, which is known to be unimodal on a given interval. Suppose that the interval is of length a and that one has function evaluations at the ends p_1, p_4 of the interval and at a single internal point p_3 which is distance b from p_1 (see Figure 9.1). If one takes a second evaluation at p_2, with coordinate $c < b$, say, then one locates the maximum of f in either the interval $p_1 p_3$ (with an internal evaluation at p_2) or the interval $p_2 p_4$ (with an internal evaluation at p_3). One wishes to take a prescribed number of evaluations sequentially so as to minimize the maximal length of the interval within which the maximum can finally be guaranteed to lie.

This is a sequential minimax problem for which the only cost-relevant aspect of past observations is (a, b): the length of the interval within which the maximum has been located and the relative coordinate b of the internal evaluation. If $F_s(a, b)$ is the minimax cost starting from (a, b) with s further function evaluations permitted, then

$$F_0(a, b) = a \tag{1}$$

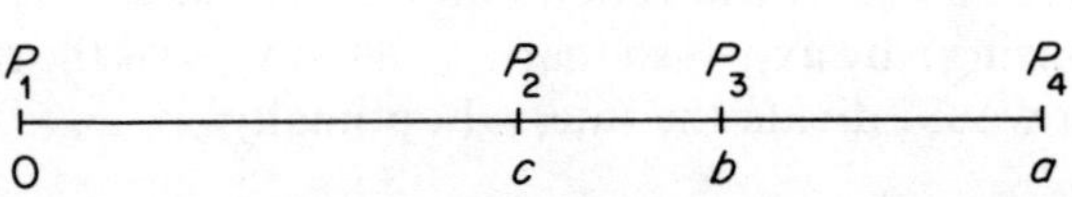

Figure 9.1

and, by the argument outlined above, the optimality equation is

$$F_s(a,b) = \min \begin{Bmatrix} \min\limits_{0 \le c \le b} \max[F_{s-1}(b,c), F_{s-1}(a-c, a-b)] \\ \\ \min\limits_{b \le c \le a} \max[F_{s-1}(c,b), F_{s-1}(a-b, a-c)] \end{Bmatrix} \quad (s > 0). \tag{2}$$

The solution of this equation, due to Kiefer (1953), is not trivial as was the case for (8.4).

Theorem 9.1. *The solution of* (1), (2) *is*

$$F_s(a,b) = \min \begin{Bmatrix} \max\left[\dfrac{b}{f_s}, \dfrac{a-b}{f_{s-1}}\right] \\ \\ \max\left[\dfrac{a-b}{f_s}, \dfrac{b}{f_{s-1}}\right] \end{Bmatrix} \tag{3}$$

where f_s is the sth Fibonacci number.

The Fibonacci numbers are those of the sequence $\{f_0, f_1, f_2, \dots\} = \{1, 1, 2, 3, 5, 8, 13, 21, \dots\}$ generated by the difference equation

$$f_s = f_{s-1} + f_{s-2}$$

with the initial conditions $f_0 = f_1 = 1$. For large s,

$$f_s \sim \frac{1}{\sqrt{5}}\left[\frac{1+\sqrt{5}}{2}\right]^s.$$

The *sth* Fibonacci number is usually denoted F_s— confusing in our context.

That (3) solves (1), (2) can be verified directly, if somewhat laboriously. However, of rather more interest is the quantity $\min_{0 \le b \le a} F_s(a,b)$, which is the minimax cost if the first given observation is located optimally. This must be proportional to a, so we shall set

$$\min_b F_s(a,b) = a\phi_{s+1}.$$

A subscript $s+1$ rather than s is appropriate, since $a\phi_s$ is the minimax cost if one is allowed a total of s observations in an initial interval of length a. Let us also denote the optimizing b by ax_{s+1}, so that $x_{s+1}/(1-x_{s+1})$ is the ratio in which the first observation would divide the interval optimally.

Theorem 9.2. $\phi_s = f_s^{-1}$ and $x_s = f_{s-1}/f_s$ *or* f_{s-2}/f_s. *The first two observations should be taken to divide the interval in the ratio* $(f_{s-1}/f_{s-2})^{\pm 1}$ *; the surviving* *observation in the sub-interval of interest at the next stage will automatically satisfy this rule.*

Proof. Assume $x_{s+1} > \frac{1}{2}$. If $c = ay_{s+1}$ is the second observation point chosen in the interval $[0, a]$ then we may assume by symmetry that $y_{s+1} = 1 - x_{s+1}$, so that then

$$F_s(a,b) = F_{s-1}(b,c) = F_{s-1}(a-c, a-b)$$
$$= bF_{s-1}\left(1, \frac{c}{b}\right)$$
$$\geq b\phi_s = ax_{s+1}\phi_s. \tag{3}$$

The inequality becomes equality if

$$\frac{c}{b} = \frac{y_{s+1}}{x_{s+1}} = x_s. \tag{4}$$

We then have the pair of relationships

$$\phi_{s+1} = x_{s+1}\phi_s, \tag{5}$$

$$y_{s+1} = 1 - x_{s+1} = x_s x_{s+1}. \tag{6}$$

Substituting for x from (5) into (6) we deduce that

$$\frac{1}{\phi_{s+1}} = \frac{1}{\phi_s} + \frac{1}{\phi_{s-1}} \tag{7}$$

Since $\phi_0 = \phi_1 = 1$ then we deduce from (7) that indeed $\phi_s = f_s^{-1}$, and from (5), (6) that $x_s = f_{s-1}/f_s$, $y_s = f_{s-2}/f_s$. ■

10. SELLING AN ASSET

Suppose one is selling an asset, for which one knows a price y_t can be realized on day t; if not then accepted, the chance to sell at that price is lost. There is a discount factor β, and one wishes to realize the maximal discounted price.

If the course of future prices is indeed known, and if one has not sold before day t, then one will sell on day r, where r is the value in $r \geq t$ maximizing $\beta^r y_r$.

In general the course of future prices will not be known, and one has a choice of minimax or stochastic approaches for this case. The minimax policy is surprisingly simple (actually, a maximin policy, since we are concerned with reward maximization rather than cost minimization).

Theorem 10.1. *If one has not sold before time t then a maximin policy is to sell at time t if and only if Min* $(\beta^r y_r | W_t)$ *is maximized in* $r \geq t$ *at* $r = t$.

Proof. Since the problem is one of reward maximization the 'pessimizing' operator $M(\cdot | W_t)$ is replaced by Min $(\cdot | W_t)$, a minimization over quantities unobservable at time t.

Define $p(r) = \mathrm{Min}(\beta^r y_r | W_t)$ and $\bar{p} = \max_{r \geq t} p(r)$. Both quantities depend upon t, but t can be regarded as fixed for the moment.

If $p(t) = \bar{p}$ and one does not sell at t, then there is a contingency (consistent with W_t) under which one realizes a discounted price no better than $\bar{p}$. There is thus nothing to be gained by not selling at t.

If $p(t) < \bar{p}$ and one sells at t, then one realizes a discounted price smaller than one could have been sure of receiving, which is certainly non-optimal.

Optimality of the proposed policy (in the maximin sense) is thus established. ■

The result may appear trivial. However, the corresponding policy in the stochastic case (to accept at t if and only if $E(\beta^r y_r | W_t)$ is maximal in $r \geq t$ at $r = t$) is a 'certainty-equivalent' policy which would scarcely ever be optimal (see Section 12.4).

11. RISK-AVERSE LQG CONTROL

Situations in which one has linear dynamics and a criterion which is a suitable function of a quadratic form in the variables are especially tractable under both the stochastic and minimax criteria. In Chapter 19 we treat a stochastic version of LQG control which, in one class of cases, reduces to a minimax problem. The criterion depends on a parameter θ which, when negative, expresses a degree of risk-aversion on the part of the optimizer. Under these circumstances the minimization of the expected cost with respect to policy becomes equivalent to the minimization of $M_\pi(\mathbb{S} | W_0)$, where $\mathbb{S}$ is a quadratic form in the variables. Solution of this minimax problem can be expressed in certainty-equivalence form (see Chapter 19).

This problem is one in unbounded variables and costs, and it is worth while nothing that

$$\inf_\pi M_\pi(\mathbb{S} | W_0) = +\infty$$

if the optimizer is sufficiently pessimistic (i.e. θ is sufficiently negative). This is the case in which the strength of his fears outweighs the reassurance of statistics.

Part II

The stochastic formulation

Chapters 10 and 11 are devoted to an exposition of the general structure. The reader interested in applications can proceed directly to Chapters 12 to 14 if willing to take the various forms of the optimality equation as self-evident.

Part II

CHAPTER 10

The general optimality equation

1. THE STOCHASTIC CRITERION

As emphasized in Chapter 9, the stochastic and minimax formulations are alternative ways of dealing with problems of incomplete information. The analogy between them is close enough that the treatment of that chapter could be transferred to the stochastic case by little more than substitution of the expectation operator E for the maximization operator M. Such a transformed treatment provides the core of this chapter and the next. However, it is appropriate to clothe this core somewhat, in view of the facts that the stochastic version has an immense literature, has been applied to a great variety of problems and has an extensively developed theory (of, for instance, the infinite-horizon case). In a word, notwithstanding the completeness of analogue between the two approaches, the stochastic approach is standard. The material of Section 9.2, on observables and policies, holds equally for the stochastic case.

2. STRUCTURE AND CAUSALITY

We are concerned with structural rather than technical questions for the moment, and so shall assume all variables discrete-valued, so that probabilities and expectations are well defined and conditional probabilities and expectations may be given their naïve evaluation. If x and y are random variables we shall write $P(x)$ for the probability of x (making no distinction between the random variable and a particular value of it), so that the expectation of $\phi(x)$ is given by

$$E\phi(x) = \sum_x \phi(x)P(x).$$

The joint distribution of x, y will be denoted $P(x, y)$ and the distribution of x conditional on y by $P(x|y)$, with the evaluation

$$P(x|y) = \frac{P(x, y)}{P(y)}.$$

The conditional expectation $E[\phi(x)|y]$ will have the evaluation

$$E[\phi(x)|y] = \sum_x \phi(x)P(x|y) = \frac{\sum_x \phi(x)P(x, y)}{\sum_x P(x, y)}.$$

Sometimes the distribution of x is affected by the value of a variable u which is not specified as a random variable jointly with x, but is simply a quantity whose value parametrizes the distribution of x. We shall speak of such a variable as a *parametrizing variable*, and denote the distribution of x for prescribed u as $P(x|;u)$. Correspondingly, $P(x|y;u)$ will denote the distribution of x as conditioned by y and parametrized by u.

Sometimes the stochastic description is completed so that x, u can be regarded jointly as random variables. In such a case it may or may not be true that $P(x|u) = P(x|;u)$ (see Exercise 2).

We return now to the structure of the temporal decision process. We assume that there is an underlying stochastic process $\{x_t\}$ whose evolution in time represents the evolution of all variables of interest, observable or not. The variable x_t is not a state variable, unless further assumptions are made; we shall refer to it as the *process variable*. The partial sequence $\{x_0, x_1, x_2, \ldots, x_t\}$ will be denoted X_t, and the infinite sequence X_∞ denoted simply X; similarly for other variables.

Specification of the dynamic model is equivalent to specification of $P(X|;U)$. Specification of a policy is equivalent to specification of the action probabilities $P_\pi(u_t|W_t)$ for $t \geq 0$, where W_t is the observation history at time t.

We now assume the structure of model and observables subject to the following conditions.

1. *Separateness of model and policy*

$$P_\pi(X, U) = P(X|;U) \prod_{t=0}^{\infty} P_\pi(u_t|W_t) \tag{1}$$

2. *Causality*

$$P(X_t|;U) = P(X_t|;U_{t-1}) \tag{2}$$

3. *Non-anticipation.* W_t is (X_t, U_{t-1}, t)-measurable.
4. *Retention of information.* W_t is W_{t+1}-measurable, and (U_{t-1}, t) are W_t-measurable.

These conditions are all natural; they could be translated verbally as: (1) The joint distribution of X and U under a policy π factorizes as relation (1) indicates. (2) Future actions cannot affect current process dynamics. (3) Process history up to the present, past actions and clock time are the most that could be observed at a given time. In particular, process future is known only insofar as this predictable. (4) Information once gained is not lost. In particular, past actions and clock time are recalled.

A convenient convention is to denote (X, U) by W, and regard knowledge of W as *complete knowledge* (of all relevant variables). Note that one is able to speak of 'partial policies'. For instance, the policy before time t would be the prescription of $P(u_j|W_j)$ for $0 \leq j \leq t$.

Theorem 2.1. Under condition (1)

(a) $E_\pi[\phi(W)|W_t]$ *is independent of policy before time t.*
(b) $E_\pi[\phi(W)|W_t, u_t]$ *is independent of policy up to and including time t.*

Proof. To establish (a), note that

$$E_\pi[\phi(W)|W_t] = \frac{\Sigma\phi(W)P_\pi(X, U)}{\Sigma P_\pi(X, U)} \tag{3}$$

where the sums cover (X, U) consistent with the prescribed value of W_t, and P_π has evaluation (1). Since $P(u_j|W_j)$ is constant in this set for $j < t$, these factors cancel out in (3),whence (a) follows. Assertion (b) follows similarly. ■

Theorem 2.2. *Under conditions* (2)–(4) $P(W_t|;U)$ *and* $P(W_t|W_j;U)$ *for* $0 \leq j \leq t$ *depend upon U only through* U_{t-1}, *and so can be written* $P(W_t|;U_{t-1})$ *and* $P(W_t|W_j;U_{t-1})$ *respectively, for* $t \geq 0$.

Proof. The first assertion follows from conditions (2) and (3). The second assertion follows from the first and the W_t-measurability of W_j. ■

Theorem 2.3. (Consequences of causality etc.) *Under conditions* (1)–(4)
(a) $P_\pi(W_t|W_0)$ *is independent of policy from time t onwards*
(b) $P_\pi(W_{t+1}|W_t, u_t)$ *is independent of policy and, if* $W_t = (Y_t, U_{t-1}, t)$, *then for arbitrary* ϕ,

$$E[\phi(W_{t+1})|W_t, u_t] = \sum_{y_{t+1}} \phi(W_{t+1})P(y_{t+1}|Y_t;U_t) \tag{4}$$

Proof. Since $P(W_t|W_0;U) = P(W_t|W_0;U_{t-1})$ then $P_\pi(W_t|W_0) = P(W_t|W_0;U_{t-1}) \prod_0^{t-1} P(u_j|W_j)$, whence (a) follows.

$P(W_{t+1}|W_t, u_t)$ is independent of policy before time $t+1$, by Theorem 2.1, and of policy from $t+1$ onwards, by (a). For the explicit expression (4), if $\sum'$ is a summation over X, U consistent with the prescribed W_t, u_t and $\sum''$ a summation over X_{t+1} consistent with the prescribed W_t, u_t, then

$$\begin{aligned} E_\pi[\phi(W_{t+1})|W_t, u_t] &= \frac{\sum'\phi(W_{t+1})P_\pi(X, U)}{\sum' P_\pi(X, U)} \\ &= \frac{\sum''\phi(W_{t+1})P(X_{t+1}|;U_t)}{\sum'' P(X_{t+1}|;U_t)}, \end{aligned} \tag{5}$$

the second equality following from (1), (3). Relation (4) is an alternative expression of (5). ■

A probabilist treating non-discrete variables would take these calculations more circumspectly. A complete stochastic model corresponds to the specification of a probability triple $(P_\pi, \mathscr{F}, \Omega)$. Here Ω is the sample space whose points are the 'elementary outcomes' W, $\mathscr{F}$ is a σ-algebra of sets on this space and P_π is a probability measure (necessarily π-dependent) defined on these sets. Events whose occurrence can be verified at time t correspond to sets in a sub-σ-algebra $\mathscr{F}_t$ of $\mathscr{F}$. The sequence $\{\mathscr{F}_t\}$ is increasing, reflecting the fact that information gained is retained. Onto this formalism one must now build the parametrizing character of U and the causality principle. The decision u_t may be subject not only to the structural measurability condition, that u_t should depend upon W_t alone, but also to a technical measurability condition that this dependence should be sufficiently regular. However, we reaffirm the opinion, expressed in Chapter 1, that the natural view of these matters has yet to emerge.

Exercises and comments

1. Causality does *not* imply that W_t is statistically independent of u_j $(j \geq t)$ under a prescribed π. The assertion is plainly false for $j = t$ (since W_t and $u_t(W_t)$ will be statistically independent only in the case when u_t is chosen constant for all W_t) and is in general false for $j > t$.

   This is the typical circularity of a feedback loop. A system may be causal in that its output (y) can only be affected by input (u) which precedes it in time. However, if one allows u to depend upon past y (i.e. allows feedback) then, despite causality, past output is not in general statistically independent of future input.
2. For such a causal system, $P(y_1, y_2 | ; u_1, u_2)$ will be functionally independent of u_2, whereas $P_\pi(y_1, y_2 | u_1, u_2)$ in general will not. Hence the two expressions are in general unequal. Construct an example.

3. COST FUNCTIONS AND RECURSIONS

There is an economy in notation which is natural if explained. It may be that for each t one is interested in a function $\phi_t(W_t)$ of W_t. One can well then drop the t subscript and write the function simply as $\phi(W_t)$, with the understanding that this is defined for all relevant t and W_t. Specification of W_t implies specification of t, of course, and the fact that W_s and W_t take values in different sets for $s \neq t$ is no bar to the use of the apparently t-independent notation $\phi(W_t)$.

Because of our convention that all relevant variables are W-measurable, we can take the cost as being a function $C(W)$ of W. The expected cost under policy π conditional on W_t is then

$$V(\pi, W_t) = E_\pi[C(W) | W_t]. \tag{1}$$

We henceforth assume conditions (1)–(4) of Section 2.

Theorem 3.1. *$V(\pi, W_t)$ is independent of policy before time t and obeys the recursion*

$$V(\pi, W_t) = E_\pi[V(\pi, W_{t+1}) | W_t]. \tag{2}$$

Proof. The first assertion is a consequence of Theorem 2.1(a). The second is a consequence of the W_{t+1}-measurability of W_t, so that for any W-measurable random variable z,

$$E_\pi(z|W_t) = E_\pi[E(z|W_{t+1})|W_t]. \blacksquare$$

Theorem 3.1 has an immediate and useful corollary.

Corollary 3.1. *Let π'' be the policy for which one follows π' up to time t and π thereafter. Then*

$$V(\pi'', W_t) = E_{\pi'}[V(\pi, W_{t+1})|W_t]. \tag{3}$$

As in Section 9.4, the problem has horizon T by definition if $V(\pi, W_T)$ is independent of policy from time T onwards, and so, by Theorem 3.1, is independent of π altogether. $V(W_T)$ can then effectively be taken as the cost function C, a function of W_T alone (see Section 6).

4. THE OPTIMALITY EQUATION

We take $t = 0$ as the standard starting time, so that the optimal policy (if one exists) is that which minimizes $V(\pi, W_0)$. Define

$$G(W_t) = \inf_\pi V(\pi, W_t). \tag{1}$$

Theorem 4.1 (The optimality equation). (a) *The infimal conditional costs $G(W_t)$ satisfy the recursion*

$$G(W_t) = \inf_\pi E_\pi[G(W_{t+1})|W_t]. \tag{2}$$

(b) *This can equivalently be written*

$$G(W_t) = \inf_{u_t} E[G(W_{t+1})|W_t, u_t]. \tag{3}$$

(c) *If π is optimal then it satisfies*

$$G(W_t) = E_\pi[G(W_{t+1})|W_t]. \tag{4}$$

Proof. Relations (2) and (3) are alternative forms of the optimality equation. That they are indeed equivalent follows from Theorem 2.3(b) and the consequent relation

$$E_\pi(G(W_{t+1})|W_t) = \sum_{u_t} P_\pi(u_t|W_t)E[G(W_{t+1})|W_t, u_t].$$

In the case of a finite horizon T the theorem follows by backward induction, with the terminal condition $G(W_T) = C(W_T)$. We simply transfer the proof of Theorem 9.5.1 with M replaced by E, and obtain the additional conclusion that (4) (for $0 \leq t < T$) is sufficient as well as necessary for optimality of π.

In the infinite-horizon case the proof needs more care. A useful further elision in notation is to suppress the argument W_t altogether in relations (3.2) and (2), and write these as

$$V(\pi) = A_\pi V(\pi)$$
$$G = \mathscr{A} G$$

The operators A_π and $\mathscr{A}$ are thus effectively defined. They share the property of *monotonicity*, so that $\phi \geq \psi$ implies $\mathscr{A}\phi \geq \mathscr{A}\psi$

Note now that $G \geq \mathscr{A}G$, because

$$G = \inf_\pi V(\pi) = \inf_\pi A_\pi V(\pi) \geq \inf_\pi A_\pi G = \mathscr{A}G.$$

To obtain the reverse inequality, note that for any $\varepsilon > 0$, finite N and prescribed t one can find a policy π such that

$$V(\pi, W_{t+1}) < N \vee G(W_{t+1}) + \varepsilon$$

where $N \vee G = \max(N, G)$. Thus

$$\begin{aligned} E_{\pi'}[N \vee G(W_{t+1}) | W_t] &> E_{\pi'}[V(W_{t+1}, \pi) | W_t] - \varepsilon \\ &= V(W_t, \pi'') - \varepsilon \end{aligned}$$

where π'' is the policy under which one follows π' up to time t, and π thereafter (see (3.3)). Since π' and ε are arbitrary, this implies that

$$\mathscr{A}[N \vee G] \geq G.$$

Letting N tend to $-\infty$ and appealing to monotone convergence one deduces that $\mathscr{A}G \geq G$, establishing (2), (3).

If an optimal policy π exists, then (4) follows by substitution of $V(\pi) = G$ into (3.2). ■

5. CONSERVATION AND IMPROVEMENT

There are a number of concepts which, in the finite-horizon case, seem to amount to no more than different views of the same simple structure. However, they turn out to mark valid distinctions for the infinite-horizon problem.

Corollary 3.1 implies that if, for two policies π and π' and given t,

$$A_{\pi'} V(\pi, W_t) \leq V(\pi, W_t) \tag{1}$$

for all W_t with strict inequality somewhere, then $V(\pi'') \leq V(\pi)$, with strict inequality for some W_t at that t value. We shall say then that π' *improves* π. If there is no π' which improves π, at any t, then π is said to be *unimprovable*. The policy π is thus unimprovable if and only if

$$V(\pi) \leq \mathscr{A} V(\pi). \tag{2}$$

It follows from the definition of $\mathscr{A}$ that $\mathscr{A}G \leq A_\pi G$ for any π. Suppose π is such that equality holds in this relation:

$$A_\pi G = \mathscr{A} G$$

(for all t and W_t, understood). Then π is said to be *conserving*. The term follows from the fact that then $G = A_\pi G$, and indeed $G = A_\pi^s G$. That is, if one used π for one step (or indeed for s steps) and then reverted to an optimal policy, then nothing would have been lost, in that the same cost would be incurred as if one had used an optimal policy throughout.

Theorem 5.1. *An optimal policy is both unimprovable and conserving.*

Proof. If π is optimal then both $V(\pi) = \mathscr{A} V(\pi)$, by (4.2), and $G = A_\pi G$, by (4.4). ∎

If (1) holds then, by virtue of Corollary 3.1, $V(\pi^*) \leq V(\pi)$, where π^* is the policy which agrees with π' at time t, and with π otherwise. Inequality is strict for some t, W_t if it is so for (1).

This expresses a *solidarity* property of the cost function $V(\pi, W_t)$: if π can be modified so that V is decreased at a given (t, W_t), then this improvement need not be at the cost of a deterioration in performance for other (t, W_t). The optimization is a vector optimization problem in that one would like to minimize $V(\pi, W_t)$ with respect to π at all W_t simultaneously. For most vector problems this is impossible; for a temporal problem, with non-decreasing information and optimizations independent at different times, it is possible.

6. FINITE HORIZONS

As for the minimax case of Section 9.4, let us define the problem as having horizon T if $V(\pi, W_T)$ is independent of policy from time T onwards.

Theorem 6.1. *If the problem has horizon T then $V(\pi, W_T)$ is independent of π. The problem can then be normalized so that the cost function is*

$$C(W_T) = E[\text{Cost} \mid W_T] \tag{1}$$

a function of W_T alone.

Proof. The first assertion follows from the fact that, by Theorem 2.2, $V(\pi, W_T)$ is independent of policy before time T. The second assertion follows from the fact that

$$V(\pi, W_t) = E_\pi[V(\pi, W_T) \mid W_t]$$

for $0 \leq t < T$; it is only over this time interval that policy can be optimized. The π-independent quantity $V(\pi, W_t)$, otherwise written as in (1), is thus the effective cost function. ■

Effectively, then, $V(\pi, W_t) = C(W_T)$ for $t \geq T$ (although see Exercise 1), whence the following assertion.

Theorem 6.2. *For a T-horizon problem*

$$V(\pi) = A_\pi^T V(\pi') = A_\pi^T G \tag{2}$$

for any π'.

Proof. The second equality is a consequence of the first, which follows from

$$\begin{aligned} A_\pi^T V(\pi', W_t) &= E_\pi[V(\pi', W_{t+T}) | W_t] \\ &= E_\pi[V(\pi, W_{t+T}) | W_t] \\ &= V(\pi, W_t). \ ■ \end{aligned}$$

Theorem 6.3. *For a finite-horizon problem a policy which is either conserving or unimprovable is optimal.*

Proof. If π is conserving then

$$G = A_\pi G = A_\pi^T G = V(\pi),$$

whence the first assertion. If π is unimprovable, then, for any π',

$$V(\pi) \leq A_{\pi'} V(\pi) \leq A_{\pi'}^T V(\pi) = V(\pi'),$$

whence the second assertion. ■

If the infimum is attained in (4.3) for $0 \leq t < T$, then it follows that there is a deterministic policy which is conserving, and so optimal. In general one has:

Theorem 6.4. *For a finite-horizon problem there is a deterministic policy which is at least ε-optimal.*

Proof. One can find a value of $u_t(W_t)$, and hence a deterministic policy π, such that, for any $\varepsilon > 0$,

$$A_\pi G \leq \mathscr{A} G + \frac{\varepsilon}{T} = G + \frac{\varepsilon}{T}.$$

Thus

$$V(\pi) = A_\pi^T G \leq G + \varepsilon.$$

We have used the fact that $A_\pi(G + c) = A_\pi G + c$ for constant c. ■

Exercises

1. Suppose the problem has horizon T. Show that $V(\pi, W_t)$ is independent of policy from time T onwards if $t \leq T$.
2. Show that the assertion of Exercise 1 holds also if $t > T$, in the sense that the set $\mathcal{N}_t$ of W_t for which this is not true makes a zero contribution to $E_\pi(V(\pi, W_t)|W_T)$. More specifically, $E_\pi(\phi(W_t)V(\pi, W_t)|W_T) = 0$ for any uniformly bounded ϕ which vanishes outside $\mathcal{N}_t$.
3. Show that the problem has horizon T if $V(\pi, W_0)$ is independent of policy from time T onwards, in the sense that the set of W_T for which $V(\pi, W_T)$ is π-dependent makes a zero contribution to $V(\pi, W_0) = E_\pi(V(\pi, W_T)|W_0)$.

7. RECURSIVELY GENERATED COST FUNCTIONS

The treatment of Section 9.6 transfers bodily, with only one point of difference.

Theorem 7.1. *Suppose that $C(W) = C^{(0)}(W)$, where $\{C^{(t)}(W)\}$ is generated by recursion (9.6.1) with ψ linear and strictly increasing in $C^{(t+1)}$ for all t, W. Define $F(W_t) = \inf_\pi E_\pi[C^{(t)}(W)|W_t]$. Then the optimality equation takes the form*

$$F(W_t) = \inf_{u_t} E[\psi(W_{t+1}, F(W_{t+1}))|W_t, u_t]. \tag{1}$$

If the horizon is finite then the minimizing u_t defines an optimal action rule $u_t(W_t)$.

Proof. We proceed as in the proof of Theorem 9.6.1, with the operator $\mathscr{E}_t = \inf_{u_t} E[\cdot|W_t, u_t]$ replacing the operator $\mathscr{M}_t$. If $\mathscr{E}_t$ is to commute with ψ_t in its second argument, then ψ_t must not only be increasing in this argument, but also linear; in this the two cases differ. Since the optimality equation (4.3) is established also for the infinite-horizon case, then so is (1). The minimizing u_t in (1) (if it exists) defines a conserving policy, and so is optimal for finite T, by Theorem 6.3. ■

The conditions on ψ imply that, for the stochastic case, the recursion (9.6.1) must in fact be of the linear form

$$C^{(t)} = g(W_{t+1}) + \beta(W_{t+1})C^{(t+1)}$$

with $\beta > 0$. If we set $g(W_{t+1}) = g_t$, $\beta(W_{t+1}) = \beta_t$, then this corresponds to a criterion of the form

$$C = g_0 + \beta_0 g_1 + \beta_0\beta_1\beta_2 g_2 + \cdots$$

with g_t and β_t interpretable as W_{t+1}-dependent instantaneous cost and discount factor respectively. This is sometimes referred to as a *separable* cost function, with $\beta_0\beta_1 \cdots \beta_{t-1} g_t$ identified as the cost-equivalent at time 0 of operations at time t.

CHAPTER 11

Markov decision processes (finite horizon)

1. STATE STRUCTURE

A temporal optimization problem with stochastic formulation and state structure is termed a *Markov decision process.*

We have already developed the concept of state structure for the deterministic and minimax cases (Sections 2.2 and 9.2). It can be characterized by the existence of a *state variable* x with value x_t at time t, which is *sufficient* in that (i) the optimal value of u_t is a function at most of x_t and t ($t \geq 0$) and (ii) x_0 is observable at $t = 0$ and x_{t+1} is calculable from x_t, u_t and the new observable at time $t+1$.

The development is an almost complete analogue of that for the minimax case in Section 9.7, except that we now base it upon the more clearly formulated dynamic structure of Section 10.2.

Suppose that the process variables $\{x_t\}$ and decision variables $\{u_t\}$ obey the structural conditions (10.2.1) and (10.2.2), and also the conditions for $t \geq 0$:

(a) $W_t = (X_t, U_{t-1}, t)$.

(b) $P(x_{t+1} | X_t, U_t) = P(x_{t+1} | x_t, u_t)$. (1)

(c) The set of values which u_t may adopt depends on x_t and t at most; $\mathcal{U}(x_t, t)$, say.

(d) $C = C^{(0)}$, where $\{C^{(t)}\}$ is generated by the backward recursion

$$C^{(t)} = \psi(x_t, x_{t+1}, u_t, t, C^{(t+1)}) \qquad (0 \leq t < T), \tag{2}$$

with the terminal condition $C^{(T)} = K(x_T)$. Here ψ is linear and strictly increasing in its last argument.

Assumption (a) states that X_t is observable at time t, which is strictly more than one needs; the assumption (ii) above is sufficient.

Assumption (b) is the stochastic version of a simply recursive plant equation. By virtue of (a), (b) and Theorem 10.2.3 we have, for any ϕ,

$$E[\phi(x_{t+1}) | W_t, u_t] = E[\phi(x_{t+1}) | x_t, u_t] \tag{3}$$

where the left-hand member is independent of π and so is well defined.

Theorem 1.1 (The optimality equation with stochastic state structure). *Assume the conditions listed above. Then the minimal future cost $F(W_t)$ defined in Theorem* 10.7.1 *is a function of x_t, t alone, $F(x_t, t)$, say, and the optimality equation* (10.7.1) *becomes*

$$F(x_t, t) = \inf_{u_t} E[\psi(x_t, x_{t+1}, u_t, t, F(x_{t+1}, t+1)) | x_t, u_t] \qquad (0 \le t < T) \qquad (4)$$

with the terminal condition

$$F(x_T, T) = K(x_T). \qquad (5)$$

The minimizing value of u_t *in* (4) *is optimal.*

Proof. This follows the familiar inductive pattern. Assuming the asserted reduction of F valid for $t+1$ we have, from (10.7.1),

$$\begin{aligned} F(W_t) &= \inf_{u_t} E[\psi(x_t, x_{t+1}, u_t, t, F(x_{t+1}, t+1)) | W_t, u_t] \\ &= \inf_{u_t} E[\psi(x_t, x_{t+1}, u_t, t, F(x_{t+1}, t+1)) | x_t, u_t], \end{aligned}$$

the second equality following from (3). All assertions now follow. ■

The condition that ψ be linear and increasing in its last argument implies the fairly considerable reduction of recursion (2): it must have the form

$$C^{(t)} = g_t + \beta_t C^{(t+1)} \qquad (6)$$

and so

$$C = \sum_{t=0}^{T} \beta_0 \beta_1 \cdots \beta_{t-1} g_t \qquad (7)$$

where

$$\beta_t = \beta(x_t, x_{t+1}, u_t, t) > 0,$$

$$g_t = \begin{cases} g(x_t, x_{t+1}, u_t, t) & (0 \le t < T), \\ K(x_T) & (t = T). \end{cases} \qquad (8)$$

As in Section 10.7. g_t and β_t have the interpretations of instantaneous cost and of discount factor at time t, functions of the arguments indicated in (8). The optimality equation (4) reduces to

$$F(x_t, t) = \inf_{u_t} E[g_t + \beta_t F(x_{t+1}, t+1) | x_t, u_t] \qquad (0 \le t < T). \qquad (9)$$

The recursion (6) can be even further reduced. The instantaneous cost can be normalized to a function of x_t, u_t and t at most by the substitution

$$g(x_t, u_t, t) = E[g(x_t, x_{t+1}, u_t, t) | x_t, u_t]. \qquad (10)$$

Further, define the functions

$$P^*(x_{t+1}|x_t, u_t) = \beta(x_t, x_{t+1}, u_t, t)P(x_{t+1}|x_t, u_t),$$

$$\beta^*(x_t, u_t, t) = \sum_{x_{t+1}} P^*(x_{t+1}|x_t, u_t). \tag{11}$$

Then, if

$$\beta^*(x_t, u_t, t) \leq 1 \tag{12}$$

for all values of the argument, one can regard P^* as the effective transition probability (without discounting) and $1 - \beta^*(x_t, u_t, t)$ as the probability of transition (conditional on x_t, u_t) into a supplementary state value x^0 which is absorbing and costless. That is, once one enters x^0 one effectively leaves the system.

If assumption (12) holds one can thus normalize β to 1 by adding the new state x^0 and defining $P(x_{t+1}|x_t, u_t)$ as P^* for $x_{t+1} \neq x^0$, and as $1 - \beta^*$ for $x_{t+1} = x^0$.

2. MARKOV POLICIES

A policy is said to be *Markov* if x_t is a state variable insofar as conditions (a) and (b) of Section 1 are satisfied, and u_t depends at most on x_t, t, so that $P(u_t|W_t) = P(u_t|x_t; t)$, for all relevant t. If the policy is also deterministic u_t must depend upon these quantities deterministically: $u_t = h(x_t, t)$. Theorem 1.1 can be construed as saying that, for a finite-horizon Markov decision process, there is an optimal policy which is deterministic Markov.

If the policy is also not explicitly dependent upon t then it is said to be *stationary*. Therefore, for a deterministic Markov policy which is also stationary $u_t = h(x_t)$ for some h. To a large extent we shall use the term 'stationary' to describe just such a policy, because we shall be working within the class of deterministic Markov policies. If we intend the wider definition then this will be indicated.

If assumption (d) of Section 1 is also satisfied then we can define

$$\phi(\pi, W_t) = E_\pi[C^{(t)}|W_t]$$

interpretable as the expected future cost at time t under policy π, conditional on current observables.

Theorem 2.1. *Suppose that conditions (a) to (d) of Section 1 hold and that π is Markov. Then $\phi(\pi, W_t)$ depends on W_t through x_t, t at most, and can be denoted $\phi(\pi, x_t, t)$, say. This function obeys the recursion*

$$\phi(\pi, x_t, t) = E_\pi[\psi(x_t, x_{t+1}, u_t, t, \phi(\pi, x_{t+1}, t+1)|x_t] \qquad (t < T). \tag{1}$$

Proof. The proof goes exactly as for Theorem 1.1, with ultimate appeal to Theorem 10.3.1 rather than to Theorem 10.4.1. ■

There is a notation for policies which is useful. Consider the policy π' under which one chooses $u_0 = h(x_0)$ and thereafter pursues policy π postponed one step in time. This is denoted $\pi' = (h, \pi)$. The deterministic Markov policy $u_t = h_t(x_t)$ $(t \geqq 0)$ is denoted $(h_0, h_1, h_2, \cdots)$ while the stationary policy $(h, h, h, \cdots)$ is denoted h^∞.

3. THE TIME-HOMOGENEOUS CASE; OPERATOR FORMULATION

Assuming state structure, we shall say that the problem is *time-homogeneous* if the instantaneous cost g, the discount factor β, the action set $\mathscr{U}(x, t)$ and the conditional distribution $P(x_{t+1} | x_t, u_t)$ do not depend on time explicitly. We shall continue to use the notation $g_t = g(x_t, x_{t+1}, u_t)$, $\beta_t = \beta(x_t, x_{t+1}, u_t)$. Time-homogeneity implies no loss of generality in a formal sense, because one can always incorporate the variable t in the state variable. However, to do so makes all values of state transient, a state of affairs one might otherwise avoid, so that the assumption of time-homogeneity does have some force.

As in Part I, for the time-homogeneous finite-horizon case it is natural to work in terms of time-to-go $s = T - t$ rather than t itself. The quantity s can be regarded as the horizon (i.e. distance of the horizon) relative to the current instant.

The cost function over horizon s corresponding to the structure (1.7) can then be written

$$C_s = \sum_{t=0}^{s-1} \beta_0 \beta_1 \cdots \beta_{t-1} g_t + \beta_0 \beta_1 \cdots \beta_{s-1} K(x_s) \tag{1}$$

and

$$V_s(\pi) = E_\pi[C_s | W_0] \tag{2}$$

is the expected s-horizon cost under policy π, a function of W_0 as well as of π. In fact, it depends upon the specification of π only for $t < s$.

The time-homogeneous version of Theorem 2.1 then states that, if π is Markov, then $V_s(\pi)$ depends upon W_0 through x_0 alone, $V_s(\pi, x_0)$, say, and that

$$V_s((h, \pi), x) = E[g_0 + \beta_0 V_{s-1}(\pi, x_1) | x_0 = x, u_0 = h(x)] \qquad (s > 0). \tag{3}$$

We shall write this equation more compactly as

$$V_s(h, \pi) = L(h) V_{s-1}(\pi) \qquad (s > 0), \tag{4}$$

with the supplementary condition

$$V_0(\pi) = K. \tag{5}$$

If $\mathscr{F}$ is the set of functions $\mathscr{X} \to R$, i.e. real scalar-valued functions of state, then $L(h)$ is an operator $\mathscr{F} \to \mathscr{F}$. It is moreover a linear operator, and has other properties to be established in Section 22.6. Note the consequence of (4), (5):

$$V_s(h^\infty) = \begin{cases} L(h) V_{s-1}(h^\infty) & (s > 0), \\ L(h)^s K & (s \geq 0). \end{cases} \tag{6}$$

The infimal s-horizon loss

$$F_s = \inf_\pi V_s(\pi)$$

is identifiable with $F(W_0, T-s)$ and so is a function of s and x_0 alone, which we shall write as $F_s(x_0)$. The optimality equation (1.9) for this time-homogeneous case becomes

$$F_s(x) = \inf_u E[g_0 + \beta_0 F_{s-1}(x_1) | x_0 = x; u_0 = u] \qquad (s > 0), \tag{7}$$

an equation which we shall write as

$$F_s = \mathscr{L} F_{s-1} \qquad (s > 0) \tag{8}$$

in analogue to (4). Since $F_0 = K$ then

$$F_s = \mathscr{L}^s K \qquad (s \geq 0). \tag{9}$$

$\mathscr{L}$ is also an operator from $\mathscr{F}$ to $\mathscr{F}$. In fact, if ϕ is an element of $\mathscr{F}$ then we can write

$$\mathscr{L}\phi = \inf_h L(h)\phi, \tag{10}$$

where the infimum is over all forms of the function $h(\cdot)$. An operator thus derived from a linear operator is called *quasi-linear*; it inherits many properties of the linear operator $L(h)$, as we shall see in Section 22.7. These properties are important, as it is apparent from (9) that study of policy optimization over a variable horizon amounts to a study of the iterates $\mathscr{L}^s$ of $\mathscr{L}$.

Exercise

1. Show that the properties of $\mathscr{L}$ and F_s established in Exercises 2.4.1 to 2.4.4 continue to hold in the present stochastic formulation.

4. THE FORMAL ACHIEVEMENT OF STATE STRUCTURE

Suppose one makes no assumptions on process structure or observability apart from the basic properties (1)–(4) of Section 10.2, but considers the observational history W_t as itself a candidate for the role of the state variable at time t. By virtue of Theorem 10.2.2, W_t inherits the properties required of the process variable; it also trivially possesses the properties (a) to (c) required of x_t in Section 1. Property (d) of that section is also formally satisfied, if one regards the problem as having the purely terminal cost

$$K(W_\infty) = C(W_\infty) = E[\text{Cost} | W_\infty].$$

The general optimality equation (10.4.3) is the analogue of recursion (1.9), and indeed appears to be a special case of it, insofar as the problem is time-homogeneous and purely terminal. If one adds the extra feature that future costs are defined and related by the recursion (9.6.1) then equation (10.7.1) holds—a formally time-homogeneous version of (1.9).

From this point of view it is reasonable to term W_t the 'observational state'. However, the catch with this trivially valid state formulation is that the space $\mathscr{W}$ which can contain all points W_t for all t is necessarily infinite. Moreover, motion through $\mathscr{W}$ is necessarily transient if information is indeed acquired, for it is then impossible that W_t and $W_{t'}$ could have the same value for $t \neq t'$. The motivation behind the definition of state structure is that there is a sufficient description which is economical, and finite in that the motion of x_t through $\mathscr{X}$ is in some sense recurrent. These are properties which the observational state cannot possess.

However, the fact that the most general formulation can be read as a state formulation may sometimes have a useful converse: results derived on the most special assumptions, of time-homogeneous Markov structure, may have some validity for the most general case.

5. CONJECTURES FOR THE INFINITE-HORIZON CASE

The infinite-horizon case presents a number of special features: some potential simplifications and some pitfalls. Part IV is devoted to these. Nevertheless, in discussing the examples of the next few chapters we shall often consider infinite-horizon versions, as is only realistic, and assume that intuitively plausible limit results hold. In this section we consider what limit behaviour might indeed be plausible for a time-homogeneous problem.

Suppose that one may expect infinite-horizon costs to be bounded in that, under any reasonable policy, C_s converges in some stochastic sense to a finite limit, as $s \to \infty$. This may happen because of discounting or because the process ultimately becomes absorbed in a state of zero cost ('transience'). In such a case one might expect that F_s tends to a limit F_∞ as $s \to \infty$, equal to $F = \inf_\pi \lim_s V_s(\pi)$, and identifiable as the minimal infinite-horizon cost. We shall rely on such expectations in the next chapters, assuming that $F_s = \mathscr{L}^s K \to F$ for all K in some appropriate class and that a property shared by all F_s for some such K is shared by F. In particular, one might expect F to depend upon W_0 only through x_0, as the F_s do, and for there to be an optimal infinite-horizon policy which is Markov—indeed, stationary Markov.

One might also expect that F satisfies the equilibrium optimality equation

$$F = \mathscr{L}F \tag{1}$$

and can be identified as the unique solution of this equation within some suitably regular class of functions. Corollaries of such a conjecture are that, if π is Markov, then $V(\pi) = \mathscr{L}V(\pi)$ implies optimality of π and $V(\pi) \neq \mathscr{L}V(\pi)$ implies non-optimality. One might also conjecture that, if

$$\mathscr{L}F = L(h)F, \tag{2}$$

so that the infimum with respect to u implicit in the definition of $\mathscr{L}$ is attained for $u = h(x)$, then the stationary policy $\pi = h^{\infty}$ is infinite-horizon optimal.

If the process incorporates neither discounting nor transience it may still converge to some kind of equilibrium behaviour as both starting point and horizon become infinitely remote, at least under an optimal policy. Under such conditions the notion of an *average cost* makes sense. This would be a cost averaged over both statistical fluctuations and over time; it would necessarily be minimized by an optimal policy.

In such a case one might expect F_s/s to have a limit γ, identifiable with this minimal average cost. A stronger conjecture is that

$$F_s(x) = \gamma s + f(x) + o(1) \tag{3}$$

for large s, where $f(x)$ is the transient expected cost ('potential' or 'differential cost') associated with starting in a given state x and the $o(1)$ represents deviations from this pattern due to finiteness of horizon. If (3) holds then one might expect that the asymptotic optimality equation, instead of taking the form (1), would rather take the form

$$\gamma + f = \mathscr{L}f. \tag{4}$$

If, moreover,

$$\mathscr{L}f = L(h)f \tag{5}$$

one might expect the stationary policy h^{∞} to be infinite-horizon optimal in some sense.

All these conjectures can rather easily be confounded by counterexample; however, they are valid under reasonable regularity conditions, considered in Part IV.

CHAPTER 12

Examples

1. THE LQ MODEL WITH PROCESS NOISE

Consider the quadratic regulation problem of Section 5.2, modified only in that the plant equation becomes

$$x_{t+1} = A_t x_t + B_t u_t + \varepsilon_{t+1}. \tag{1}$$

Here ε_t is a sequence of random variables, to be regarded as random disturbances driving the plant equation, or 'process noise'.

As for all state models hitherto, we suppose that $W_t = (X_t, U_{t-1}, t)$. Relation (1) would indeed imply the conditional Markov property (b) of Section 11.1, making x_t a state variable, if we demanded the property

$$P(\varepsilon_t | W_t; u_t) = P(\varepsilon_t | x_t, u_t). \tag{2}$$

In fact, we shall demand that

$$E(\varepsilon_t | W_t, u_t) = 0, \tag{3}$$

$$E(\varepsilon_t \varepsilon_t' | W_t, u_t) = N_t, \tag{4}$$

where N_t depends on W_t only in its possible dependence on t. Relations (3) and (4) ask both more and less than (2). The demand less, in that they place conditions only on first and second conditional moments. They demand more, in that these moments are required to be functionally independent even of x_t, u_t.

Theorem 1.1. *Assume the conditions of Theorem* 5.2.1, *except that the plant equation is now specified by* (1), (3), (4). *Then*

$$F(x_t, t) = x' P_t x_t + d_t \tag{5}$$

and the optimal control is

$$u_t = K_t x_t \tag{6}$$

where P_t, K_t have the same evaluations as in Theorem 5.2.1, *and*

$$d_t = d_{t+1} + \operatorname{tr}(N_t P_{t+1}) \qquad (t < T) \tag{7}$$

with $d_T = 0$.

That is, the optimal control, when specified in feedback form, is the same as in the case of zero process noise. The only effect of process noise is to add the purely t-dependent term d_t to $F(x_t, t)$.

Proof. Much as for Theorem 5.2.1. Under the inductive hypothesis relation (5.2.7) holds, except that the last term in the brackets is replaced by

$$E[(Ax + Bu + \varepsilon_{t+1})_t P_{t+1}(Ax + Bu + \varepsilon_{t+1})_t | W_t, u_t] + d_{t+1}$$

$$= (Ax + Bu)_t' P_{t+1}(Ax + Bu)_t + d_{t+1} + \mathrm{tr}(N_t P_{t+1})$$

whence the theorem follows. ■

The disturbed plant equations (1) and (5.4.1) correspond to opposite extremes, in that the α_t of (5.4.1) is known, and so perfectly predictable, while conditions (3), (4) imply that ε_t is perfectly unpredictable, at least in the linear least-square sense. Chapter 17 includes a study of the intermediate case, when disturbing sequences are partially predictable.

Note that all assertions of Section 5.4 will survive if the term ε_{t+1} is added to the driving term α_{t+1} in the plant equation (5.4.1), except that this term should also be added to the later forms (5.4.5), (5.4.13) of the plant equation, and the term d_t to $F(x_t, t)$. In particular, the control rules (5.4.8) and (5.4.12) are unchanged.

However, it is only when control rules are expressed in feedback form that they remain unaffected by white process noise (i.e. process noise satisfying (3), (4); see Exercise 2).

Exercises

1. *Stability.* The plant equation under optimal control now becomes

$$x_{t+1} = \Gamma_t x_t + \varepsilon_{t+1}.$$

'Stability' is now reflected by the fact that x_t deviates not excessively from 0 rather than it converges to 0. In the time-homogeneous case Γ_t will converge to the stability matrix Γ as $T \to \infty$, and $\{x_t\}$ will converge to stationarity (at times remote both from initial and horizon points). Show that the covariance matrix Λ of x_t in equilibrium satisfies

$$\Lambda - \Gamma \Lambda \Gamma' = N.$$

2. *Open- and closed-loop control* (cf. Section 4.7). Consider for simplicity the infinite-horizon time-homogeneous case, when the optimal control is $u_t = Kx_t$. This could also be written $u_t = K\Gamma^t x_0$ in the deterministic case ($\varepsilon \equiv 0$). Demonstrate that this second form of control is not equivalent to the first when process noise is present, and is not optimal.

The first form of control is *closed-loop* or *feedback* control, in that control is exhibited as a function of current state. The second is *open-loop* control, exhibiting control as a function of initial state and time. It can scarcely be optimal if current state is not determined by these quantities.

The dynamic programming approach automatically yields control in feedback form. However, even this form of the control rule will in general differ for deterministic and stochastic versions of the same problem: that there is no difference in the LQ case is a consequence of the special structure.

2. THE SECRETARY PROBLEM. (A) MAXIMIZATION OF EXPECTED SCORE

The 'secretary problem', sometimes referred to as the 'marriage problem', or, more neutrally, simply as the 'optimal choice problem', is a classic decision problem with many versions. It is the archetype of problems in which one must make an irrevocable choice from a number of opportunities whose values are revealed only sequentially. It is also a classic example of a problem amenable to a dynamic programming treatment, since Cayley gave in effect just such an analysis in 1875 (see Cayley, 1874, 1875).

The situation is that in which an employer interviews N candidates for a post in turn, and must decide immediately after each interview whether or not to appoint the current candidate (at most one to be appointed). Appointment or non-appointment is thus binding—past candidates cannot be recalled. Candidates can be evaluated, or at least ranked, as they are interviewed, and one wishes to appoint a candidate who ranks highly. One may even insist on appointing the best, or none. The question is, when to take the positive decision of appointment. If made too early in the sequence, one is neglecting the possibility of good candidates in the remaining field; if made too late, the field remaining and available is small.

In the simplest version of the problem, each candidate can be characterized by a scalar-valued 'score' y, which is revealed upon interview, and one seeks a procedure that maximizes the expected score of the candidate appointed. We suppose that scores are independently distributed over candidates, with a fixed, known distribution. (If the distribution is not known then one has a dynamic decision problem with an important new component—an element of inference as well as of decision. In effect, one must go through a period of 'learning' before being in a position to make a decision; see Part VI.)

The optimality equation (7) below is easily deduced immediately, but the reasoning is kept clearer if one reduces the problem to the formal framework of Chapters 10 and 11. If $C^{(t)}$ is the reward incurred from the moment of interviewing the tth candidate, then

$$C^{(t)} = \begin{cases} y_t & (u_t = A), \\ C^{(t+1)} & (u_t = R), \end{cases} \qquad (t < N) \tag{1}$$

where the decisions of acceptance or rejection are denoted by $u = A$, R

respectively. The information W_t at this point consists of Y_t and the fact that no candidate has hitherto been accepted. If the post must be filled, then

$$C^{(N)} = y_N. \tag{2}$$

Since the quantities y are independent one sees readily that y_t is a state variable, and that, by virtue of (1), the optimality equation (11.1.4) becomes

$$F(y_t, t) = \max [y_t, EF(y_{t+1}, t+1)] \tag{3}$$

where the expectation in (3) is over y_{t+1}. In fact, if we define

$$\phi(t-1) = EF(y, t), \tag{4}$$

then (3) implies that

$$\phi(t-1) = E \max [y_t, \phi(t)] \qquad (t < N). \tag{5}$$

The quantity $\phi(t)$ is the maximal expected score if t candidates have been interviewed and not appointed. It is a maximal future reward defined just after rather than just before decision. This change of definition is often helpful; in the present case it reduces relevant history to the single fact: that no appointment has hitherto been made.

Recursion (5) is better written in terms of $\phi_s = \phi(N-s)$, the maximal expected score if no appointment has been made and s candidates remain.

Theorem 2.1. *Define ϕ_s, the maximal expected score from a field of s candidates. Then*

$$\phi_1 = E(y), \tag{6}$$

$$\phi_s = E \max [y, \phi_{s-1}] \qquad (s > 1), \tag{7}$$

and the optimal rule is: accept the first of the s candidates if her score exceeds ϕ_{s-1}.

Proof. Equation (6) is a consequence of (2), (7) of (5), and the action rule is that implicit in the optimality equation (7). ■

The recursion (7) determining the sequence of critical thresholds ϕ_s can be written in various ways. If y has distribution function $H(y)$ then (7) can be written

$$\phi_s = \phi_{s-1} H(\phi_{s-1}) + \int_{\phi_{s-1}}^{\infty} y \, \mathrm{d}H(y) \tag{8}$$

or

$$\phi_s - \phi_{s-1} = \int_{\phi_{s-1}}^{\infty} [1 - H(y)] \, \mathrm{d}y.$$

Some particular cases are considered in the exercises.

Exercises

1. Suppose y to be the standard exponential. Recursion (8) then becomes

$$\phi_s = \phi_{s-1} + e^{-\phi_{s-1}}.$$

Approximating this by the differential equation

$$\frac{\partial \phi}{\partial s} = e^{-\phi}$$

one deduces the approximation

$$\phi_s \sim \log(c + s)$$

for some c and large s.

2. Suppose y distributed uniformly on $(0, 1)$. Deduce the recursion

$$\phi_s = \tfrac{1}{2}[1 + \phi_{s-1}^2]$$

and approximate solution

$$\phi_s \sim 1 - \frac{2}{c + s}.$$

3. THE SECRETARY PROBLEM. (B) MAXIMIZATION OF THE PROBABILITY OF APPOINTING THE BEST

Suppose the interviewer takes the perfectionist criterion, that he wishes to maximize the probability of appointing the best of the field, rather than that of simply maximizing the expected score of the candidate appointed. He then runs the risk of appointing no candidate at all.

The problem occurs in differing versions, according to whether candidates observed can only be ranked, or absolute scores (of known distribution) can be attached to them. We consider these in order.

Let W_t be the observation history up to and including the tth candidate (the 'current' candidate out of a total field of N). Then the only relevant aspects of this history are t itself, and whether the current candidate outranks her predecessors. This event can be labelled by the indicator variable

$$z_t = \begin{cases} 1, & \text{if } t\text{th candidate is best hitherto,} \\ 0, & \text{otherwise.} \end{cases} \quad (1)$$

We can then write $F(W_t)$ as $F(z_t, t)$.

Define

$$\phi(t) = E[F(z_{t+1}, t+1) | W_t; u_t = R] \quad (2)$$

the maximal probability of appointing the best if t candidates have been interviewed and not appointed. Since the z_t are independent, with $P(z_t = 1) = 1/t$; then

$$\phi(t) = \frac{t}{t+1} F(0, t+1) + \frac{1}{t+1} F(1, t+1) \qquad (t < N). \tag{3}$$

The terminal condition is

$$\phi(N) = 0. \tag{4}$$

One has also

$$F(z,t) = \begin{cases} \phi(t) & (z = 0), \\ \max\left(\frac{t}{N}, \phi(t)\right) & (z = 1), \end{cases} \tag{5}$$

since in the first case one has no option but to continue, while in the second one can appoint or continue, with respective probabilities t/N or $\phi(t)$ of appointing the best.

Relations (3) and (5) relate maximal probabilities before and after interview; one derives an optimality equation by eliminating either F or ϕ. Taking the first course we deduce the optimality equation

$$\phi(t-1) = \frac{t-1}{t}\phi(t) + \frac{1}{t}\max\left(\frac{t}{N}, \phi(t)\right) \tag{6}$$

or

$$\phi(t-1) - \phi(t) = \left[\frac{1}{N} - \frac{\phi(t)}{t}\right]_+ \tag{6'}$$

with terminal condition (4).

Theorem 3.1. *Let t_0 be the smallest integer satisfying*

$$\sum_{t}^{N-1} \frac{1}{r} \leq 1. \tag{7}$$

Then the optimal policy is to interview the first $t_0 - 1$ candidates without making an appointment, and then appoint the first candidate superior to all predecessors, if one such occurs. The quantity $\phi(t)$ defined in (2) has the evaluation

$$\phi(t) = \begin{cases} \phi(t_0) & (t \leq t_0), \\ \frac{t}{N} \sum_{t}^{N-1} \frac{1}{r} & (t \geq t_0). \end{cases} \tag{8}$$

Proof. One sees from (6′) that $\phi(t)$ is non-increasing. If t_0 is the smallest value of t such that $t \geq N\phi(t)$ then the first or second options hold in (6) according to whether $t \geq t_0$ or $t < t_0$—from which evaluation (8) and characterization (7) follow. One will accept the tth candidate if she is the best hitherto and if $t/N \geq \phi(t)$, whence the action rule asserted. ■

For large N approximate equality in (7) amounts to $\log(N/t) \sim 1$, so that $t_0 \sim N/e$. The initial probability of selecting the best is then $\phi(0) = \phi(t_0) \sim t_0/N \sim 1/e$.

Suppose now that scores y are observed which are independent, with the known distribution function $H(y)$. Let $\phi(x, t)$ denote the maximal probability of appointing the best, conditional on the fact that t candidates have been interviewed and not appointed and that x is the highest score observed, the *record*. It is intuitively clear that x and t are sufficient at this stage.

The relation analogous to (6) is, then, by a similar argument,

$$\phi(x, t-1) = H(x)\phi(x, t) + \int_x^\infty \max\,[H(y)^{N-t}, \phi(y, t)]\,\mathrm{d}H(y) \qquad (t \leq N) \quad (9)$$

with

$$\phi(x, N) = 0,$$

and the tth candidate will be accepted if she is the best hitherto and her score y satisfies

$$H(y)^{N-t} \geq \phi(y, t). \qquad (10)$$

That is, the probability that she is the best also of those remaining must exceed the probability of success if one continues.

If the y-distribution possesses a density then one can normalize to the case when y is uniformly distributed on (0, 1) by a non-linear change of scale. Relations (9), (10) then become

$$\phi(x, t-1) = x\phi(x, t) + \int_x^1 \max\,[y^{N-t}, \phi(y, t)]\,\mathrm{d}y, \qquad (11)$$

$$y^{N-t} \geq \phi(y, t). \qquad (12)$$

Relation (11) is to be solved with the terminal condition $\phi(x, N) = 0$. However, evaluation seems much easier by a direct argument, due to Gilbert and Mosteller (1966).

Theorem 3.2. *Define $\phi_s(x) = \phi(x, N-s)$—the maximal probability of appointing the best if s candidates remain and if x is the record. Let ξ_s denote the solution in* (0, 1) *of*

$$\xi^s = \phi_s(\xi). \qquad (13)$$

Then

$$\phi_s(x) = \sum_{r=1}^{s} \binom{s}{r} \frac{x^{r-s}(1-x)^r}{r} \qquad (x \geq \xi_s) \qquad (14)$$

and $\xi_s = (1 + \zeta_s)^{-1}$, *where* ζ_s *is the positive root of*

$$\sum_{r=1}^{s} \binom{s}{r} \frac{\zeta^r}{r} = 1. \tag{15}$$

If the candidate being interviewed has s other candidates to follow her, then she should be appointed if she is the best hitherto, and if her score exceeds ξ_s.

Proof. The final criterion is just a rephrasing of that associated with inequality (12). Suppose $x \geq \xi_s$ and s candidates remain. Since $\phi_r(x)$ decreases as r does, then so does ξ_r, and one will appoint the first candidate whose score exceeds x, if there is one, and no other. The probability that there are r such candidates is $\binom{s}{r} x^{r-s}(1-x)^r$; the probability that the first of these is indeed the best is $1/r$. Hence expression (14). Equation (15) follows from (13), (14). It is plain that (15) has a unique positive solution and hence (13) a unique solution in $(0, 1)$. ■

4. SELLING AN ASSET

Consider the secretary problem in the score-optimization version of Section 2, but suppose the future discounted by a factor β. Suppose also that the pool of candidates is infinite; i.e. that the horizon is infinite. Plausibly the maximal expected score ϕ_s will now have a limit ϕ, obeying the analogue of (2.7):

$$\phi = E \max(y, \beta\phi). \tag{1}$$

The optimal rule is to accept the first candidate whose score exceeds $\beta\phi$.

The problem as thus formulated can also be viewed as a stochastic version of the asset-selling problem considered in Section 9.10. The scores y are regarded as prices bid sequentially for an asset, and the question is; when to accept the current bid? Relation (1) corresponds to the infinite-horizon case, with bids statistically independent.

In a general stochastic formulation one would take account of the observational history W_t at the time when acceptance or rejection of the tth bid is being considered. The analogue of the minimax solution would be: accept the current bid at time t if $E[\beta^r y_r | W_t]$ is maximal in $r \geq t$ at $r = t$. This 'certainty-equivalent' rule is not optimal, however, as the reader can convince himself for the case of independent bids, considered above.

The independent bid case can be varied in a number of ways. If a cost c per unit time is incurred for maintaining the asset (or continuing with an unfilled vacancy) then relation (1) becomes

$$\phi = E \max(y, \beta\phi - c). \tag{2}$$

Another generalization is to allow past bids to remain open. This would seem to make no difference to the policy; the optimal policy will be to accept the current bid if it exceeds the critical value $\beta\phi - c$ determined by (2), or else to forget it. The intuitive argument is that, if a bid is not accepted at a given time, then there are equally no grounds for accepting it later, whether or not it has been superseded in the interim. In fact, one can make an assertion valid also for the finite-horizon case.

Theorem 4.1. *Consider the situation in which existing bids, statistically independent, remain open. Let the current best bid be x and suppose that s bids remain. Then the optimal policy is to accept x if $s = 0$ and to accept x when first it exceeds the value $\beta\phi - c$ determined by* (2) *if $s > 0$.*

So, if the best bid at time $t - 1$ is not acceptable, and that at time t is, and t is not the horizon point, then it must be the current bid y_t which is accepted.

Proof. Let $F_s(x)$ denote the maximal expected reward if the current best bid is x and s bids remain. The optimality equation is then

$$F_s(x) = \max\,[x, \beta E F_{s-1}(x^*) - c] \tag{3}$$

where $x^* = \max(x, y)$ and the expectation in (3) is with respect to y. The terminal condition is

$$F_0(x) = x \tag{4}$$

if some bid must be accepted. It follows from (3), (4) that one will accept the best bid x at $s = 1$ if

$$x \geq \beta E \max(x, y) - c. \tag{5}$$

Denote the critical value of x for which equality holds in (5) by $\beta\phi - c$; inserting this value into the equality form of (5) we deduce relation (2). That is, ϕ can be identified with the ϕ of (2), and the assertion of the theorem is verified for $s = 1$.

Now the region $x \geq \beta\phi - c$, once entered, can never be left, because x_t is increasing in time. It follows, then, by a general theorem (Theorem 15.3.1), that the optimal stopping region at $s = 1$ determined by (5) is the optimal stopping region for all $s > 0$. ■

The reference to Theorem 15.3.1 places this problem in a more general context. It is a *stopping problem*: one can either stop (i.e. accept a bid or candidate) or else continue. Criterion (5) is a *one-step look-ahead* (OSLA) *rule*. One compares the options of immediate stopping and of continuation for one step followed by stopping. Theorem 15.3.1 states that the stopping region determined by the OSLA rule, optimal for $s = 1$, is also optimal for all $s > 0$ if one can never leave it, i.e. if it is an absorbing set for the state variable.

Exercises

1. Demonstrate uniqueness of the solution of (2) under suitable hypotheses.
2. *Parking.* This is another optimal stopping problem. Parking places are

arranged linearly along the route to one's destination. Each parking place is free with probability p, these events being independent; one cannot observe whether a place is free until one reaches it. If one parks s places from one's destination one incurs a cost of s; if one reaches one's destination without having parked one incurs a cost of K.

Let F_s be the minimal expected cost if one is distant s places from one's destination and has not yet parked. Show that

$$F_0 = K,$$

$$F_s = p \min(s, F_{s-1}) + qF_{s-1} \qquad (s = 1, 2, \ldots)$$

where $q = 1 - p$. If r is the smallest integer satisfying

$$q^{r-1} < (pK + q)^{-1}$$

show that the optimal policy is: never park if $s \geq r$, but take the first available place if $s < r$.

5. REPLACEMENT, WITH COMPLETE INFORMATION ON COMPONENT STATE

There is a whole category of problems characterized by the fact that some component of a system deteriorates with time and must at some stage be replaced or reconditioned. The component might, for example, be an electric lamp, the cutting edge of a tool, the refractory lining of a furnace or the lubricating oil in the sump of an engine. In some cases one may carry out an occasional inspection to determine the 'state' of the component, in some cases not. One may replace or repair the component before such action is strictly necessary. This may be because failure and replacement during operation is particularly costly or simply because it is convenient to replace or repair at planned intervals rather than at random failure times.

One then has the decisions: when and how to inspect, when and how to carry out remedial action. This formulation also includes the problem of *quality control*, when a machine must occasionally be reset for satisfactory performance.

Let x denote the state of the component, which we assume to be immediately observable, and to follow a Markov process if left to itself. The state of a new component will have a standard value, $x = 0$, say. 'Failure' could be regarded as another state (or states), but we shall regard it simply as an event which makes immediate replacement obligatory. Assume the cost of a voluntary replacement to be c and the cost of a replacement forced by failure in state x to be $c + d(x)$. Let $g(x)$ be the cost of operating a functioning component in state x for one unit of time and let $p(x) = 1 - q(x)$ be the probability that a component which has reached state x without failure fails during the next step if one attempts to continue. The equilibrium optimality equation for the discounted problem is then

$$F(x) = min\,[c + \beta F(0), g(x) + \beta p(x)(c + d(x) + \beta F(0)) + \beta q(x)E[F(x')|x]] \qquad (1)$$

where x, x' are generic values of x_t, x_{t+1}. Replacement is then to be advised in all states x for which the second expression in the square brackets exceeds the first. This criterion can be simplified under some circumstances.

Theorem 5.1. *Suppose that $g(x_t) + p(x_t)d(x_t)$ increases with t with probability one and that the problem is such that ultimate replacement is certain. The optimal rule is then to replace when first*

$$g(x) + \beta p(x)d(x) \geq (1 - \beta)(c + \beta F(0)). \tag{2}$$

The plausibility of the result is clear, since $g(x) + p(x)d(x)$ is the effective one-step cost incurred in state x. The threshold value in the right-hand member of (2) depends upon $F(0)$; to determine this one must in principle solve the optimality equation, although, having determined the general form of the optimal rule, one may be satisfied to fix the threshold value by ad hoc methods.

Proof. Equation (1) exhibits the problem as a stopping problem with the stopping cost $c + \beta F(0)$; criterion (2) is just the OSLA rule for such a problem. The hypothesis of the theorem ensures the optimality of the OSLA rule (see Section 4 and Theorem 15.3.1). ■

For an OSLA rule to be valid it is not necessary that the state should revert to a particular value 0 on replacement, but merely that it should regenerate, i.e. become independent of the state x just quitted. For example, a replacement may not be new, but drawn at random from a population with the state distribution $\pi(x)$. The quantity $F(0)$ in (1), (2) would then be replaced by $\sum \pi(x)F(x)$.

The hypothesis that one-step costs increase strictly with time can be weakened (see Exercise 1).

If one equates state with age then

$$E[F(x')|x] = F(x + 1)$$

and the problem can be treated fairly explicitly. This formulation of the problem corresponds to a situation where actual state is not observable, and so one assumes costs, failure rates, etc., to be simply age-dependent. The only observables then are age and whether the component functions or fails.

The optimality equation (1) now determines an *age replacement policy* (ARP): a decision on replacement which depends on the component's age. An ARP is conventionally assumed to have the simple form of a *mandatory* ARP: that one replaces a component when it fails, or when its age reaches or exceeds a critical value ξ. The optimal policy need not necessarily have this simple form. However, if ξ is the smallest value of x for which the optimal policy recommends replacement, then one might as well assume the mandatory policy, because, once the first component has been replaced, later components will not be allowed to continue in service beyond age ξ in any case. Under a mandatory ARP one can solve (1) to obtain

$$F(0) = \frac{\sum_{x=0}^{\xi-1} \beta^x q_0 q_1 \dots q_{x-1}[g_x + p_x(c + d_x)] + \beta^\xi q_0 q_1 \dots q_{\xi-1} c}{1 - \sum_{x=0}^{\xi-1} \beta^{x+1} q_0 q_1 \dots q_{x-1} p_x - \beta^{\xi+1} q_0 q_1 \dots q_{\xi-1}} \tag{3}$$

where we have written $q(x)$ as q_x, etc. The cut-off ξ will be chosen to minimize this quantity.

Exercises

1. Suppose that equation (1) is simplified to

$$F_s(x) = \min\,[c + \beta F_{s-1}(0), g(x) + \beta E(F_{s-1}(x')|x)]$$

in the finite-horizon case, so that the event of failure has effectively been incorporated in the state description. Suppose that x takes the values $x = 0, 1, 2, \dots$, with increasing x corresponding to increasing degeneration in that (a) $g(x)$ is an increasing function of x and (b) $P(x' \geqq z|x)$ is a non-decreasing function of x for all z. Show that $F_s(x)$ is an increasing function of x if $F_{s-1}(x)$ is, and that the optimal rule is to replace if x exceeds some s-dependent threshold ξ_s.
2. Consider evaluation (3) in the case $g(x) \equiv 0, d(x) \equiv d$. Show that in the limit $\beta \uparrow 1$ this evaluation corresponds to an average cost per unit time of

$$\gamma = \frac{c + d(1 - q_0 q_1 \dots q_{\xi-1})}{\sum_0^\xi q_0 q_1 \dots q_{x-1}}.$$

3. For the general problem with the optimality equation (1), suppose that $p(x)$, $g(x)$ and $d(x)$ are all independent of x. Show that it is then never optimal to replace a functioning component. Interpret.

6. REPLACEMENT, WITH OPTIONAL INSPECTION

To carry over the notation of the last section, suppose that the component can only adopt the two states $x = 0, 1$ which we shall regard as 'good' and 'faulty' respectively. The costs of using the component per unit time in its good and faulty states are taken as being 0 and d respectively. The transition matrix is taken as

$$(p_{xx'}) = \begin{pmatrix} 1-\alpha & \alpha \\ \cdot & 1 \end{pmatrix}$$

so that a fault develops with probability α and then remains. The cost of an inspection (which tells one the state exactly) is b; the cost of restoring a faulty component to state 0 is c.

The dynamics of degeneration are thus the simplest non-trivial version of those formulated in the last section, but inspection is now both costly and optional.

Since the component state is not necessarily observed at all times, it cannot be the state variable of the decision process. We shall see in Chapter 39 that, in such cases, the effective state variable of the decision process is the *posterior distribution* of the state variable of the underlying Markov process; i.e. the actual distribution over process states, conditional on current observables. In this case, this is simply the posterior probability, p, say, that the system is in the fault-free state, conditional on current observations.

If one does not carry out an inspection, then p suffers the transformation

$$p \to (1-\alpha)p. \tag{1}$$

If one carries out an inspection, then one either confirms the component as satisfactory or else finds it faulty and replaces it. So, under either contingency, the component reverts to state 0, that is, $p \to 1$.

The equilibrium optimality equation for the discounted problem is then

$$F(p) = \min\begin{bmatrix}(1-p)d + \beta F((1-\alpha)p) \\ b + (1-p)(c+d) + \beta F(1)\end{bmatrix}. \tag{2}$$

Theorem 6.1. *$F(p)$ is a decreasing function of p. The optimal strategy thus takes the form: inspect when p falls below a critical value p^*. Inspection thus takes place at intervals of r units of time, where r is the smallest integer such that*

$$(1-\alpha)^r < p^*. \tag{3}$$

Proof. The second two statements are immediate consequences of the first, since $p = (1-\alpha)^r$ at time r if $p = 1$ at time 0 and no inspections have since taken place. The first assertion follows from the usual induction proof. Write (2) as $F = \mathscr{L}F$. If $\phi(p)$ is decreasing then so is $-pd + \beta\phi((1-\alpha)p)$, and hence so is $\mathscr{L}\phi$, and so also is $\mathscr{L}^s\phi$ and $F = \lim \mathscr{L}^s\phi$. That the limit exists and equals F follows from the general theory of Chapter 23. ■

Of course, one should allow more component states, for realism. If one also allows variations of the inspection option, then one almost has a formulation of the general experimentation/action situation of Part VI. A relatively realistic and simple treatment is available in some cases (see Section 41.12).

7. OPTIMAL ALLOCATION BETWEEN CONSUMPTION AND INVESTMENT

Consider the model of Section 3.1 again, but suppose now that the factor by which capital grows at time t is a random function of time, a_t. Suppose that the operations of the investor are small enough that his actions do not affect interest rates, so that a_t is independent of the investor's past actions and capital. One then finds, as in (3.1.3), (3.1.4), that

$$F_s(x, \tilde{W}_t) = A_s(\tilde{W}_t)x^{1-\nu} \tag{1}$$

$$u(x, s, \tilde{W}_t) = \frac{x}{[A_s(\tilde{W}_t)]^{1/\nu}}, \tag{2}$$

where the coefficient A_s is now a function also of $\tilde{W}_t$—the information available at time t on all economic variables affecting future growth rates. (Our assumption is that X_t and U_{t-1} belong to W_t, but not to $\tilde{W}_t$.) In place of recursion (3.1.5) we now have

$$A_s(\tilde{W}_t)^{1/\nu} = 1 + \{\beta E[A_{s-1}(\tilde{W}_{t+1})a_t^{1-\nu} | \tilde{W}_t]\}^{1/\nu}. \tag{3}$$

If the model is time-homogeneous then the utility growth rate γ of (3.1.6) will now be the maximal solution of

$$\gamma^\nu \phi(\tilde{W}_t) = \beta E[\phi(\tilde{W}_{t+1})a_t^{1-\nu} | \tilde{W}_t] \tag{4}$$

for non-zero ϕ. Suppose, for example, that

$$\tilde{W}_t = \{a_r; r < t\}$$

and

$$a_t = a_{t-1}^\alpha \zeta_t \tag{5}$$

where $0 < \alpha < 1$ and $\{\zeta_t\}$ is a sequence of independent and identically distributed positive random variables. The process $\{a_t\}$ is then Markov and (4) reduces to

$$\gamma^\nu \phi(a) = \beta E[\phi(a^\alpha \zeta)(a^\alpha \zeta)^{1-\nu}]$$

where the expectation is over ζ. One finds a solution:

$$\phi(a) = a^{\alpha(1-\nu)/(1-\alpha)},$$

$$\gamma = [\beta E(\zeta^{(1-\nu)/(1-\alpha)})]^{1/\nu}. \tag{6}$$

Since this ϕ is strictly positive, then, by Perron-Frobenius theory, evaluation (6) is correct, in that it is the maximal solution of (4).

Exercises

1. Obtain the reduced (Markov) form of recursion (3) in the special case associated with (5).
2. Verify relations (1) to (3). Note that these must be consequences of the general optimality equation, as Markov structure is not postulated. The only dynamic assumptions are that $x_{t+1} = a_t(x_t - u_t)$ and that $\{a_t\}$ is causally independent of the investor's actions or state.

8. PORTFOLIO SELECTION

Suppose that one holds capital x_t at time t, which can then be distributed between a number of investments. An amount x invested in stock j at time t will grow to $a_{jt}x$ by time $t+1$, say, where the a_{jt} are random variables.

We suppose that rewards are purely terminal (so that there is no consumption of assets) and that capital may be switched between investments without cost. Current capital x_t is supposed known, and so is a component of W_t. It will obey the recursion

$$x_{t+1} = \left(\sum_j u_{jt} a_{jt}\right) x_t \tag{1}$$

where u_{jt} is the fraction of x_t invested in stock j. The optimality equation will then be

$$F_s(W_t) = \max_{u_t} E[F_{s-1}(W_{t+1}) \mid W_t, u_t] \tag{2}$$

where $u_t = \{u_{jt}\}$ is constrained to be a distribution over the values of j.

Suppose the investor is small enough that his actions do not affect the future course of growth rates. Suppose also that the terminal reward takes the form

$$F_0(W_t) = x_t^{1-v} \tag{3}$$

where v takes a value in $[0, 1)$. If $v = 0$ then one is simply maximizing expected terminal capital; if $v > 0$ then the terminal reward is strictly concave in terminal capital, which implies a degree of risk-aversion (i.e. that random variation about the expected value is penalized).

With these assumptions one finds that F_s has the form

$$F_s(W_t) = A_s(\tilde{W}_t) x_t^{1-v} \tag{4}$$

where $\tilde{W}_t$ is the market information—the component of W_t that actually conditions future rates of growth. The $\tilde{W}_t$-dependent coefficient A_s obeys the optimality equation

$$A_s(\tilde{W}_t) = \max_{u_t} E\left[A_{s-1}(\tilde{W}_{t+1}) \left(\sum_j u_{jt} a_{jt}\right)^{1-v} \middle| \tilde{W}_t \right]. \tag{5}$$

If one is maximizing a quantity such as $(\sum u_j a_j)^{1-v}$ for known non-negative a_j one of course concentrates the distribution u on the values of j for which a_j is largest—on the fastest-growing stocks. If the a_j were unknown and one were maximizing $E(\sum u_j a_j)^{1-v}$, where the expectation is over the values of the a_j, then, in the case of risk-indifference, $v = 0$, one would again concentrate on those stocks of the greatest expected growth factor $E(a_j)$. In the case $v > 0$ the optimal distribution would be spread more widely, however, in an effort to hedge against risk.

For the dynamic case (5) one might imagine that, because actions do not affect future growth rates and capital can be transferred between investments without cost, the optimal rule would be a one-step look-ahead rule. That is, one would simply choose u_t to maximize the factor by which utility increased at the next step, i.e. maximize $E[(\sum_j u_{jt} a_{jt})^{1-\nu} | \tilde{W}_t]$. This is not so, is general, because the term under the expectation in (5) is weighted multiplicatively by the subsequent growth factor. In the case $\nu = 1-$, however, the conjecture is true (see Exercise 2).

Under appropriate assumptions recursion (5) will, of course, reduce. If we suppose that $\tilde{W}_t = \{a_r; r < t\}$, where a_r is the set of growth factors a_{jr} at time r, and suppose also that $\{a_r\}$ is Markov, then $\tilde{W}_t$, $\tilde{W}_{t+1}$ can be replaced in (5) by a_{t-1}, a_t respectively.

Exercises

1. Consider the special case $F_0(W_t) = x_t$. Show that there is an optimal policy which concentrates investment on a single option at every stage.
2. Consider the special case $F_0(W_t) = \log x_t$. Prove optimality of the one-step look-ahead rule: choose u_t to maximize $E[\log(\sum_j u_{jt} a_{jt}) | \tilde{W}_t]$.
3. Analyse, as far as you can, a consumption/investment model, combining the features of the models of this and the preceding section.

9. OPTIMAL GAMBLING. (A) INSTABILITY OF THE STOPPING PROBLEM

The theory of probability received much of its founding impetus from investigation of games of chance. Dubins and Savage (1965) showed that this impetus continues and that substantial further work was required to develop a theory of *optimal* gambling. These authors did not use dynamic programming arguments explicitly, but it seems likely that a largely equivalent dynamic programming treatment is possible.

One of their principal results is that, if one trying to maximize the probability of winning a given amount, then cautious play is optimal if the game is favourable and bold play is optimal if the game is unfavourable. That is, that one should stake lightly or heavily (in senses to be made explicit in Sections 10 and 11) according as the odds are in one's favour or against one. This is intuitively reasonable: in the first case one will want to give the law of large numbers time to operate; in the second case one will not.

In this section we consider a simpler problem: the decision of when to retire from a fair game with fixed stakes. Nevertheless, this, the simplest of all gambling problems and of *stopping problems* (see Chapter 29), shows a classic and interesting pathology.

Suppose that money is held in multiples of a unit amount, which may be negative (so that debts are permitted); that at each step a unit amount is staked; and that the game is fair, so that one makes a net gain of $+1$ or -1, each with probability $\frac{1}{2}$. If the amount of one's current assets (i.e. amount of money held) is x, then the question is whether to continue playing or to retire. The criterion is the maximization of expected assets at termination.

For the finite-horizon case current assets x fairly obviously constitute a state variable, and one will have the optimality equation

$$F_s(x) = \max\left[x, \tfrac{1}{2}F_{s-1}(x-1) + \tfrac{1}{2}F_{s-1}(x+1)\right] \qquad (s > 0) \tag{1}$$

with the terminal condition

$$F_0(x) = x. \tag{2}$$

Here $F_s(x)$ is construed as the maximal expected reward rather than the minimal expected loss, and relations (1), (2) hold for $x \in Z$.

One readily determines that (1) has the solution

$$F_s(x) = x \tag{3}$$

so that the optimal expected reward does not differ from that obtainable by terminating immediately. In fact, if we define the operator T by

$$T\phi(x) = \tfrac{1}{2}[\phi(x-1) + \phi(x+1)]$$

then

$$T^s x = x$$

for integral s, so that one gets the same expected reward if one plays a predetermined number of bets.

However, over an infinite horizon, one can do better. The sequence $\{x_t\}$ generated by continued play is a simple random walk, which can attain any given value in Z, with probability one, if one waits long enough. One can then terminate with an arbitrarily high level of assets if one only continues until that level is reached. Thus $F(x) = +\infty$.

The difference between F and F_∞ reveals a fundamental difference between finite- and infinite-horizon optimization, referred to as *instability*. Almost any modification that encourages or forces play to terminate in a finite time will remove instability.

For example, suppose that one discounts future rewards, so that (1) is modified to

$$F_s(x) = \max\left[x, \beta T F_{s-1}(x)\right]. \tag{4}$$

One conjectures that the continuation region will have the form $x < a$, so that

$$F(x) = \begin{cases} x & (x \geq a), \\ \beta/2[F(x-1) + F(x+1)] & (x < a). \end{cases}$$

The second of these relations has a solution

$$F(x) = c\alpha^x + c'\alpha^{-x} \qquad (x \leq a) \tag{5}$$

where

$$\alpha = \frac{1 + \sqrt{(1 - \beta^2)}}{\beta}$$

Since $F(x) \geqq x$ and $F(x)$ is monotone non-decreasing we must have $c' = 0$. Determining c by matching of the two solution forms at $x = a$ we deduce that

$$F(x) = \begin{cases} x & (x \geq a), \\ a\alpha^{x-a} & (x < a). \end{cases} \tag{6}$$

Application of the inequalities implicit in (4) at $x = a - 1, a$ yields

$$(a - 1)\alpha < a,$$
$$a\alpha \geq a + 1,$$

whence we deduce that a is the smallest integer satisfying

$$a \geq (\alpha - 1)^{-1},$$

a positive quantity. F has the form indicated in Figure 9.1.

Note that, as $\beta \uparrow 1$, then $\alpha \downarrow 1$ and $F(x) \uparrow \infty$. The discounted version of the process is stable (because discounting discourages continuation if one is in a favourable position), but the limit of the discounted process gives the correct evaluation for the unstable undiscounted process.

Another way to stabilize the process is to penalize or prohibit debts. If debts are

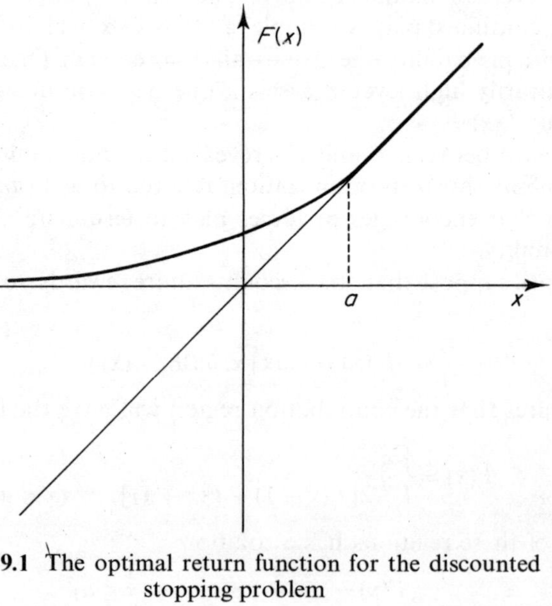

Figure **9.1** The optimal return function for the discounted stopping problem

prohibited, then relation (1) holds for $x = 1, 2, 3, \ldots$ only, and one has the extra boundary condition

$$F_s(0) = 0.$$

Then (3) is still the solution of this equation, but indefinite play can yield no better reward than $F(x) = x$. Indeed, since ultimate absorption at the origin is certain, infinite play is impossible.

Another way to view the distinction between the two cases is to note that

$$T\phi = \phi$$

has the solution

$$\phi(x) = c_1 + c_2 x$$

but that one necessarily has $c_1 = 0$ if the boundary condition $\phi(0) = 0$ is imposed (see also Exercise 2).

Exercises

1. Show that the discounted problem is stable, in that $F_s(x)$ converges to solution (6) as s increases.
2. Solve the discounted version of the problem in which no debts are permitted and show that $F(x)\uparrow x$ as the discount factor tends to unity. This is to be compared with $F(x)\uparrow\infty$ for the case when debts are permitted and unpenalized.

10. OPTIMAL GAMBLING. (B) OPTIMALITY OF CAUTIOUS PLAY UNDER CERTAIN CIRCUMSTANCES

We consider the simplest version of *rouge et noir*. Money is handled in multiples of a unit and debts are not permitted. Subject to these constraints bets of any positive size can be placed (so that the amount staked must be integral and must not exceed one's current assets). Zero bets are prohibited, to force action. Trials are independent and the player has a success probability of p at each trial. A win or a loss implies that one's stake is doubled or lost respectively. So, if one has assets x and bets an amount u, then x changes to $x + u$ or $x - u$ with probabilities p and $q = 1 - p$ respectively. The game is *favourable* or *unfavourable* according to whether $p > \frac{1}{2}$ or $p < \frac{1}{2}$. The case $p = \frac{1}{2}$ is termed *fair*.

The element of choice is then the size of the bet one places under different circumstances. In this section we shall show that, under a variety of criteria, the optimal policy is to bet the minimal amount, of unity. This is *cautious play*.

Theorem 10.1. *Cautious play is optimal if the game is favourable and one wishes to maximize the probability of attaining a given level of assets.*

Proof. Suppose the desired level is N; the game will then terminate if $x = 0$ or if $x \geq N$, and we wish to maximize the probability of the second contingency. If $G(x)$ is the maximal expected reward starting from x, then G will satisfy the optimality equation

$$G(x) = \max_u [pG(x+u) + qG(x-u)] \qquad (0 < x < N) \tag{1}$$

with the boundary conditions

$$\begin{aligned} G(x) &= -\infty \qquad (x < 0), \\ G(0) &= 0 \\ G(x) &= 1 \qquad (x \geq N). \end{aligned} \tag{2}$$

The stake u in (1) is restricted to the set of values $1, 2, \ldots, x$. We shall write (1) as

$$G = \mathscr{M}G. \tag{3}$$

Let $R(x)$ be the expected reward from a cautious policy. This then obeys

$$R(x) = pR(x+1) + qR(x-1) \qquad (0 < x < N)$$

with the same boundary conditions as G, so that

$$R(x) = \frac{1 - \theta^x}{1 - \theta^N} \tag{4}$$

where

$$\theta = \frac{p}{q} < 1.$$

If we can show that $R = \mathscr{M}R$ then it follows from a later theorem (Theorem 27.4.2) that the policy upon which R is calculated is optimal. Note now that

$$R(x) - pR(x+u) - qR(x-u) = \frac{\theta^x[H(\theta^u) - 1]}{1 - \theta^N} \tag{5}$$

where

$$H(\alpha) = p\alpha + q\alpha^{-1}.$$

However, as α decreases from θ, then $H(\alpha)$ increases from 1. Expression (5) is thus zero for $u = 1$, positive for $u > 1$, whence $R = \mathscr{M}R$. ■

Suppose the aim is simply one of survival—to keep playing as long as possible before ruin ensues. Then it turns out that cautious play maximizes survival time (in a number of senses) whether the game is favourable or unfavourable.

We define τ as the time until ruin, when x first becomes zero, with $\tau = +\infty$ if ruin never occurs.

Theorem 10.2. (*a*) *Suppose the game is unfavourable. Then cautious play maximizes the expected survival time* $E(\tau)$.

(*b*) *Suppose the game is favourable. Then cautious play maximizes the probability of indefinite survival,* $P(\tau = +\infty)$. *It also maximizes* τ *stochastically, i.e. maximizes* $P(\tau > k)$ *for any* k.

Proof. (a) Quantities such as $E(\tau)$ and $P(\tau > k)$ are calculated conditional on initial x. With cautious play $E(\tau) = x/(q-p)$. One finds that this satisfies the optimality equation if $q > p$.

(b) With cautious play the probability of indefinite survival is $1 - \theta^x$. As before, this satisfies the optimality equation if $p > q$.

To establish the stochastic maximization of τ, denote $P(\tau > k \mid x_0 = x)$, calculated on the assumption of cautious play, by $R_k(x)$. Note then that $R_k(x)$ is decreasing in k and increasing in x. Let r be the (random) number of trials required for assets to reach one of the values $x + u$ or $x - u$, and denote the (random) value of assets at this stage by X. Then, by (4),

$$P(X = x + u) = \frac{1 - \theta^u}{1 - \theta^{2u}} = \frac{1}{1 + \theta^u} \geq p.$$

Appealing to these facts we then have

$$\begin{aligned} R_k(x) &= E[R_{k-r}(X)] \\ &\geq E[R_{k-1}(X)] \\ &= P(X = x + u)R_{k-1}(x + u) + P(X = x - u)R_{k-1}(x - u) \\ &\geq pR_{k-1}(x + u) + qR_{k-1}(x - u). \end{aligned} \tag{6}$$

Inequality (6) demonstrates, recursively for increasing k, that the probability of surviving k stages of play cannot exceed $R_k(x)$, the value attained by cautious play. ■

Exercise

1. Consider a discounted version of the problem of Theorem 10.1. Show that cautious play is optimal for a favourable game only if discounting is weak enough.

11. OPTIMAL GAMBLING. (C) OPTIMALITY OF BOLD PLAY UNDER CERTAIN CIRCUMSTANCES

We return to the problem of choosing stakes so as to maximize the probability that a level of assets N is attained. However, we shall no longer assume that stakes and assets must be integer-valued; arbitrary non-negative values are permitted.

Bold play is characterized by the policy

$$u = \min(x, N - x). \tag{1}$$

That is, one stakes the amount $N - x$ required to reach level N with a single successful play, if one has it. Otherwise, one stakes all one has. The reward function $R(x)$ for bold play thus satisfies

$$R(x) = \begin{cases} pR(2x) & (0 < x \leq N/2), \\ p + qR(2x - N) & (N/2 \leq x < N), \end{cases} \tag{2}$$

with the boundary conditions

$$\begin{aligned} R(x) &= -\infty \qquad (x < 0), \\ R(0) &= 0 \\ R(x) &= 1 \qquad (x \geq N). \end{aligned} \tag{3}$$

Let us write (2) in the operator form

$$R = MR$$

and also define the operators M_u and $\mathscr{M}$ by

$$M_u\phi(x) = p\phi(x+u) + q\phi(x-u),$$

$$\mathscr{M}\phi = \sup_{u \geq 0} M_u\phi.$$

Let us define as *risk-encouraging* those monotone non-decreasing functions $\phi(x)$ with boundary specification (3) for which

$$\mathscr{M}\phi = M\phi.$$

That is, if ϕ were to be the reward function after one stage of play, then bold play would be optimal for that stage.

Lemma 11.1. *Suppose ϕ is risk-encouraging and the game unfavourable, and let a, b be a pair of values in $[0, N]$. Then $\phi(b+u) + \phi(a-u)$ is maximal in $u \geq 0$ at $u = \min(a, N-b)$.*

Proof. We shall prove the result first for $a = b = x$. Setting $p = \frac{1}{2} - \delta$ where $\delta > 0$ we have

$$\tfrac{1}{2}[\phi(x+u) + \phi(x-u)] = M_u\phi(x) + \delta[\phi(x+u) - \phi(x-u)].$$

The right-hand member is certainly maximal at $u = \min(x, N-x)$ in the range $0 \leq u \leq \min(x, N-x)$, because $M_u\phi(x)$ is maximal at that value and the final form is increasing in u. The left-hand member cannot increase if u is increased further, in view of the boundary specifications (3).

The result is thus established for $a = b$. The assertion then follows for $a \leq b$, since we can set $a = x - v, b = x + v$ and apply the result just proved with $u' =$

$u+v$ as an effective stake. This in turn implies the result for $a \geq b$, for any value of $\phi(b+u)+\phi(a-u)$ attainable for a value of u such that $b+u \leq a-u$ is also attainable for an admissible value u' such that $b+u' > a-u'$: namely $u' = a-b-u$. ■

Lemma 11.2. *Suppose the game is unfavourable. Then $M\phi$ is risk-encouraging if ϕ is.*

Proof. $M\phi$ is plainly non-decreasing with the boundary evaluations (3). To prove the other property, suppose that $0 \leq x \leq N/2$, and note the evaluations

$$\begin{aligned} M_u M\phi(x) &= p^2\phi(2x+2u) + qp\phi(2x-2u) \\ &= pM_{2u}\phi(2x) \qquad (x+u \leq N/2), \end{aligned} \tag{4}$$

$$\begin{aligned} M_u M\phi(x) &= p[p + q\phi(2x+2u-N)] + qp\phi(2x-2u) \\ &= p^2 + pq[\phi(2x-N+2u) + \phi(2x-2u)] \\ &\qquad (x-u \leq N/2 \leq x+u). \end{aligned} \tag{5}$$

By the risk-encouraging property and Lemma 11.1 these expressions are formally maximized at $u = \min(x, N/2 - x)$ and $u = \min(x, N-x) = x$ respectively. If $x \leq N/4$ then evaluation (4) necessarily holds, and $u = x$ is the optimizing value. If $N/4 \leq x \leq N/2$ then the optimizing value of u in $u \leq N/2 - x$ is $N/2 - x$; the optimizing value in $u \geq N/2 - x$ is x. The optimizing value is thus $u = x$, which it is then for all $x \leq N/2$.

In the same way, the value $u = N - x$ is proved optimizing for all x in $[N/2, N]$. ■

Theorem 11.1. *Suppose the game is unfavourable and that one wishes to maximize the probability of attaining a level of assets N, solvency being maintained at all stages. Then bold play is optimal at all horizons.*

Proof. Let $G_s(x)$ be the s-horizon expected reward under optimal play. Then

$$G_0(x) = 0 \qquad (0 \leq x < N)$$

and $G_s(x)$ has the boundary specification (3) for all x. But then G_0 is risk-encouraging, so that, by Lemma 11.2,

$$M^s G_0 = \mathcal{M}^s G_0 = G_s.$$

Thus bold play is optimal over any finite horizon s. By a later result (Theorem 27.4.3) this assertion also holds for an infinite horizon. ■

12. SCHEDULING TASKS ON SIMULTANEOUS PROCESSORS

This analysis (due to Weiss and Pinedo, 1980) is a nice example of dynamic programming arguments which appeal to structure rather than to explicit solution.

N tasks may be allocated to m processors. Each processor can process one task at a time and each task can be processed by one processor at a time. Interruptions in processing and switches of a task between processors are allowed. Task j is of size z_j and processor i works at rate s_i, in that if task j is allocated to processor i then it will be completed in time $T = z_j/s_i$. In fact, z_j is not known; it is only known that z_j is distributed exponentially with parameter $\lambda(j)$ (expectation $\lambda(j)^{-1}$). The time to completion T would thus be exponentially distributed with parameter $\lambda(j)s_i$. We assume the processors numbered in order of rate, so that processor 1 is the fastest. We can normalize to the case when there are always processors available by assuming that processors $N+1, N+2, \ldots$ exist for which $s_i = 0$. The state of affairs x at a given time will be the set of taks currently uncompleted. The aim is to choose an allocation of tasks to processors sequentially to minimize $\int_0^\infty g(x(t))\,dt$. We assume the cost function g such that

$$g(\varnothing) = 0, \tag{1}$$

$$g(x) > g(x') \text{ if } x \supset x'. \tag{2}$$

That is, the aim of the optimizer is to despatch the tasks rather than to maintain employment. The minimal cost function $F(x)$ will then evidently share properties (1), (2). The decision at any time will be the allocation of tasks. Plainly, we will use the $|x|$ fastest processors; let j_i be the label j of the task (of those in x) allocated to processor i ($i = 1, 2, \ldots, |x|$). This allocation $(j_1, j_2, \ldots)$ defines a decision u.

The form of the optimality equation over an infinitesimal time interval Δ will then be

$$F(x) = \min_u \Big[g(x) + \Delta \sum_i s_i \lambda(j_i) F(x - j_i) + \Big(1 - \Delta \sum_i s_i \lambda(j_i)\Big) F(x) \Big] + o(\Delta)$$

which yields in the limit

$$g(x) = \max_u \sum_i s_i \lambda(j_i) [F(x) - F(x - j_i)]. \tag{3}$$

We thus deduce:

Theorem 12.1. *The allocation u will be optimal if and only if the quantities*

$$\delta(j_i) = \lambda(j_i)[F(x) - F(x - j_i)] \tag{4}$$

are non-increasing in i.

Proof. The sum $\sum_i s_i \delta(j_i)$ is maximal when $\{j_i\}$ is such that $\{s_i\}$ and $\{\delta(j_i)\}$ are similarly ordered. ■

There are now two extreme allocations defining stationary policies which are candidates for optimality under an appropriate g. One is that in which the smallest jobs are always assigned to the fastest processor, so that $\lambda(j_i)$ is non-increasing in i. This defines the SEPT (shortest expected processing time) rule. The other is that in which the largest jobs are always assigned to the fastest processors, so that $\lambda(j_i)$ is non-decreasing in i. This defines the LEPT (longest expected processing time) rule.

Theorem 12.2. *Assume, in addition to* (1) *and* (2), *that* $\{g(x-(k))\}$ *has the opposite ordering to* $\{\lambda(k)\}$ *and that* g *is convex in that*

$$g(x) - g(x-(k)) - g(x-(l)) + g(x-(k,l)) \geqq 0 \tag{5}$$

for k, l *members of* x. *Then the LEPT rule is optimal.*

Proof. One chooses $\{j_i\}$ so that $\lambda(j_i)$ is non-increasing in i and demonstrates by induction on $|x|$ that $F(x)$ is convex in the sense (5) and that $\delta(i_j)$ is non-increasing in i (see the paper quoted for details). This latter fact implies optimality of the policy, by Theorem 12.1. ■

Theorem 12.3. *Assume, in addition to* (1) *and* (2), *that* $g(x)$ *is a function* $h(|x|)$ *of* $|x|$ *and that* $[h(i) - h(i-1)]/s_i$ *is non-increasing in* i ($0/0$ *being interpreted as* 0). *Then the SEPT rule is optimal.*

Proof. One chooses $\{j_i\}$ so that $\lambda(j_i)$ is non-decreasing in i and demonstrates by induction on $|x|$ that

$$\delta(j_i) \geq \frac{h(i) - h(i-1)}{s_i}$$

and that $\delta(j_i)$ is non-decreasing in i (see the paper quoted for details). ■

The SEPT rule is thus seen to be optimal if

$$g(x) = |x|$$

when it minimizes *expected flow-time*. More generally, it is optimal if

$$g(x) = \sum_{j \in x} w_j$$

where $w_j \geq 0$ and w_j has the same ordering as $\lambda(j)$.

The LEPT rule is seen to be optimal if

$$g(x) = \begin{cases} 0 & (x = \varnothing), \\ 1 & (x \neq \varnothing), \end{cases}$$

when it minimizes *expected makespan*, the expected time to completion of all tasks.

The LEPT rule is also optimal if

$$g(x) = \sum_{i=1}^{|x|} s_i$$

and so minimizes the amount of work done by the fast processors among the 'fast' policies (the policies which always use the $|x|$ fastest processors).

More generally, the LEPT rule is optimal among the fast policies if

$$g(x) = \sum_{i=1}^{|x|} c_i$$

where c_i/s_i is non-increasing in i. If c_i can be regarded as the cost per unit time of employing processor i, then the LEPT rule minimizes total processing costs.

Exercise

1. Suppose N interchangeable although non-identical components are available for m locations in a system ($N \geq m$). Component j in location i has a constant failure rate $\lambda(j)s_i$. Thus s_i is interpretable as a wear rate in location i. Show that the policy which maximizes the expected time elapsing until fewer than m unfailed components are available is the LEPT one: to allocate the best components to the worst locations.

13. CONTROL OF QUEUES

This is a subject which has been considered in extensively varied versions by a number of authors (see, for example, Serfozo, 1981, and references quoted therein). We shall consider just the one specimen problem: that of an $M/M/l$ queue in continuous time with a fixed arrival rate λ and a serving rate μ which can be varied arbitrarily. The aim is to vary μ with the queue size x so as to minimize the average value of $x + c(\mu)$, where $c(\mu)$ reflects the cost of serving at rate μ. We suppose $c(\mu)$ to be monotone increasing and strictly convex. In Section 43.3 we shall consider the corresponding $M/G/l$ problem as a semi-Markov decision problem.

We consider average cost optimality for the undiscounted case. The equilibrium optimality equation is

$$\gamma = \begin{cases} \min_{\mu} [x + c(\mu) + \lambda(f(x+1) - f(x)) + \mu(f(x-1) - f(x))] & (x > 0), \\ \min_{\mu} [x + c(\mu) + \lambda(f(x+1) - f(x))] & (x = 0), \end{cases} \tag{1}$$

where γ is the minimal average cost and $f(x)$ the differential cost; see (11.5.4).

Let the minimizing value of μ for the queue length x be denoted as μ_x, Relation (1) and our assumptions on $c(\mu)$ imply that $\mu_0 = 0$ and that, if μ_x is positive, then it is the unique solution of

$$c'(\mu_x) = f(x) - f(x-1) \tag{2}$$

where $c'(\mu) = \partial c(\mu)/\partial \mu$. From (1), (2) we thus deduce:

Theorem 13.1. *The optimal service rate μ_x for a queue of size x satisfies the first-order recursion*

$$\mu_x c'(\mu_x) - c(\mu_x) - \lambda c'(\mu_{x+1}) = x - \gamma \qquad (x \geq 0), \tag{3}$$

with the boundary condition $\mu_0 = 0$. The constant γ has a value such that the sequence $\{\mu_x\}$ thus generated increases monotonically.

If $\mu c'(\mu) - c(\mu)$ is of greater order than $c'(\mu)$ for μ large, then for x large μ_x is asymptotically equal to the root of

$$\mu c'(\mu) - c(\mu) = x. \tag{4}$$

Proof. We know that $\mu_0 = 0$ and that relation (3) follows from (1), (2) if $\mu_x > 0$ for $x > 0$. To prove that this latter supposition holds, suppose that $\mu_x = 0$ $(x \leq r)$ and that $\mu_{r+1} > 0$, for some $r > 0$. It follows then from (1) that

$$\Delta_x := f(x) - f(x-1) = \frac{\gamma - x + 1}{\lambda} \qquad (1 \leq x \leq r+1) \tag{5}$$

and from the optimality conditions that

$$c'(0) - \Delta_r \geq 0,$$
$$c'(0) - \Delta_{r+1} < 0. \tag{6}$$

But we see from (5) that $\Delta_{r+1} < \Delta_r$, which is incompatible with (6). Thus either $\mu_x > 0$ for all $x > 0$ or $\mu_x = 0$ for all x. We can exclude the latter alternative, so the first holds. Note a consequence of this—that necessarily $c'(0) - \Delta_1 < 0$ or

$$\gamma > \lambda c'(0). \tag{7}$$

The final assertion of the theorem is evident from (3), if μ_x and μ_{x+1} are of the same order in x. In this and in the characterization of γ in the theorem we are appealing to the intuitively evident fact that μ_x increases smoothly from $\mu_0 = 0$ to the asymptotic evaluation implied by (4). The analysis is to that extent incomplete, in that this behaviour should be demonstrated and its implications for the evaluation of γ made explicit. This is a non-trivial task. ■

The approximation derived for μ_x from (4) is independent of γ. For example, if we choose $c(\mu) = c\mu^{1+\alpha}$ with $\alpha > 0$, then we find from (4) that

$$\mu_x \sim \left(\frac{x}{\alpha c}\right)^{1/(1+\alpha)}. \tag{8}$$

Calculating the second-order term in this asymptotic evaluation from (3) we find that

$$\mu_x \sim \left(\frac{x}{\alpha c}\right)^{1/(1+\alpha)} + \frac{\lambda}{\alpha}, \tag{9}$$

which is still independent of γ to this degree of approximation.

Exercises

1. If μ_x is the service rate when the queue is of length x, then the equilibrium probability $\rho(x)$ that the queue has size x is proportional to $\lambda^x/\mu_1\mu_2\cdots\mu_x$ (see, for example, Cox and Miller, 1965, p. 176). The quantities μ_x must be such that $\sum_x \rho(x)(x + c(\mu_x))$ is minimal. Show that the corresponding stationarity conditions imply that

$$\rho(x)\mu_x c'(\mu_x) = \sum_x^{\infty} \rho(y)(y + c(\mu_y) - \gamma)$$

and that these imply (3).

2. Suppose the problem is modified in that there is probability intensity λp_j that a group of j persons may join the queue ($j = 1, 2, \ldots$). Show that if $c(\mu) = c\mu^{1+\alpha}$, then the analogue of formula (9) is

$$\mu_x \sim \left(\frac{x}{\alpha c}\right)^{1/(1+\alpha)} + \frac{\lambda}{\alpha}\Sigma j p_j.$$

Note that the second term is proportional to the expected number of persons arriving during a complete service at the unit service rate.

This variation could be regarded as representing the situation in which the arrivals are indeed single individuals, who declare the 'work-load' j they are bringing to the queue on arrival. The variable x would then represent the total work outstanding rather than the total number of customers outstanding, and 'completion of a service' represents the despatch of a quantum of work rather than the despatch of a customer.

CHAPTER 13

Inventory and storage problems

1. STORAGE MODELS

There is a class of models which might be called 'storage models', whose dynamics are expressible by the simple plant equation

$$x_{t+1} = x_t + u_t + \zeta_t. \tag{1}$$

The state variable x_t will be a scalar in the simplest cases, representing the amount of a particular commodity in storage at time t. The term u_t represents the planned or controlled component of change in x; the term ζ_t represents the uncontrolled component of change.

For example, in inventory problems one is concerned with stocking a commodity for retail sale; u_t then represents the planned addition to stocks (an order placed with a wholesale supplier) and ζ_t is a negative quantity, representing sales. If one sells as much as one can (the constraints are discussed in Sections 2 and 7) then sales are governed by demand, and demand will be unpredictable. For an inventory problem, then, input to stocks is controlled and output is uncontrolled.

In dam problems x_t is to be regarded as the amount of water held in a dam. The flow of water into the dam is determined by rainfall, and so is outside one's control. However, the release of water from the dam is within one's control. This case is then opposite to that of the inventory, in that input to stocks is uncontrolled and output is controlled.

The problem of warehousing (or 'commodity management', as it has come to be known, when dealings are on a large enough scale) is different yet again. An entrepreneur who is in business for profit will hold an amount x_t of a commodity in his warehouse. One can imagine a situation in which he controls both input and output, in that decisions on how much to buy in and how much to sell out are his. What is probably outside his control, and only partially predictable, are market prices for the commodity—both buying prices and selling prices. It is by buying cheap and selling dear that he makes his living. Of course, prices reflect availability, being low in times of plenty and high in times of scarcity. If the entrepreneur is dealing on large enough a scale that his actions affect availability, then they will also affect prices.

A limit case in this direction is that of a single entrepreneur, a monopoly. If his motive is social good rather than individual profit, then his role is virtually that of an agency charged with the stabilization of the supply of the commodity concerned. He will accept all of the commodity that comes on the market and will release it in such a way as to achieve the greatest utility for the community. One is then virtually back in the dam situation.

In all the situations described the stock has the role of a reserve which acts as a buffer or stabilizer, to even out variations in either demand or supply. In the inventory problem one smoothes out a variable demand. In the dam situation one smoothes out a variable supply. The intention of a warehouse (or stabilizing agency) is to match demand variations and supply variations. The motive for doing so is either profit or social concern. Note, of course, that there may be many complications. One may have to hold inventories of several different commodities and so operate stores in parallel. Dams may feed into each other, so that one must operate stores 'in series'—or at the nodes of a more complicated graph than a linear one. If the warehouseman operates on a large enough scale then his actions may affect supply, demand and prices.

Two of the simplest storage models are the diffusion versions of dam and inventory control (see Sections 38.3 and 38.4). To some extent the diffusion approximation is unrealistic, but it also both captures some essential features and allows simple solution.

2. INVENTORY CONTROL; FORMULATION

Let x_t be the stock of a particular commodity held at the end of day t. The decision made is to then order a further amount u_t from the supplier, which is to be delivered by the beginning of day $t+1$, in the simplest case. If demand for the item on day $t+1$ is ε_{t+1} and all demand is satisfied, then the recurrence relation for x_t is

$$x_{t+1} = x_t + u_t - \varepsilon_{t+1}. \tag{1}$$

By assuming that demand is always satisfied, we are admitting the possibility that stocks may run negative. 'Negative stocks' are to be regarded as a state of debt to some supplier—the practice of 'backlogging'.

As far as costs are concerned, one imagines that there is a cost $A(x)$ for holding stock x overnight, a cost $B(u)$ for placing an order of size u and receipts $R(\varepsilon)$ on sales ε. For x positive, $A(x)$ represents the costs of storage; for x negative, it represents the costs of running into deficit. The two aspects of this cost component are really quite different, one discouraging the holding of an excessive stock, the other discouraging the state of deficit.

The random element is the demand ε. If this is supposed independently and identically distributed on different days, with a known distribution, then the process is a Markov decision process, with the state variable x, and the equilibrium optimality equation for the discounted problem is

$$F(x) = \min_u [A(x) + B(u) - R(\varepsilon) + \beta EF(x + u - \varepsilon)]. \tag{2}$$

The expectation is with respect to ε.

The prime constraint on u is that it should be non-negative: $u \geq 0$. If one has limited storage capacity C then there is the second constraint $u \leq C - x$. There will be cases where these are the only constraints and others where the commodity naturally occurs in integral amounts, and then u will necessarily be integral, as will x and C. These constraints of boundedness and integrality do not really change the character of the problem and reduce the process to one with finite state and action spaces.

A natural choice for the order cost is

$$B(u) = \begin{cases} 0 & (u = 0), \\ b + cu & (u \geq 0). \end{cases} \tag{3}$$

Here c is the unit cost of the commodity to the retailer and b the *transaction cost*, the flat cost of placing any order at all. The constant b is what would be referred to in other contexts as a *set-up cost*. It plays a critical role in the sequel, because it is this that discourages too frequent ordering. We shall sometimes write the cost function (3) as $bi(u) + cu$, where $i(u)$ is the *order indicator*

$$i(u) = \begin{cases} 0 & (u = 0), \\ 1 & (u > 0). \end{cases} \tag{4}$$

If we settle on form (3) for the order cost, then it is convenient to work in terms of the normalized cost functions

$$F^*(x) = F(x) - \frac{1}{1-\beta}[ER(\varepsilon) - cE(\varepsilon)] + \frac{cx}{\beta}, \tag{5}$$

$$A^*(x) = A(x) + c\left(\frac{1-\beta}{\beta}\right)x, \tag{6}$$

in terms of which the optimality equation reduces to

$$F^*(x) = \min_u [A^*(x) + bi(u) + \beta EF^*(x + u - \varepsilon)] \tag{7}$$

The additional terms in (5) and (6) all have an interpretation. The constant term subtracted from F is simply the expected discounted net income from sales if no storage, deficit or transaction costs were incurred. The term cx/β is the cost value (on the current day) of the stock held. The additional term in (6) is the overnight change in cost value of this stock due to discounting.

3. INVENTORY CONTROL; OPTIMALITY OF THE ONE-BIN POLICY WITH ZERO TRANSACTION COST

Theorem 3.1. *Suppose that $A^*(x) = A(x) + c((1-\beta)/\beta)x$ is convex, with the limit $+\infty$ at $x = \pm\infty$, and that $B(u) = cu$. Then the optimal ordering policy is of the form*

$$u = \begin{cases} 0 & (x > S), \\ S - x & (x \le S). \end{cases} \tag{1}$$

This is the *'one-bin' policy*, in which S is the size of the 'bin' and one brings stock up to S as soon as it falls below.

Proof. The function of u which is being minimized in (2.6) is

$$\theta(x+u) = EF^*(x+u-\varepsilon). \tag{2}$$

Let the class of functions which are convex, with limit $+\infty$ for any infinite value of the argument, be denoted $\mathscr{C}$. Suppose that $\theta(x)$ belongs to $\mathscr{C}$ and has its minimum at $z = S$.

In minimizing $\theta(x+u)$ over $u \ge 0$ it is plain that rule (1) is the maximizing rule. The maximized function is

$$\max_u \theta(x+u) = \begin{cases} \theta(S) & (x \le S), \\ \theta(x) & (x > S), \end{cases}$$

which would also belong to $\mathscr{C}$, but that it has zero gradient for $x < S$.

We have now only to show that $F^* \in \mathscr{C}$. Let the equation (2.7) be written $F^* = \mathscr{L}F^*$. It is plain from the argument of the previous paragraph that, in the case $b = 0$, $\mathscr{L}\phi$ belongs to $\mathscr{C}$ if ϕ does. So then also does $F^* = \lim (\mathscr{L})^r \phi$. ■

The determination of the optimal value of S is not difficult. We have

$$F^*(x) = A^*(x) + \beta EF^*(S-\varepsilon) = A^*(x) + \beta\theta(S) \qquad (x \le S) \tag{3}$$

by the optimality of the one-bin rule. Setting $x = S - \varepsilon$ in (3) and taking expectations, we find that

$$\theta(S) = \frac{EA^*(S-\varepsilon)}{1-\beta} \tag{4}$$

and consequently deduce:

Theorem 3.2. *The normalized minimal cost function is given for $x \le S$ by equation* (3), *where $\theta(S)$ has the evaluation* (4) *and S is the value minimizing $EA^*(S-\varepsilon)$.*

Exercises

1. Suppose that $A(x)$ equals ax $(x \ge 0)$ or $-a'x$ $(x \le 0)$, where $a, a' > 0$, and

suppose that ε has the exponential density $\lambda e^{-\lambda\varepsilon}$. Deduce the evaluation

$$S = \frac{1}{\lambda}\log\left[\frac{a+a'}{a+c(1-\beta)/\beta}\right].$$

2. Show directly from the optimality equation that
   (a) $\theta(x) > \theta(S)$ for $x < S$.
   (b) $\theta(x)$ is non-decreasing for $x > S$. It is plain then that S is a global minimum of θ.

4. INVENTORY CONTROL; OPTIMALITY OF THE TWO-BIN POLICY

Theorem 4.1. *Suppose that $A^*(x) = A(x) + c((1-\beta)/\beta)x$ is convex, with the limit $+\infty$ at $x = \pm\infty$, and that $B(u) = b + cu$. Then the optimal ordering policy is of the form*

$$u = \begin{cases} 0 & (x > s), \\ S - x & (x \le s), \end{cases} \tag{1}$$

where $S > s$ if $b > 0$.

This is the two-bin or (s, S) policy. As soon as stock falls below a critical value s one brings it up to a level S. (This use of s to denote the lower critical value conflicts with our use of it to denote time-to-go. The notation is hallowed in this context, however, and s will have only the former interpretation in Sections 4 to 7.) The effect of a transaction cost is to make one order less frequently, but to order rather more when one does. We shall take the rest of the section over the proof of the theorem, which requires the introduction of some new ideas.

The relevant optimality equation is (2.7). The quantity to be maximized is thus

$$bi(u) + \beta\theta(x+u) = \begin{cases} \beta\theta(x) & (u = 0), \\ b + \beta\theta(x+u) & (u > 0), \end{cases} \tag{2}$$

where θ is the function defined in (3.2). If it were true that $\theta \in \mathscr{C}$, as previously, with a minimum at S, then one sees from Figure 4.1 that the maximizing value would be that expressed in (1), with s determined as the smaller root of

$$\beta\theta(s) = b + \beta\theta(S). \tag{3}$$

In fact, however, it turns out that, although these conclusions are correct, the functions F^* and θ do not necessarily belong to $\mathscr{C}$. To support these conclusions, one must relax the requirement of convexity to that of b-convexity.

A function $\phi(x)$ of a scalar variable is termed *b-convex* if

$$b + \phi(x+u) - \phi(x) - u\phi'(x) \ge 0 \tag{4}$$

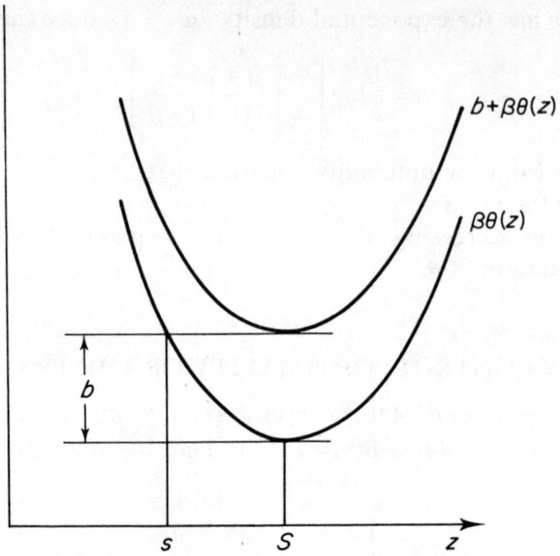

Figure **4.1** Determination of the bin sizes s,S for the optimal inventory policy

for all $u > 0$ and all x. The derivative is the right-derivative

$$\phi'(x) = \lim_{u \downarrow 0} \frac{\phi(x+u) - \phi(x)}{u}.$$

Note the following properties:

(a) b-convexity implies b'-convexity for $b' \geq b$.
(b) 0-convexity is ordinary convexity (which is known to imply the existence of one-sided derivatives; see Whittle, 1971, p. 56).
(c) If $\phi(x)$ is b-convex, then so is $\phi(x - \varepsilon)$.
(d) If ϕ_1 and ϕ_2 are b_1-convex and b_2-convex respectively, then $\alpha_1 \phi_1 + \alpha_2 \phi_2$ is $(\alpha_1 b_1 + \alpha_2 b_2)$-convex for $\alpha_1, \alpha_2 > 0$.
(e) If ϕ is b-convex, then so is $E\phi(x - \varepsilon)$.

These assertions all follow fairly directly; the last is a consequence of the two before it. We shall denote the class of b-convex functions which have the limit $+\infty$ at $x = \pm\infty$ by $\mathscr{C}_b$. The assertions (a) to (e) clearly imply:

Lemma 4.1.
(a) $\mathscr{C}_b$ is increasing in b.
(b) $\mathscr{C}_0 = \mathscr{C}$.
(c) $\mathscr{C}_b$ is closed under translation of the argument.

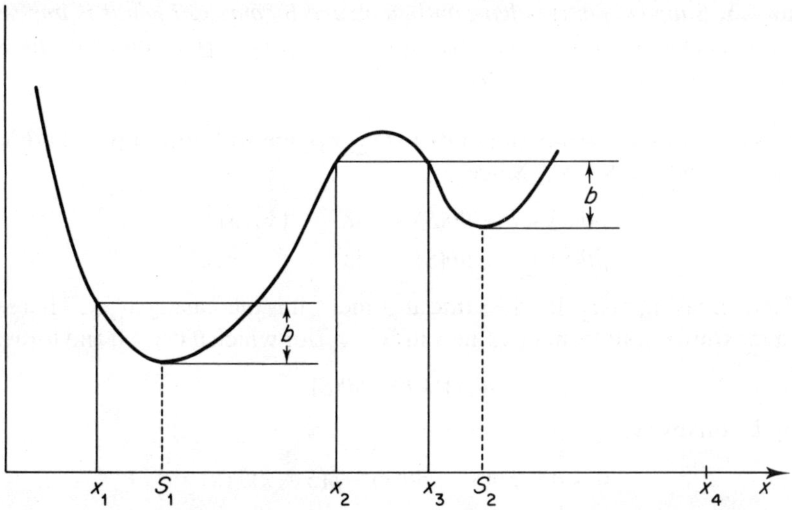

Figure **4.2** Minimization of $bi(u) + \beta\theta(x + u)$ for general θ

(d) $\phi_i \in \mathscr{C}_{b_i} \Rightarrow \sum \alpha_i \phi_i \in \mathscr{C}_b$ with $b = \sum \alpha_i b_i \quad (\alpha_i \geqq 0;\ i = 1, 2, \ldots)$.
(e) $\mathscr{C}_b$ is closed under averaging.

The possibilities for the maximization of $bi(u) + \theta(x + u)$ with a non-convex θ are illustrated in Figure 4.2. From x one will order up to a stock S if $S > x$ and $b + \beta\theta(S) < \beta\theta(x)$; the value S will necessarily be a local minimum of $\beta\theta$, otherwise it would be advantageous to vary S. In the diagram the function $\theta(x)$ has local minima at $x = S_1$ and $x = S_2$. One will order to S_2 in the interval (x_2, x_3), but one will not order at all for values of x in (x_1, x_2) or in the interval beginning at x_3. In the interval terminating at x_1 one will order to a stock level of S_1.

We shall prove a number of lemmas, referring to Figure 4.2 for clarification.

Lemma 4.2. *Suppose θ is left-continuous. If one orders for all x in some interval, then one orders up to a common stock value for all such x.*

So, for all x in (x_2, x_3), one orders up to the common value S_2.

Proof. Let $x' < x''$ be two values in the interval, from which one orders up to distinct levels S' and S'' respectively. Then one must have $\theta(S') \leq \theta(S'')$, otherwise it would be preferable to order to S'' from x'. One must also have $S' < x''$, otherwise it would be possible (and so preferable) to order to S' from x''. Thus S' belongs to the interval. Since S' is the value of z minimizing $\theta(z)$ for $z \geq x$ for all x in (x', S'), then for x in (x', s') one will order up to level S' if one orders at all. But for x near enough to S' the inequality $b + \beta\theta(S') < \beta\theta(x)$ is violated, and one will not order. But this contradicts the supposition that one orders for all x in the interval. ∎

Lemma 4.3. *Suppose $\beta\theta(x)$ is left-continuous and b-convex. Then it is impossible that in three consecutive sub-intervals $(x_1 x_2)$, $(x_2 x_3)$, $(x_3 x_4)$ the optimal rule is not to order, to order and not to order respectively.*

Proof. By Lemma 4.2, from all points in (x_2, x_3) one will order up to a common stock level S, where $S > x_3$. Since

$$\begin{aligned} \beta\theta(x) &\le b + \beta\theta(S) \qquad \text{in} \qquad (x_3, S), \\ \beta\theta(x) &> b + \beta\theta(S) \qquad \text{in} \qquad (x_2, x_3), \end{aligned}$$

then θ is decreasing at x_3. By the same argument, it is increasing at x_2. There thus exists a maximizing stationary value $\bar{x}$ in (x_2, x_3) for which $\theta'(\bar{x}) = 0$ and for which

$$\beta\theta(\bar{x}) > b + \beta\theta(S). \tag{5}$$

But, by b-convexity,

$$\begin{aligned} 0 &\le b + \beta\theta(S) - \beta\theta(\bar{x}) - \beta(S - \bar{x})\theta'(\bar{x}) \\ &= b + \beta\theta(S) - \beta\theta(\bar{x}). \end{aligned} \tag{6}$$

The contradiction of (5), (6) proves the lemma. ■

Lemma 4.4. *If $\beta\theta \in \mathscr{C}_b$ then an (s, S) policy is optimal.*

Proof. By Lemma 4.3 the optimal rule must either be of the form

$$\begin{aligned} &\text{order} && \text{if} \quad x \le s, \\ &\text{do not order} && \text{if} \quad x > s, \end{aligned}$$

or

$$\begin{aligned} &\text{do not order} && \text{if} \quad x \le s, \\ &\text{order} && \text{if} \quad x > s, \end{aligned}$$

for some s. By Lemma 4.2, in the first case one will order (when one does) up to a fixed value S; in the second case one will order up to $+\infty$. The first alternative is an (s, S) policy, the second an unacceptable policy incurring infinite immediate cost. ■

Finally, we need:

Lemma 4.5. *If $\beta\theta(x)$ is b-convex, then so is $\psi(x) = \min_{u \ge 0} [bi(u) + \beta\theta(x + u)]$.*

Proof. We know, from Lemma 4.4, that

$$\psi(x) = \begin{cases} \beta\theta(x) & (x > s), \\ b + \beta\theta(S) = \beta\theta(s) & (x \le s). \end{cases}$$

We must establish the inequality (4) for ψ. For $x, x + u \ge s$ it follows from the fact that $\psi = \beta\theta$. For $x, x+u \leq s$ it follows from the fact that ψ is constant. For $x < s \leq x+u$ the left-hand member of (4) equals

$$b + \beta\theta(x+u) - \beta\theta(s) \geq b + \beta\theta(S) - \beta\theta(s) = 0. \blacksquare$$

The way to the proof of Theorem 4.1 is now clear. It follows from Lemma 4.5 and the assumption on A^* that $\mathscr{L}\phi$ is b-convex if ϕ is. This can easily be strengthened to the assertion that $\mathscr{L}\phi$ belongs to $\mathscr{C}_b$ if ϕ does. Thus $F^* = \lim \mathscr{L}^r\phi$ belongs to $\mathscr{C}_b$, whence optimality of the (s, S) policy follows from Lemma 4.4.

5. INVENTORY CONTROL; OPTIMAL DETERMINATION OF THE (s, S) PARAMETERS

As a consequence of the optimality of an (s, S) policy, equation (2.7) will reduce to

$$F^*(x) = A^*(x) + b + \beta\theta(S) \qquad (x \leq s) \tag{1}$$

and

$$F^*(x) = A^*(x) + \beta\theta(x) \qquad (x > s) \tag{2}$$

where θ is defined again by (3.2). The optimality condition (4.3) implies that (1) can be rewritten as

$$F^*(x) = A^*(x) + \beta\theta(s) \qquad (x \leq s), \tag{3}$$

whence it follows, precisely as for $\theta(S)$ in the one-bin case, that

$$\theta(s) = \frac{EA^*(s-\varepsilon)}{1-\beta}. \tag{4}$$

It is not true, however, that s takes the value minimizing $\theta(\cdot)$, although it might appear from expression (3) that this would be the case. In appealing to (4.3) we have invoked a relation between the optimal values of s and S; relation (4) would not have held for an arbitrary value of s.

It is, in fact, true that S is a minimizing value of θ, but the functional form for θ near $x = S$ will not be given by (4). Indeed, one must take considerably more account of the distribution of demand ε before one can evaluate θ for $x > s$. One amenable case, treated below, is that in which demand is exponentially distributed (see Chap. 14 of Arrow, Karlin and Scarf, 1958). Another, treated in Section 38.4, is a continuous time diffusion version, due to Bather (1966).

Theorem 5.1. *Suppose that demand ε has the probability density $\lambda e^{-\lambda\varepsilon}$ $(\varepsilon \geq 0)$ Then the optimal values of s and S are determined by the relations*

$$\beta[\alpha^*(s) - A^*(S)] = b(1-\beta), \tag{5}$$

$$[e^{\nu S}A^*(S) - e^{\nu s}\alpha^*(s)] = \nu\int_s^S A^*(x)e^{\nu x}\,dx, \tag{6}$$

where

$$\alpha^*(s) = EA^*(s - \varepsilon),$$
$$\nu = \lambda(1 - \beta).$$

Proof. For notational simplicity we drop the asterisks during the proof. Differentiating (2) with respect to x and eliminating θ we deduce that

$$F' + \nu F = A' + \lambda A \qquad (x > s). \tag{7}$$

We know that S minimizes the expression $\beta\theta(x) = F(x) - A(x)$, so that $F'(S) = A'(S)$, and so, from (7),

$$F(S) = \frac{A(S)}{1 - \beta}. \tag{8}$$

Substituting evaluations (4) and (8) into the relation

$$F(S) = A(S) + \beta\theta(S) = A(S) + \beta\theta(s) - b \tag{9}$$

we deduce (5). Integration of (7) from s to S yields

$$e^{\nu S}[F(S) - A(S)] - e^{\nu s}[F(s) - A(s)] = \lambda\beta \int_s^S A(x)e^{\nu x}\,dx.$$

Substituting again evaluations (8) and (4) for $F(S)$ and $F(s) = A(s) + \beta\theta(s)$ we deduce (6). ■

These results simplify in the undiscounted case. In the undiscounted limit relations (5) and (6) become

$$\alpha^*(s) = A^*(S) = \frac{1}{S - s}\left[\frac{b}{\lambda} + \int_s^S A^*(x)\,dx\right].$$

Exercise

1. Write out the determining equations for the case when $A(x)$ has the piecewise linear form of Exercise 3.1. Show that $s > 0$ if

$$a' > \sqrt{(2ab\lambda)}.$$

6. INVENTORY CONTROL; DELAY IN THE DELIVERY OF ORDERS

Suppose that there is a delay of m in the fulfilment or orders, so that relation (2.1) is replaced by

$$x_{t+1} = x_t + u_{t-m} - \varepsilon_{t+1} \tag{1}$$

where u_t is the order placed with the supplier at the end of day t. It is plain that the sufficient statistics at time t are x_t and the string of orders which are placed but not yet delivered: $u_{t-m}, u_{t-m+1}, \ldots, u_{t-1}$. The equilibrium optimality equation (in unnormalized form) will be

$$F(x_t, u_{t-m}, \ldots, u_{t-1})$$
$$= \min_{u_t} [A(x_t) + B(u_t) - ER(\varepsilon) + \beta EF(x_t + u_{t-m} - \varepsilon, u_{t-m+1}, \ldots, u_t)] \quad (2)$$

where the expectation is with respect to ε. Despite the fact that one is in the situation of having to store information from a block of time of length m, the problem reduces essentially to the previous case, $m = 0$.

Theorem 6.1. *F has the form*

$$F(x_t, u_{t-m}, \ldots, u_{t-1}) = \sum_{j=0}^{m} H_j\left(x_t + \sum_{k=0}^{j-1} u_{t-m+k}\right) \quad (3)$$

where the H_j satisfy

$$H_0(x) = A(x) \quad (4)$$
$$H_j(x) = \beta EH_{j-1}(x - \varepsilon) \qquad (j = 1, 2, \ldots, m-1), \quad (5)$$
$$H_m(x) = \min_u [\beta EH_{m-1}(x - \varepsilon) + B(u) - ER(\varepsilon) + \beta EH_m(x + u - \varepsilon)]. \quad (6)$$

The optimal value of u_t is thus determined from the recursion

$$H_m(\xi_t) = \min_{u_t} E[A_m(\xi_t) + B(u_t) - R(\varepsilon_t) + \beta H_m(\xi_t + u_t - \varepsilon_t)] \quad (7)$$

where

$$\xi_t = x_t + u_{t-m} + u_{t-m+1} + \cdots + u_{t-1} \quad (8)$$

and

$$A_m(\xi) = \beta^m EA(\xi - \zeta_m), \quad (9)$$

ζ_m being a sum of m independent demands, $\sum_{k=0}^{m-1} \varepsilon_{t+k}$.

Once one sees the result, interpretation is fairly immediate. The quantity $\xi_t - \zeta_{mt} = \xi_t - \sum_{k=0}^{m-1} \varepsilon_{t+k}$ is the stock one will hold m time units in the future—at the time when the order u_t currently being placed will be delivered. The component ξ_t is current stock plus outstanding orders and the component ζ_m is the demand incurred between the placing of the order and its delivery—necessarily a random variable. The quantity $A_m(\xi)$ is the expected discounted value of storage costs for the period immediately before u_t is delivered. The decomposition (3) is a representation of F as expected discounted storage costs up to the delivery of u_t, plus $H_m(\xi_t)$. This last component comprises all costs from time t onwards, with the exception of storage costs from t to $t+m$.

The quantity ξ_t is a sufficient statistic, in that it obeys the simply recursive updating

$$\xi_{t+1} = \xi_t + u_t - \varepsilon_{t+1} \tag{10}$$

and in that both the optimal u_t and the normalized loss function H_m are functions of ξ_t alone. Recall of the outstanding order u_{t-m} is effectively supplied by arrival of that order.

Proof. This follows the usual inductive pattern. If $\phi(W_t)$ has the form (3), then so does $\mathscr{L}\phi$, and so plausibly does $F = \lim \mathscr{L}^r\phi$. The equations (4), (5), (6) are then consequences of $F = \mathscr{L}F$, and themselves, with (2), imply (7). ■

Equation (7) has the implication that the problem reduces essentially to the case of zero order delay, with ξ_t replacing x_t and $A_m(\xi)$ replacing $A(x)$. Theorem 4.1 then has an immediate analogue:

Theorem 6.1. *Suppose that $B(u) = bi(u) + cu$. If the function*

$$A^*(\xi) = \beta^m EA(\xi - \zeta_m) + \frac{c(1-\beta)}{\beta}\xi \tag{11}$$

is convex and has the limit $+\infty$ at $\xi = \pm\infty$, then the optimal ordering policy is of the (s, S) form

$$u_t = \begin{cases} 0 & (x_t + \sum_{k=1}^m u_{t-k} > s), \\ S - x_t - \sum_{k=1}^m u_{t-k} & (x_t + \sum_{k=1}^m u_{t-k} \le s). \end{cases} \tag{12}$$

Note that, because of the factor β^m in (11), the storage cost must now be more powerfully increasing for large positive x (as compared with the case $m = 0$) if expression (11) is to have the properties required. That is, because an order cost incurred now must be weighed against a possible storage cost incurred at a time m in the future, which will be discounted by a factor β^m, one is more prone to place large orders as m increases. Storage costs for large stock must be thus stiffer if they are to supply an effective constraint.

7. INVENTORY CONTROL WITHOUT BACK-LOGGING

A common constraint will be that inventory is simply not permitted to run negative: that demand can be satisfied only as long as stocks are available. Instead of incurring a cost from negative inventory, caused by satisfying excess demand, one will now incur a cost because of unsatisfied excess demand (representing customer disappointment, etc.).

If backlogging is not permitted then the stock relation (2.1) is modified to

$$x_{t+1} = (x_t + u_t - \varepsilon_{t+1})_+. \tag{1}$$

There will be an unsatisfied demand of $(\varepsilon_{t+1} - x_t - u_t)_+$, and actual sales on day $t + 1$ will be

$$\min(x_t + u_t, \varepsilon_{t+1}) = \varepsilon_{t+1} - (\varepsilon_{t+1} - x_t - u_t)_+. \tag{2}$$

The equation corresponding to (2.2) will be

$$F(x) = \min_u E[A(x) + B(u) - R(\min(x + u, \varepsilon)) + D((\varepsilon - x - u)_+) + \beta EF((x + u - \varepsilon)_+)]. \tag{3}$$

The new component D expresses the cost of unsatisfied demand. All functions have non-negative argument only.

If we assume the linear or near-linear cost forms

$$B(u) = bi(u) + cu, \tag{4}$$

$$R(s) = rs, \tag{5}$$

$$D(v) = dv, \tag{6}$$

then equation (3) simplifies with a normalization

$$F^*(x) = F(x) + \frac{(d + r)x}{\beta} + \frac{dE(\varepsilon)}{1 - \beta}, \tag{7}$$

$$A^*(x) = A(x) + \frac{(d + r)(1 - \beta)x}{\beta}, \tag{8}$$

$$c^* = c - (d + r) \tag{9}$$

to the form

$$F^*(x) = \min_u E[A^*(x) + bi(u) + c^*u + \beta F^*((x + u - \varepsilon)_+)]. \tag{10}$$

That is, the terms R and D are, in this linear case, essentially normalized to zero. As previously, we shall use $\mathscr{L}$ to denote the operator defined in this normalized form (10) of the optimality equation.

The term $d + r$ occurs in (8), (9) as a combination of direct receipts r and disappointment cost d avoided if the unit amount of a commodity in demand is sold rather than not sold. Thus $-c^* = d + r - c$ is the net gain for the unit amount of commodity sold, if a sale is immediate. The quantity $A^*(x)$ represents the cost of storing x for one night, plus the cost due to deferring a certain sale of stock for one day.

We shall assume that the operation is intrinsically profitable, so that

$$-c^* = d + r - c > 0. \tag{11}$$

We shall also assume that it is never profitable to order before goods are needed, so that

$$\frac{\partial}{\partial x}\left[A(x)+\frac{c(1-\beta)x}{\beta}\right]>0. \tag{12}$$

Expression (12) is the marginal cost of buying stock one day before one need have done. The expression in square brackets is the A^* previously defined for backlogged demand. In view of (11), inequality (12) implies that the A^* defined in (8) is increasing. In fact, it will usually be true that A itself is increasing, for $x \geq 0$.

If $A(x)$ is convex, then the proofs of one-bin optimality, two-bin optimality, etc., go through very much as in the backlogged case. The one new factor in the situation is that, while it is true that if $\psi(x)$ is b-convex than so is its translate $\psi(x-\varepsilon)$, this is not necessarily true of the 'stopped translate' $\psi((x-\varepsilon)_+)$. However, one verifies readily that the stopped translate will also be b-convex if only

$$b+\psi(x) \geq \psi(0) \qquad (x>0). \tag{13}$$

For the case $b=0$ (which is the most demanding) we then require that ψ be non-decreasing for the non-negative argument.

Theorem 7.1. *Suppose $B(u)=bi(u)+cu$. If $A(x)+c((1-\beta)/\beta)x$ is convex and strictly increasing, and $c<d+r$, then an (s, S) policy with S finite is optimal.*

Proof. The proof is much the same as that of Theorem 4.1, except that we must prove $F^*(x)$ non-decreasing for $x \geq 0$ if $EF^*((x-\varepsilon)_+)$ is to be b-convex. One verifies readily that if ϕ has a non-negative derivative, then the derivative of $\mathscr{L}\phi$ exceeds that of

$$A^*(x)=A(x)+c\left(\frac{1-\beta}{\beta}\right)x+\frac{1-\beta}{\beta}(d+r-c)x,$$

which is positive, by the hypotheses of the theorem. The loss function F^* is then also strictly increasing, by the usual inductive proof. ■

One knows that the strictly increasing nature of $A(x)+c((1-\beta)/\beta)x$ precludes indefinitely large orders, and so forces S to be finite. However, it is useful to see how this conclusion would follow from dynamic programming arguments. In forming $\mathscr{L}\phi$, we will find a finite minimizing value for u if $\psi(x)= c^*x+\beta E\phi((x-\varepsilon)_+)$ is strictly increasing for large enough x. Let $\phi_r=\mathscr{L}^r\phi$ and let d_r be the limit of the derivative of $\psi_r(x)$ as $x \rightarrow +\infty$. One finds from the optimality equation that

$$d_{r+1}-c^* \geq \beta a^*+\beta(d_r-c^*)$$

where

$$a^*=a+\frac{1-\beta}{\beta}(d+r)$$

and a is $\lim_{x \to \infty} A'(x)$. Thus

$$\lim_{r \to \infty} d_r \geq \frac{\beta a^*}{1-\beta} + c^*$$

$$= \frac{\beta a}{1-\beta} + c,$$

and this quantity is positive, by the hypotheses of the theorem.

The detailed results of Sections 3 to 5 have analogues in this case, indicated in the exercises.

Exercises

1. We assume the cost forms (4) to (6). Show that

$$F^*(x) = M + A^*(x) - c^* x \qquad (x \leq s)$$

where

$$M = \frac{\beta E A((s-\varepsilon)_+) + (d + r - \beta c) E(s-\varepsilon)_+ + (c - d - r)s}{1-\beta}.$$

2. Show that, in the case $b = 0$, s takes the value minimizing M. Show that if ε has a density $\rho(\varepsilon)$, then the minimizing value is determined by

$$\beta \int_0^s A'(s-\varepsilon)\rho(\varepsilon)\,d\varepsilon + (d + r - \beta c)P(\varepsilon \leq s) = d + r - c.$$

In the case $A(x) = ax$ this reduces to

$$P(\varepsilon \leq s) = \frac{d + r - c}{\beta(a-c) + d + r}$$

3. Obtain relations determining s and S in the case of exponentially distributed ε in both the discounted and undiscounted cases.

8. THE REGULATION OF A DAM WITH INDEPENDENT INPUTS

We return to the dam regulation problem of Section 6.4, except that the inflows v_t are assumed stochastic. In fact, we shall write the plant equation (6.4.2) as

$$x_{t+1} = x_t - u_t + \varepsilon_{t+1} \tag{1}$$

where the ε are the random inflows. In the simplest case, treated by Bather (1962), the ε are supposed to be independently and identically distributed. (See also Section 38.3 or Bather, 1968, for a very tractable continuous time model.)

As in Section 6.4 we suppose a utility $g(u)$ for draw-off, whose concave character encourages uniformity of draw-off. Thus does the dam fulfil its buffering role of smoothing inflow variation.

If $G(x)$ is the maximal discounted utility over an infinite horizon for initial water stocks of x, then this will obey the optimality equation

$$G(x) = \max_u [g(u) + \beta E G(x - u + \varepsilon)] \tag{2}$$

where the expectation is over ε. The choice of u is subject to the constraints

$$(x - C)_+ \leq u \leq x. \tag{3}$$

That is, u is non-negative and must be such that $0 \leq x - u \leq C$. One cannot abstract more water than is available and any unstorable water $x - C$ must be immediately released. We are thus assuming that water released by overflow is included in u and contributes to the utility $g(u)$. One establishes the following results:

Theorem 8.1. *Suppose $g(u)$ concave. Then:*

(*a*) *$G(x)$ is also concave.*

(*b*) *The optimal release rule $u(x)$ is such that both $u(x)$ and $x - u(x)$ are non-decreasing functions of x, and $u(x)$ is determined by the relations*

$$u(x) = x \qquad (x \leq a), \tag{4}$$

$$g'(u(x)) = EG'(x - u(x) + \varepsilon) \qquad (a \leq x \leq b), \tag{5}$$

$$u(x) = x - C \qquad (x \geq b), \tag{6}$$

with a, b determined by continuity of u at these points.

(*c*) $$G'(x) = g'(u(x)). \tag{7}$$

That is, the release has the pattern depicted in Figure 8.1. If the amount stored is less than a one uses all the water there is; if it exceeds b then one releases just the excess to capacity and no more. In between the amount increases with x, as does also the amount *not* released, $x - u$.

The continuity conditions determining a and b will be obtained by saying that rules (4) and (5) must agree at a; rules (5) and (6) at b. That is,

$$g'(a) = \beta EG'(\varepsilon),$$

$$g'(b - C) = \beta EG'(C + \varepsilon).$$

The assertions of the theorem are very much what one would expect; we refer to Bather (1962) for a careful argument.

The fact that the release rate never falls to zero if $x > 0$ is a reflection of the fact that one naturally uses all the dam; if $u(x)$ were zero for x less then some value C_0, then one would effectively have a dam only of size $C - C_0$. However, it is somewhat counter-intuitive that, when water is low, one does not eke it out, but,

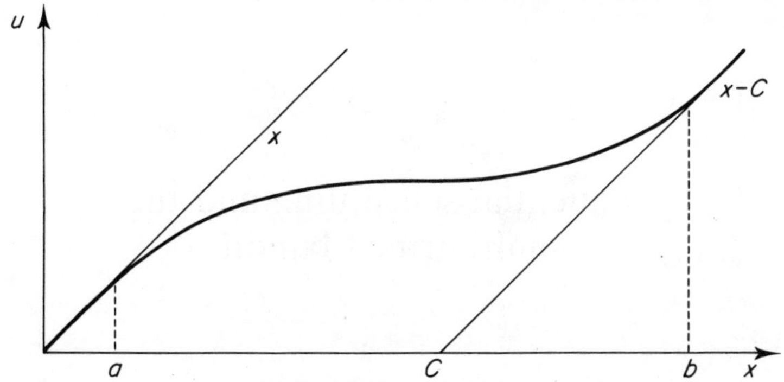

Figure **8.1** The optimal release rule for a dam with independent inputs

rather, uses all of it. This demonstrates itself in the fact that

$$G'(x) < g'(0),$$

i.e. that the utility gained from keeping a small amount is less than that gained from immediate consumption. The reason for this apparent lack of foresight is the fact that the ε_t are assumed independent. If one could expect that there would be low inflow for some rather definite period, then one would husband resources over that period. However, such an expectation is inconsistent with independent inflows. Results for dams with Markov inflow are obtained in a later paper by Bather (1963); solution is complete in some special cases.

Exercises

1. Show that, in the undiscounted case, the optimal release rule is unaffected if $g(u)$ is modified to $g(u) + cu$.
2. Show that, in the discounted case, the addition of a storage cost hx has the same effect as the modification of $g(u)$ to $g(u) - [\beta hu/(1-\beta)]$.

CHAPTER 14

Sequential scheduling and the multi-armed bandit

1. INTRODUCTION

The subject of this chapter can be regarded as an important special case of both sequential experimentation and stochastic scheduling. It also embodies in essential form a conflict evident in all human action. This is the conflict between taking those actions which yield immediate reward and those (such as acquiring information or skill, or preparing the ground) whose benefit will come only later.

In a basic version of the problem one has N parallel projects, labelled $i = 1, 2, \ldots, N$, and at each instant of discrete time can work on only one project. Let the state of project i at time t be denoted $x_i(t)$ and the label of the project engaged at time t be denoted $i(t)$. States up to the present and all past actions are supposed observable:

$$W_t = \{x_i(s), i = 1,2,\ldots,N, s \leq t; i(r), r < t; t\}.$$

If project i is engaged at time t (that is, $i(t) = i$) then one acquires an immediate expected reward of $R_i(x_i(t))$. (This is a problem for which it is traditional and natural to work in terms of rewards rather than costs.) Rewards are additive and discounted in time by a factor β. Given that $i(t) = i$, then $x_j(t+1) = x_j(t)\,(j \neq i)$ and $P(x_i(t+1)|W_t) = P(x_i(t+1)|x_i(t); i)$. That is, the states of unengaged projects do not change, while the state of the engaged project changes by a Markov transition rule, in that the value of $x_i(t+1)$ is conditioned only by $x_i(t)$ and i. The problem is the choice of $i(t)$: of how to allocate one's effort over projects sequentially in time so as to maximize expected total discounted reward. It is generally considered over an infinite horizon, although the finite-horizon version of course has interest.

Such, a process is sometimes called a *bandit process*, because an archetype of such problems is the 'multi-armed bandit', in which the projects represent the arms of a gambling machine. At each instant of time one can place a unit stake and try one of the arms. The different arms have differing (and unknown) probabilities θ_i of pay-off, but as one acquires experience of an arm one builds up an estimate, improving in precision, of its pay-off probability. The state $x_i(t)$ is then a summary of one's relevant statistical experience of arm i. (In fact, it is $P(\theta_i|W_t)$; see Chapter 39.) In choosing between an arm which one knows to be good (i.e. to have a high θ) and an arm of which one has little experience (and so a poor estimate of θ) one has the conflict mentioned above. From the point of view of immediate expected yield, it would be preferable to use the first arm. However, if long-term performance is important (i.e. if discounting is not too strong) then it may be preferable to test the second arm, which has a chance of being better than the first. That is, from the long-term point of view the gaining of additional information may be worth more than the gaining of guaranteed immediate yield; experimentation may be justified.

This is a case where development of the project (i.e. the evolution of its state) takes the form that one learns more about some important unobservable. The same is true in medical experimentation, in the situation where one is choosing between alternative treatments for a given condition to be applied to a number of patients presented sequentially. The projects are the treatments; their state is one's state of knowledge of their efficacy. To go for short-term yield is to maximize the probability of success for the current patient. To go for the extremest form of long-term yield is to experiment until one is confident one has found the best treatment. The conflict between these two alternatives is usually regarded as an ethical one; we would regard it as a question of goal, i.e. of choice of utility function. As β varies from 0 to 1 one progresses from the first extreme to the second. By solving the optimization problem for intermediate values of β one may hope to be striking the best balance between the present and posterity. In general, evaluational investigations (e.g. the evaluation of different medical treatments, different strains of wheat, different building materials) represent situations in which 'project development' corresponds to *learning*. However, there are other problems in which 'development' has a much more evident and predictable character: e.g. the stages of a building project or the stages of bringing agricultural land to a productive state.

If one considers the general scheduling problem, then what we have referred to as 'projects' would more usually be called 'activities', and the problem will have a more general structure than the one we have supposed. For example, it will usually be possible to carry out more than one activity at a given time (although there will normally be resource constraints on the total rate of activity). It may also be that some activities have to be completed before others can start, so that there are precedence constraints. Progress on some activities may imply progress on others. These are special cases of the phenomenon that activities may interact, and the reward structure may not be the simple one we have supposed.

The theory of the present chapter has been extended in several directions to meet these points (see the references to Gittins, Glazebrook and Nash). However, we shall keep to the problem first enunciated, which is general enough to be useful and, as the multi-armed bandit problem, has offered a classic challenge.

The Gittins index and the proof of its optimality are indeed due to Gittins (see Gittins and Jones, 1974; Gittins, 1979), as is also the material of Sections 6 to 8. The material of Sections 4, 5 and 11 is due to Whittle (1980, 1981a).

2. FORMULATION

The problem defined in the last section is a Markov decision problem with the state variable $x = (x_1, x_2, \ldots, x_N)$. (Note that the subscript refers to the project, not to time). In order to keep within the classic framework of the discounted case we shall suppose rewards to be uniformly bounded:

$$-\infty < A(1-\beta) \leq R_i(x) \leq B(1-\beta) < \infty \tag{1}$$

(although see Exercise 4.1). The constants A and B are then lower and upper bounds on total discounted reward respectively.

For simplicity, let us denote the reward $R_{i(t)}(x_{i(t)}(t))$ realized at time t simply by $R(t)$. The total discounted reward is then $\sum_{t=0}^{\infty} \beta^t R(t)$ and its maximal expected value

$$G(x(0)) = \sup_{\pi} E_{\pi}\left[\sum_{t=0}^{\infty} \beta^t R(t) | W_0\right] \tag{2}$$

is a function only of the initial state $x(0)$, as indicated. The reward function G will obey the dynamic programming equation

$$G = \max_i L_i G \tag{3}$$

and will be the unique bounded solution of this equation. Here L_i is the one-step operator if project i is engaged, so that

$$L_i G(x) = R_i(x_i) + \beta E[G(x(t+1)) | x(t) = x, i(t) = i]. \tag{4}$$

Of course, $x(t)$ and $x(t+1)$ differ only in the ith component.

Consider the modification of the process which allows the additional option of retirement for a lump reward of M. That is, one can either continue with selection from the projects or abandon operations permanently and settle for a retirement payment of M. The additional parameter M thus introduced turns out to be significant.

Let $\Phi(x, M)$ be the maximal expected conditional reward for the modified process, conditional on $x(0) = x$. Then Φ will be the unique bounded solution of the modified version of (3):

$$\Phi = \max\left[M, \max_i L_i \Phi\right]. \tag{5}$$

Let us refer to the original version of the process as the 'continuing process' and the modified version as the 'M process'. There is an immediate relation between the two.

Theorem 2.1. *$\Phi(x, M)$ is a non-decreasing convex function of M for which*

$$\Phi(x, M) = \begin{cases} G(x) & (M \leq A), \\ M & (M \geq B) \cdot \end{cases} \tag{6}$$

The optimal policies of the continuing process and of the M-process are identical for $M < A$.

Proof. To increase M cannot decrease reward, so the non-decreasing character is evident. To retire is always optimal for $M \geq B$ and to continue is always optimal for $M \leq A$ (for the M-process), whence (6) follows. To retire is never optimal if $M < A$, whence the last assertion.

Let V be the expected return from a policy whose prescription is independent of M. Then

$$V = V_c + ME(\beta^\tau) \tag{7}$$

where V_c is the expected reward before retirement (independent of M) and τ is the time of retirement. All expectations in (7) are those determined by the policy and are conditional on W_0. The event of non-retirement can be identified with the event $\tau = +\infty$, since, because $|\beta| < 1$, either convention will cause this contingency to yield a zero contribution to $E(\beta^\tau)$.

Since G is the supremum over policies of expression (7), linear in M, it is convex in M. ■

3. THE GITTINS INDEX

Suppose that to each project i can be attached an *index*, by which we mean a quantity $M_i(x_i)$ which is a function only of the label i and the current state x_i of that project. For simplicity we shall often denote the index for project i simply by M_i, but the dependence also on project state is understood.

A policy in which one always engages a project of currently greatest index will be referred to as an *index policy*. The conjecture that there might be an index policy which is optimal (among all policies) is an attractive one, since it asserts that optimal project choice reduces simply to a comparison of the figures-of-merit M_i. It is not obvious, however, that the conjecture is true.

In fact, Gittins (Gittins and Jones, 1974; Gittins, 1979) demonstrated that, with an appropriate choice of index, an index policy is optimal. The Gittins index is defined as follows. Let $\Phi_i(x_i, M)$ be the analogue of $\Phi(x, M)$ for the situation where, instead of the N projects, one has only the single project i. The sole options are then to continue with project i or to retire with reward M. The expected reward ϕ_i will then obey the reduced version of (2.5):

$$\phi_i = \max\,[M, L_i\phi_i]. \tag{1}$$

(Since L_i changes only the value of x_i its action on ϕ_i is defined.)

In analogy with the assertions of Theorem 2.1, $\phi_i(x_i, M)$ is a non-decreasing function of M which equals M for M large enough—certainly for $M \geq B$. The Gittins index $M_i(x_i)$ is defined as the infimal value of M for which $\phi_i(x_i, M) = M$. This is exactly the break-point at which one has $M = L_i\phi_i$. The index $M_i(x_i)$ can be regarded as the value of M which makes the options of continuation (with project i) or of retirement (with reward M) equally attractive.

$M_i(x_i)$ thus represents an equitable surrender price, or reserve price, for project i in its current state x_i. It is an equitable price under a particular circumstance, however. It is a price equivalent, not to the expected reward from indefinite operation of project i, but to the expected reward when one may continue with project i, but the option of accepting the price initially offered remains open at all times.

We shall refer to the index policy which uses the Gittins index as the *Gittins index policy*. Optimality of this policy will be proved in the next section. The index which Gittins actually used was

$$v_i = (1-\beta)M_i$$

because he visualized the option of retiring at reward M as the option of adopting a *standard project* with a fixed reward rate $R = v$. The definitions differ only by a factor, and we shall find it convenient to work in terms of M rather than v.

Gittins has a second characterization of his index.

Theorem 3.1. *The Gittins index also has the characterization*

$$M_i(x_i) = \sup_\pi \frac{E_\pi[\sum_0^{\tau-1} \beta^t R_i(x_i(t)) \mid x_i(0) = x_i]}{1 - E_\pi(\beta^\tau \mid x_i(0) = x_i)} \tag{2}$$

or

$$v_i(x_i) = \sup_\pi \frac{E_\pi[\sum_0^{\tau-1} \beta^t R_i(x_i(t)) \mid x_i(0) = x_i]}{E_\pi[\sum_0^{\tau-1} \beta^t \mid x_i(0) = x_i]} \tag{3}$$

where the policy π determines a stopping time τ.

That is, v_i can be regarded as an average reward (with both reward and time discounted) over a stopping time τ, with the rule determining the stopping time chosen to maximize this average.

Proof. Since ϕ_i is the maximal reward under sequential choice between project i and retirement, then

$$\phi_i(x_i, M) \geq E_\pi\left[\sum_0^{\tau-1} \beta^t R_i(x_i(t)) + \beta^\tau M \mid x_i(0) = x_i\right]$$

for any π, with equality for some π, where τ is the instant of retirement. Giving M

the value $M_i(x_i)$ we have then

$$M_i \geq E_\pi\left[\sum_0^{\tau-1} \beta^t R_i(x_i(t)) + \beta^\tau M_i \,|\, x_i(0) = x_i\right],$$

with equality for some π, whence (2) follows. ■

4. OPTIMALITY OF THE GITTINS INDEX POLICY

Define the function

$$\hat{\Phi}(x, M) = B - \int_M^B \prod_i \frac{\partial \phi_i(x_i, m)}{\partial m} \, dm. \tag{1}$$

The motivation for this definition will transpire in the next section. Note that, since $\phi_i(x_i, m)$ is convex as a function of m, the derivative $\partial \phi_i / \partial m$ exists almost everywhere. The value assigned to it at the points of ambiguity will not affect the value of the integral (1), but for consistency we shall give it the value of the right-derivative.

It is now relatively easy to show that $\hat{\Phi} = \Phi$ and that this maximal reward is realized by the Gittins index policy, which is consequently optimal.

Lemma 4.1.

$$\hat{\Phi}(x, M) = \phi_i(x_i, M) P_i(x, M) + \int_M^\infty \phi_i(x_i, m) d_m P_i(x, m) \tag{2}$$

where

$$P_i(x, M) := \prod_{j \neq i} \frac{\partial \phi_j(x_j, M)}{\partial M} \tag{3}$$

is non-negative, non-decreasing in M and equal to unity for

$$M > M_{(i)} := \max_{j \neq i} M_j. \tag{4}$$

Proof. Equation (2) follows from (1) by partial integration. Since ϕ_i, as a function of M, is non-decreasing, convex and equal to M for $M \geq M_i$, then $\partial \phi_i / \partial M$ is non-negative, non-decreasing and equal to unity for $M \geq M_i$. The properties asserted for P_i thus follow. ■

Consider the quantity

$$\delta_i(x_i, M) = \phi_i(x_i, M) - L_i \phi_i(x_i, M)$$

and note that $\delta_i \geq 0$, with equality for $M \leq M_i$. We are interested for the moment in the dependence of the various quantities on M for fixed x, and so shall for simplicity suppress the x-argument.

Lemma 4.2.

$$\hat{\Phi} \geq M, \tag{5}$$

with equality if $M \geq \max M_j$, *and*

$$\hat{\Phi}(M) - L_i\hat{\Phi}(M) = \delta_i(M)P_i(M) + \int_M^\infty \delta_i(m) d_m P_i(m) \geq 0, \tag{6}$$

with equality if $M_i = \max M_j \geq M$.

Proof. Inequality (5) and the characterization of the equality case follow from (2) and the properties of P_i.

The first relation of (6) follows immediately from (2). The non-negativity of the expression follows from $\delta_i(M) \geq 0$ and the non-negative and non-decreasing nature of P_i. We know that $\delta_i(M) = 0$ for $M \leq M_i$ and that $d_m P_i(m) = 0$ for $m \geq M_{(i)}$, so that expression (6) will be zero if $M \leq M_i$ and $M_{(i)} \leq M_i$. This pair of conditions is equivalent to those asserted in the Lemma. ■

Theorem 4.1. $\hat{\Phi} = \Phi$ *and the Gittins index policy is optimal.*

Proof. We shall have demonstrated that $\hat{\Phi} = \Phi$ if we can show that $\hat{\Phi}$ satisfies (2.5), because Φ is the unique bounded solution of this equation (Theorem 23.3.1). We shall have demonstrated that the Gittins index policy is optimal if we can show that the maximizing option in the right-hand member of (2.5) is just that option which would be recommended by the Gittins index policy (Theorem 23.3.2). But the truth of both of these statements follows from the inequalities and cases of equality proved in Lemma 4.2. ■

Exercises

1. The results of this section do not in general extend to the finite-horizon case. They do, however, extend to the case where there is a finite horizon on each individual project, by the device of replacing x_i by (x_i, s_i), where s_i is the permitted remaining process time for project i. Complete the details. Note that this is in fact a finite-horizon problem in the conventional sense and there is no need for discounting: for a finite-horizon problem the optimality equation has a unique solution and a policy is optimal if and only if it provides the 'equalizer' in this equation.
2. A generalization of the bandit process considered by Gittins is the *bandit super-process* in which, once one chooses $i(t) = i$, there is an additional decision

variable $u_i(t)$ which affects both reward and x_i-transition. The operator L_i then generalizes to the operator L_{iu} with the action

$$L_{iu}\Phi(x) = R_i(x_i, u) + \beta E[\Phi(x(t+1))|x(t) = x, i(t) = i, u_i(t) = u]$$

where only x_i changes in the transition. The variable $u_i(t)$ can be regarded as a subsidiary decision to be made (optimally) at time t if one has decided to engage project i at that time.

One can define ϕ_i and Φ as before, with the only difference that their definition incorporates an optimization over u as well as over other action options. One can define the Gittins index policy as before, but cannot now be sure that it is optimal.

Show that Theorem 4.1 still holds if, in the continuation region for the one-project problems defining ϕ_i, the optimal u-decisions are independent of M for all i. The assertion follows from an appropriate generalization of the proof of Theorem 4.1 (see Whittle, 1980). This condition has been shown to be necessary as well as sufficient (Glazebrook, 1981).

5. MOTIVATION FOR THE DEFINITION OF $\hat{\Phi}$

Our proof that $\hat{\Phi} = \Phi$ establishes the truth of the identity

$$\Phi(x, M) = B - \int_M^B \prod_i \frac{\partial \phi_i(x_i, m)}{\partial m} dm, \tag{1}$$

and the definition (4.1) of $\hat{\Phi}$ has its origin in the conjecture that identity (1) holds. We now outline the motivation for this conjecture.

Lemma 5.1. *Let V be the expected reward, conditional on W_0, for an arbitrary policy whose prescription is independent of M. Then*

$$\frac{\partial V}{\partial M} = E(\beta^\tau | W_0) \tag{2}$$

where τ is the time of retirement under this policy.

Proof. This is an immediate consequence of (2.7). ∎

Let us now define a *write-off policy* as a policy in which project i is written off (i.e. abandoned) when first its state x_i enters a *write-off set* S_i $(i = 1, 2, \ldots, N)$. One continues as long as there are projects which have not been written off, working only on those projects; one retires as soon as all projects are written off. That is, one continues until all projects have been driven into their write-off sets.

Note that a write-off policy is not fully specified by prescription of the write-off sets S_i, because no rule has been given for the order in which active projects are to be engaged. This is something the write-off policy does not specify; moreover, a write-off policy need be neither Markov nor stationary. Note that the Gittins index policy is a write-off policy, with S_i the set of x_i for which $M_i(x_i) \leq M$.

Lemma 5.2. *Let τ be the time to retirement for an N-project policy, and τ_i the time to retirement when only project i is available. If the policy is a write-off policy, then*

$$E[\beta^{\tau}|W_0] = \prod_i E[\beta^{\tau_i}|W_0]. \tag{3}$$

Proof. We have $\tau = \sum \tau_i'$ where τ_i' is the total *process time* for project i, the total time for which project i is engaged before it is written off. Equation (3) states effectively that the τ_i' are independently distributed and that τ_i, τ_i' have the same distribution (conditional on W_0). But since evolution of the projects (in process time) is independent, the process time needed to take x_i into S_i is independent of the states of other projects; hence the assertion. ■

Lemma 5.3. *Let V and V_i be the expected rewards, conditional on W_0, for an arbitrary write-off policy whose prescription is independent of M when N projects are available and when only project i is available respectively. Then*

$$\frac{\partial V}{\partial M} = \prod_i \frac{\partial V_i}{\partial M}. \tag{4}$$

Proof. Relation (4) follows from (2), (3). Note Exercise 1. ■

Suppose now that relation (4) also holds for the optimal policies, so that

$$\frac{\partial \Phi}{\partial M} = \prod_i \frac{\partial \phi_i}{\partial M}. \tag{5}$$

Identity (5) would then follow immediately by integration of (5) and appeal to $\Phi(x, B) = B$. To establish (5) we would need analogues of Lemmas 5.1 and 5.2 for optimal policies. Lemma 5.1 indeed has an analogue:

Lemma 5.4. *Let E_M denote expectation under a policy optimal when retirement reward has the value M. Then $E_M(\beta^{\tau}|x(0) = x)$ is a sub-gradient of $\Phi(x, M)$, as a function of M, and*

$$\frac{\partial \Phi(x, M)}{\partial M} = E_M[\beta^{\tau}|x(0) = x] \tag{6}$$

for almost all M.

Proof. Since in each state there are at most $N + 1$ possible actions, optimal policies always exist. If one applies an M-optimal policy to the $(M + \delta)$-process one achieves an expected reward not exceeding $\Phi(x, M + \delta)$. This assertion and relation (2.7) imply that, for any M, δ,

$$\Phi(x, M + \delta) - \Phi(x, M) \geq \delta E_M(\beta^\tau | x(0) = x). \tag{7}$$

(The expectation is conditional on W_0, but $x(0)$ will be the only effective conditioner for an optimal policy.) Relation (7) characterizes expression (6) as a sub-gradient to Φ. It will then coincide with the gradient of Φ, wherever this exists. But Φ, being convex, will have a gradient for almost all M, whence the assertion of the Lemma. ■

However, there will be no analogue of Lemma 5.2 unless one can assert that there is an optimal policy (for fixed M) which is a write-off policy. This is the real conjecture, which is plausible, but for which there seems to be no direct proof. Its value is that it leads to the conjecture $\hat{\Phi} = \Phi$, whose truth we established in Theorem 4.1. This theorem also established optimality of the Gittins index policy, so that there is indeed a write-off policy which is optimal.

Exercises

1. Show that for a write-off policy
$$E[\beta^\tau | W_0] = E[\beta^\tau | x(0)],$$
and hence that $\partial V / \partial M$ depends on W_0 only through $x(0)$.
2. Show that relation (3) is valid even if $P(\tau_i = +\infty) > 0$ for some i.
3. Show that $\partial \Phi / \partial M$ fails to exist just at those values of M for which the optimal policy changes and at which there is more than one optimal policy.

6. EXAMPLES: THE DETERIORATING CASE

To complete the treatment for individual cases one must now calculate the index $M_i(x_i)$ from one of the two characterizations of Section 3. Relatively few cases are known in which an explicit formula for the index can be deduced, so one will often have to resort to numerical solution of the optimality equation (3.1) for an appropriate range of values of M. However, even if one is driven to numerical solution at this point, the index result still represents an enormous reduction of the original problem, as well as an essential insight.

In this and the following sections we review a number of cases for which the index can be determined explicitly. The material of Sections 6 to 9 is due to Gittins (1979).

Theorem 6.1 (The deteriorating case). *Suppose that $\phi_i(x_i(t+1), M) \leq \phi_i(x_i(t), M)$ for all M and for all $(x_i(t), x_i(t+1))$ which occur with positive probability. Then*

$$M_i(x_i) = \frac{R_i(x_i)}{1 - \beta}. \tag{1}$$

If this holds for all i then the Gittins rule is a one-step look-ahead rule.

Proof. Equation (3.1), written out explicitly, is

$$\phi_i(x_i, M) = \max[M, R_i(x_i) + \beta E[\phi_i(x_i', M)|x_i]] \tag{2}$$

where x_i and x_i' are generic values of $x_i(t)$ and $x_i(t+1)$. Now, if $M = M_i(x_i)$, so that $\phi_i(x_i, M) = M$, and stopping is optimal at x_i, then stopping will certainly be optimal at x_i', by the hypothesis of the theorem. The condition that continuation or retirement should give the same expected reward at $M = M_i(x_i)$ then yields the relation

$$M = R_i(x_i) + \beta M,$$

whence evaluation (1) follows. To choose the project maximizing M_i is then to choose the project maximizing $R_i(x_i)$; i.e. that which maximizes immediate return. ∎

A condition sufficient for the hypothesis of the theorem is that $R_i(x_i(t+1)) \leq R_i(x_i(t))$. It is also necessary (see Exercise 1). This is a rather strong condition, but one that holds in the following example.

Suppose the N projects correspond to N sites, at each of which there may be treasure. Let r_i be the expected value of the treasure at site i, P_i the probability that there is such a treasure there and c_i the cost of examining the site for unit time. It is supposed that, conditional on there being a treasure at the site, the probability that it will be found in a search over unit time is α_i. The expected immediate reward from an examination of site i is thus

$$R_i = \alpha_i P_i r_i - c_i \tag{3}$$

which is an increasing function of P_i. Now P_i can be taken as the state variable x_i for project i, and its value decreases under all circumstances. If unit search does not reveal treasure at site i then P_i drops to

$$P_i^* = \frac{P_i(1-\alpha_i)}{P_i(1-\alpha_i) + 1 - P_i}$$

by Bayes' theorem. If treasure is found then P_i increases to 1, but, since the treasure is then removed, P_i effectively falls to zero at the next step. Thus R_i decreases under all circumstances, and one is in the deteriorating case. The optimal policy is then to examine the site i for which expression (3) is maximal, where P_i is the current *posterior* probability that treasure exists at site i (see Chapter 39).

There are a number of natural variations of this problem which would alter its character. An important variant of the problem is that in which there is known to be only a single treasure, which has been placed at one of the sites. P_i is then the probability that the treasure is at site i, rather than the probability that site i contains a treasure. This produces the feature that an examination of site i will now affect P_j for all j. The basic property of a bandit process seems thus to be lost.

Nevertheless, a simple index result holds for this problem (see Black, 1965, Ross, 1969, and Exercise 15.4.2) which makes it probable that the problem can be reformulated as a bandit problem. This turns out to be the case (see Section 10).

The first version of the treasure-hunt problem is a special case of what one might term a 'break-through' process. Suppose that a project yields reward only at the moment of a 'success' and there is at most one such success in the lifetime of a project. One then has

$$R_i = r_i p_i - c_i$$

where r_i is the reward if project i succeeds, p_i is the probability of success at the next instant and c_i the cost of operating the project for unit time. The quantities p_i and c_i will generally be functions of current observables W_t. Let us suppose that they are functions only of t_i, the elapsed process time for project i, so that t_i can be taken as the state variable for project i, and we have

$$R_i(t_i) = r_i p_i(t_i) - c_i(t_i), \tag{4}$$

say. Strictly, the state variable should also record whether or not success has taken place. However, it is generally taken for granted that one is considering only those projects which have not yet succeeded. One can thus regard $p_i(t_i)$ as the probability of success at process time $t_i + 1$ given that success has not been attained previously.

If $R_i(t_i)$ is non-increasing as a function of t_i (e.g. if $c_i(t_i)$ is constant and $p_i(t_i)$ is decreasing) then one is in the deteriorating case, and one chooses i to maximize expression (4).

Exercise

1. Show that, for the deteriorating case, $M_i(x_i(t))$ is a non-increasing function of t. Hence show that one is in the deteriorating case if and only if $R_i(x_i(t))$ is a non-increasing function of t.

7. EXAMPLE: THE IMPROVING AND CLIMACTIC CASES

The deteriorating case has an obvious opposite.

Theorem 7.1 (The improving case). *Suppose that* $\phi_i(x_i(t+1), M) \geq \phi_i(x_i(t), M)$ *for all* M *and for all* $(x_i(t), x_i(t+1))$ *which occur with positive probability. Then*

$$M_i(x_i) = E\left[\sum_0^\infty \beta^t R_i(x_i(t)) | x_i(0) = x_i\right] \tag{1}$$

and, once it is optimal to engage project i, it is optimal to continue with that project indefinitely.

Proof. Suppose that $M = M_i(x_i)$, so that

$$\phi_i(x_i, M) = M = L_i\phi_i(x_i, M). \tag{2}$$

It follows from the hypothesis of the theorem that $L_i\phi_i(x_i(t), M)$ is non-decreasing in t, so that

$$L_i\phi_i(x_i(t), M) \geq M \tag{3}$$

for $t \geq 0$, so that continuation is always optimal. It follows that $\phi_i(x_i, M)$, and so $M_i(x_i)$, has the evaluation (1).

If the fact that $\phi_i(x_i(t), M)$ is non-decreasing in t implies that $M_i(x_i(t))$ is also non-decreasing in t, then the final assertion follows. This is true, however, because the characterization of the index implies that

$$\phi_i(x_i(t), M_i(x_i(t)) - \delta) \geq M_i(x_i(t)) - \delta$$

for any $\delta > 0$. Then

$$\phi_i(x_i(t+1), M_i(x_i(t)) - \delta) \geq M_i(x_i(t)) - \delta$$

which implies that $M_i(x_i(t+1)) \geq M_i(x_i(t)) - \delta$. Since this inequality holds for any $\delta > 0$, the non-decreasing character of $M_i(x_i(t))$ follows. ■

A condition obviously implying the hypothesis of the theorem is that $R_i(x_i(t))$ should increase with t.

However, the case where prospects continue to improve indefinitely as a project is pursued is likely to be rare in practice. What is more likely to occur is that the prospects of success may improve with process time, but at the moment of 'success' the project yields all its benefits, making further investigation pointless. This would be the situation for the break-through problem defined at the end of the previous section, if $R_i(t)$ were an increasing function of t: the increase would be followed by a drop at the moment of success, when there would be nothing further to pursue. We shall refer to such a case as a 'climactic case', for obvious reasons.

Theorem 7.2 (The climactic case). *Suppose there is a state $x_i = 0$ for which project i is regarded as lapsed; it is absorbing and continuation is pointless, so that, $\phi_i(0, M) = M$ for all relevant M. Suppose that $\phi_i(x_i(t+1), M) \geq \phi_i(x_i(t), M)$ for all M and for all $(x_i(t), x_i(t+1))$ of positive probability such that neither $x_i(t)$ nor $x_i(t+1)$ equal 0. Then*

$$M_i(x_i) = \frac{E[\sum_0^{\sigma-1} \beta^t R_i(x_i(t)) | x_i(0) = x_i]}{1 - E[\beta^\sigma | x_i(0) = x_i]} \tag{4}$$

where σ is the smallest t for which $x_i(t) = 0$. Once it is optimal to engage project i, it is optimal to continue with that project until it lapses.

Proof. Suppose that $M = M_i(x_i)$. It follows, then, as in Theorem 7.1, that continuation will be optimal in any state which can be reached subsequently, except for state 0, in which retirement is optimal. It thus follows that

$$M = \phi_i(x_i, M) = E\left[\sum_0^{\sigma-1} \beta^t R_i(x_i(t)) + \beta^\sigma M | x_i(0) = x_i\right], \tag{5}$$

whence (4) follows.

As before, $M_i(x_i(t))$ increases up to the moment of lapse, whence the final assertion. ■

8. EXAMPLE: PROBLEMS WITH PREDICTABLE REWARD

Suppose that the value of $R_i(x_i(t))$ is completely determined by the value of $x_i(0)$. A particular example of this would be that in which R_i was a function only of process time t_i and there were no random events such as 'lapse' or 'success' which changed the rules. One then finds that evaluation (3.2) of the index reduces to

$$M_i(x_i) = \sup_{s>0} \frac{\sum_0^{s-1} \beta^t R_i(x_i(t))}{1-\beta^s}$$

where x_i is the value of $x_i(0)$. The maximization is with respect to the number s; i.e. one is considering optimization of a deterministic stopping time for retirement.

We shall prove a rather more general result, which allows for the possibility of lapse of the project as well.

Theorem 8.1. *Suppose there is a lapse state $x_i = 0$, as defined in Theorem 7.2, and let σ denote the random time of lapse if the project is allowed to continue that far. Suppose the reward to be predictable up to lapse, in that*

$$E[R_i(x_i(t)) | W_t] = E[R_i(x_i(t)) | x_i(0)]$$

if W_t specifies that lapse has not occurred by time t. Then

$$M_i(x_i) = \max_{s>0} \frac{E[\sum_0^{\tau-1} \beta^t R_i(x_i(t)) | x_i(0) = x_i]}{1 - E[\beta^\tau | x_i(0) = x_i]}$$

where

$$\tau = \min(\sigma, s).$$

Proof. Consider the characterization (3.2) of the index. The time τ is the time of first entry to a stopping set S_i, which is to be chosen to maximize the ratio in (3.2). In the present case S_i will certainly include the state $x_i = 0$. In the contingency that $x_i(t)$ first enters S_i at the value 0, the time taken will be σ. Conditional on the fact that it enters first at some other value, the time s to entry is deterministic and determined wholly by the starting point $x_i(0)$. To optimize $S_i - \{0\}$ is equivalent to optimizing the deterministic value s, whence the assertion of the theorem. ■

9. VARIABLE PROJECT STAGES

We have hitherto assumed that any project may be engaged for unit time and then broken off. This may be unrealistic; it may be that when one engages project i in state x_i then one is committed to following it for a time z which may be random, with a distribution conditioned by x_i. We shall speak of such natural segments in the development of the project as *stages*. The dynamic programming equations will then become recursions between consecutive stage-completions rather than between consecutive units of time. If project i is engaged at time t in state $x_i(t) = x_i$, then we shall suppose that z and $x_i' = x_i(t+z)$ are conditioned by i and x_i alone, so that the definition (2.4) of the action of L_i becomes modified to

$$L_i G(x) = R_i(x_i) + E[\beta^z G(x') | x(t) = x, i(t) = i]. \tag{2}$$

Here $x_j' = x_j$ for $j \neq i$ and, as stated, z and $x_i' = x_i(t+z)$ are conditioned by x_i and i alone. $R_i(x_i)$ is now interpreted as the expected reward from the *stage* starting from state x_i.

The one-project return $\phi_i(x_i, M)$ will continue to obey equation (3.1), with L_i modified as in (1), and the first characterization of the Gittins index given in Section 3 still holds. The characterization (3.2) is modified only in that the stopping time τ is now restricted to being a time at which a stage is completed.

With these modifications, Theorem 4.1 continues to hold; that is, identity (5.1) still holds between Φ and the ϕ_i, and the Gittins index policy is optimal. The proof is a direct analogue of that given in Section 4, and will be left to the reader.

10. FURTHER EXAMPLES

The generalization of the previous section not only allows one to treat problems generalized in just that way but also to frame certain problems as bandit problems which previously did not appear to be so. The examples of this section are due to F. P. Kelly (contribution to the discussion of Gittins, 1979).

Consider the second version of the search example of Section 6, in which one is looking for a single object placed at one of N sites; an inspection for unit time of site i costs c_i and reveals the object, if it is there, with probability α_i. If P_i is the current posterior probability that the object is at site i, then $P = \{P_i\}$ is the natural state variable of the decision process (see Chapter 39). The problem is not a bandit problem, because an examination of site i changes all elements of P. Nevertheless, an index result holds; an optimal rule is to examine the site i for which $\alpha_i P_i / c_i$ is currently maximal (see Black, 1965, and Exercise 15.4.2). This is for the undiscounted case, in which one minimizes the expected time until finding. The discounted version is not optimized by an index policy.

To achieve a bandit version of this problem, let us nevertheless consider the discounted case and suppose that site-dependent costs c_i are expressed by the assumption that an examination of site i takes a time c_i. Such an examination constitutes a stage. The incentive to examine will be supplied by assuming that a reward r_i is given when the object is found in site i. If site i has been examined k_i times then the expected reward from another examination would be

$$R_i = r_i a_i (1 - \alpha_i)^{k_i} \alpha_i$$

where a_i is the value of P_i at time 0. Thus R_i depends upon events only through the process time k_i, which may be taken as the state variable for project i, and one has achieved a bandit formulation.

Obviously one is in the deteriorating case, so that

$$M_i = \frac{r_i a_i (1 - \alpha_i)^{k_i} \alpha_i}{1 - \beta^{c_i}}$$

(cf. (6.1)). Since $P_i \propto a_i(1 - \alpha_i)^{k_i}$ then

$$M_i \propto \frac{r_i \alpha_i P_i}{1 - \beta^{c_i}} \tag{1}$$

and an optimal rule is to examine the site i for which expression (1) is maximal. In the undiscounted limit case $\beta \uparrow 1$ this leads to the recommendation: examine the site for which $r_i \alpha_i P_i / c_i$ is maximal. This policy differs only from that deduced in Exercise 15.4.2 by the occurrence of the factor r_i.

In this formulation of the problem we are maximizing the expected reward

$$V = E[\beta^c r_\imath]$$

where the two random variables $\imath$ and c represent respectively the site at which the object was found and the total time-cost $\sum c_{i(t)}$ incurred up to the moment the object was found. If $\beta = \mathrm{e}^{-\gamma}$ then

$$V = E(r_\imath) - \gamma E(r_\imath c) + o(\gamma).$$

If the policy is such that the object is found with probability one then $E(r_\imath) = \sum a_i r_i$, which is independent of policy. The optimization of policy in the undiscounted limit is thus equivalent to the minimization of $E(r_\imath c)$. If we wish just to minimize $E(c)$ then we should choose r_i independent of i, and the policy deduced above reduces to that deduced in Exercise 15.4.2.

A second example is Bellman's 'gold-mining' problem. This is an example of a case in which all activities cease at 'first failure', which is a feature apparently inconsistent with a bandit formulation. Suppose that one is working N mines, one at a time, and extracts an expected amount of gold $R_i(k)$ when one works mine i for the $(k+1)$th unit of time. However, there is a probability $P_i(k)$ that one's extraction machinery may then break down, making further mining impossible. One wishes to move between mines in such a way as to maximize the expected amount extracted before breakdown.

If we set

$$P_i(k) = \beta^{z_{ik}}$$

then the probability of breakdown can be incorporated formally into the discounting, with z_{ik} being formally the time needed for the $(k+1)$th stage of working of mine i. If mine i has at a given time been worked k_i times then $\{k_i\}$ is the state vector and

$$\begin{aligned} L_i G(k) &= R_i(k_i) + \beta^{z_{ik_i}} G(k') \\ &= R_i(k_i) + P_i(k_i) G(k') \end{aligned}$$

where

$$k'_j = \begin{cases} k_j + 1 & (j = i), \\ k_j & (j \neq i). \end{cases}$$

The evolution of a project is deterministic, and the Gittins index has the evaluation

$$M_i(k_i) = \max_s \frac{\sum_{h=0}^{s-1} R_i(k_i+h) \prod_{v=0}^{h-1} P_i(k_i+v)}{1 - \prod_{v=0}^{s-1} P_i(k_i+v)}.$$

Exercises

1. Formulate the quiz problem of Exercise 15.4.1 as a bandit problem.
2. Treat the search problem of this section in the case when costs and success probability depend on k_i.

11. THE OPEN PROCESS

A radical but rather necessary generalization is to allow the process to be *open*, in that new projects become available in a statistically homogeneous Poisson stream. With appropriate modification of definitions, relation (5.1) remains valid and the Gittins index policy remains optimal (see Whittle, 1981a).

A project was previously described by its individual label i and its state x_i. With indefinitely many projects it is better to combine these two indicators into a single label x, the state of a project. If $n(x)$ is the number of projects of state x available at a given time, then the vector $n = \{n(x)\}$ describes the state of the decision process, under appropriate statistical assumptions. The quantities previously denoted $F(x)$ and $\Phi(x, M)$ will now be denoted $F(n)$ and $\Phi(n, M)$; the analogue of Theorem 2.1 plainly holds.

Let $e(u) = \{e(x,u)\}$ denote the value of the vector n which is zero except for a unit in the uth place:

$$e(x,u) = \begin{cases} 1 & (x = u), \\ 0 & (x \neq u). \end{cases}$$

The quantity

$$\phi(u, M) = \Phi(e(u), M)$$

is then the analogue of $\phi_i(x_i, M)$, the maximal expected discounted reward conditional on starting with a single project, of state u. However, in contrast to the closed problem previously considered, a one-project situation does not now remain so, and the analogue of the one-project optimality equation (3.1) is bound to be more complicated. In fact, if we also allow random stage length, then the optimality equation will be

$$\Phi(n, M) = \max\left[M, \max_{u \in U(n)} L_u \Phi(n, M)\right] \tag{1}$$

where now

$$L_u \Phi(n, M) = R(u) + E[\beta^z \Phi(n - e(u) + \mathrm{e}(u') + w, M)|u]. \tag{2}$$

The conditioning even u indicates that a project of stage u is engaged, so that expression (2) is defined only if $n \geq e(u)$. The set $U(n)$ in (1) is the set of x for which $n(x) > 0$, i.e. the set of states occupied by some currently available project. $R(u)$ is the expected stage reward if a project of state u is engaged. The random variables covered by the expectation in (2) are z, the stage length; u', the state to which the project of initial state u is brought by the end of the stage; and $w = \{w(x)\}$, where $w(x)$ is the number of projects of state x which enter the system during the stage.

Suppose that the vector number of projects $q = \{q(x)\}$ entering the system in unit time is independent of previous entries and of the state of the system (which justifies our assumption that n is indeed a state variable for the decision process). It is then shown in Whittle (1981) that relation (5.1) still holds, in the form

$$\Phi(n, M) = B - \int_M^B \prod_x \left[\frac{\partial \phi(x, m)}{\partial m}\right]^{n(x)} \mathrm{d}m \tag{3}$$

and that the Gittins index policy is again optimal, with the index $M(x)$ for a project of state x being determined as the infinal value of M for which

$$\phi(x, M) = M. \tag{4}$$

We see from (2), (3) that the relations determining the one-project rewards $\phi(x, M)$ are now

$$\phi(u, M) = \max\left[M, R(u) + E\left[\beta^z B - \int_M^B \frac{\partial \phi(u', m)}{\partial m}\left[\beta A\left(\frac{\partial \phi}{\partial m}\right)\right]^z \mathrm{d}m \middle| u\right]\right] \tag{5}$$

where

$$A(\zeta) = E\left[\prod_x \zeta(x)^{q(x)}\right] \tag{6}$$

and $A(\partial\phi/\partial m)$ is obtained by setting $\zeta(x)$ equal to $\partial\phi(x, m)/\partial m$.

Relation (5) is the open equivalent of (3.1), a good deal more complicated, even in the case of fixed unit stage-length: $z = 1$. It can be solved for decreasing M: one has $\phi(u, M) = M$, certainly for $M \geq B$ (in fact for $M \geq M(u)$), and

$$\frac{\partial\phi(u, M)}{\partial M} = E\left[\frac{\partial\phi(u', M)}{\partial M}\left[\beta A\left(\frac{\partial\phi}{\partial M}\right)\right]^z \middle| u\right]. \tag{7}$$

The right-hand member of (7) changes only when M passes through the index value $M(x)$ for some x; $\phi(u, M)$ is piecewise linear in M.

12. SCHEDULING OF PRIORITIES IN A QUEUE

Suppose that customers of different types $i = 1, 2, \ldots$ arrive in a single-server queue in independent Poisson streams. They may be served in any order, but service, once begun, is completed. One seeks for an optimal order in which to serve customers, under cost and service-time assumptions to be specified below.

This can be regarded as an open-project selection problem, a customer of type-i constituting a 'project' of type i. The simplifying feature of this particular problem is that service is always completed. That is, a project, once begun, is continued until completion. Since the project is never left in an intermediate state, there is no need to specify the state of project development. On the other hand, the time needed to complete the project will in general be random. This problem was solved by Harrison (1975a, b, 1976); his results, obtained by direct methods, can be regarded as special cases of those of Section 11.

We shall phrase the problem in continuous time. Let us also refer to a customer of type i simply as an 'i-customer'. Suppose that customer types arrive in independent Poisson streams—type i at rate λ_i. Suppose that a reward r_i is paid when the service of an i-customer is completed, but that a holding charge of h_i per unit time is charged while he is waiting. Costs are discounted by a factor $e^{-\alpha t}$ over time t.

As far as choice of customer is concerned, one can imagine that there is zero holding cost, but that the reward collected after service is

$$r_i + h_i \int_0^\infty e^{-\alpha t}\,dt = r_i + \frac{h_i}{\alpha} \tag{1}$$

since one is then saved a payment at rate h_i in perpetuity. If the service time z of an i-customer has moment generating function

$$E(e^{-\theta z}) = \psi_i(\theta) \tag{2}$$

then the reward expected from serving an i-customer is effectively

$$R_i = \left(r_i + \frac{h_i}{\alpha}\right)E(e^{-\alpha z}) = \left(r_i + \frac{h_i}{\alpha}\right)\psi_i(\alpha). \tag{3}$$

The analogue of the one-project reward will be the maximal expected reward if the queue contains only a single customer, of type i. Let us denote this $\phi_i(M)$.

Lemma 12.1. *The equations* (11.5) *for the one-project reward reduce in the present case to*

$$\phi_i(M) = \max(M, R_i + Q_i(M)) \tag{4}$$

where

$$Q_i(M) = B\psi_i(\alpha) - \int_M^B \psi_i\left[\alpha + \sum_k \lambda_k\left(1 - \frac{\partial\phi_k(m)}{\partial m}\right)\right]\mathrm{d}m \tag{5}$$

Proof. If q_k is the number of k-customers arriving in time t then

$$E[\prod_k \zeta_k^{qk}] = \exp[t\sum_k \lambda_k(\zeta_{k-1})] \tag{6}$$

by the assumption of independent Poisson streams. Also the term equivalent to the quantity $\partial\phi(u', m)/\partial m$ of (11.5) is unity, because, when service is completed, the customer is lost to the system, and $\phi(u', m) = m$, effectively. In view of (6) and this fact, we see that (11.5) indeed reduces to (4), (5). ■

Let M_i denote the Gittins index for an i-customer. If we suppose the customers ordered so that $M_1 \geq M_2 \geq M_3 \geq \ldots$, then we shall have numbered them in order of decreasing priority, since the optimal priority rule will be to serve customers of highest index first. This ordering is just what we have to determine. For simplicity of treatment we shall exclude confluent cases and suppose that $M_1 > M_2 > M_3 > \ldots$.

Define the quantities α_j determined by the equations

$$\alpha_0 = \alpha,$$

$$\alpha_j = \alpha + \sum_{k=1}^{j} \lambda_k(1 - \psi_k(\alpha_j)) \qquad (j > 0). \tag{7}$$

Since the final expression increases concavely from a positive value to a limit value as α_j increases from 0 to $+\infty$, these equations have unique positive solutions.

Theorem 12.1 (Determination of customer priorities). *Consider a relabelling of the customers at each stage, so that the j customer types of highest priority are labelled $i = 1, 2, \ldots, j$, and the others $i > j$. Suppose $M_1, M_2, \ldots, M_j$ already determined. Then*

$$M_{j+1} = \max_{i \geq j} \frac{R_i + \sum_{k=1}^{j} M_k(\psi_i(\alpha_{k-1}) - \psi_i(\alpha_k))}{1 - \psi_i(\alpha_j)} \tag{8}$$

The maximizing i determines the $(j+1)$*th customer type in labelling of decreasing priority and* α_{j+1} *is then determined from* (7).

Proof. The functions $\phi_i(M)$ are piecewise linear, with changes in derivative only at the values M_j. Define

$$\psi_{ij} = \frac{\partial \phi_i(M)}{\partial M} \qquad (M_j > M > M_{j+1}). \tag{9}$$

This would be assigned the value $\psi_{i0} = 1$ for $M > M_1$. From formula (5) we deduce that

$$\frac{\partial Q_i(M)}{\partial M} = \psi_i(\alpha_j^*) \qquad (M_j > M > M_{j+1}) \tag{10}$$

where

$$\alpha_j^* = \alpha + \sum_{k=1}^{j} \lambda_k(1 - \psi_{kj}). \tag{11}$$

On the other hand, since $\phi_i(M) = R_i + Q_i(M)$ for $i \leq j$, we deduce from (9), (10) that

$$\psi_{ij} = \psi_i(\alpha_j^*) \qquad (i \leq j). \tag{12}$$

From (11), (12) and the uniqueness of solution of equations (7) it then follows that

$$\alpha_j^* = \alpha_j, \qquad \psi_{ij} = \psi_i(\alpha_j).$$

From formula (5) we also deduce that

$$\begin{aligned} Q_i(M_{j+1}) &= \sum_{k=1}^{j} (M_{k+1} - M_k)\psi_i(\alpha_k) + M_i\psi_1(\alpha_0) \\ &= M_{j+1}\psi_i(\alpha_j) + \sum_{k=1}^{j} M_k[\psi_i(\alpha_{k-1}) - \psi_i(\alpha_k)] \end{aligned} \tag{13}$$

for all i. Now M_{j+1} is characterized by

$$M_{j+1} \geq R_i + Q_i(M_{j+1}) \qquad (i > j) \tag{14}$$

with equality for the i which is to be assigned the label $j+1$ in the new labelling. In view of (13), this is exactly the assertion of the theorem. ■

We see that the customer type i of highest priority is that which maximizes $R_i/(1 - \psi_i(\alpha))$. In the undiscounted case it turns out (Theorem 12.2) that this quantity provides an effective index for all customer types. In the strictly discounted case the more complicated characterization of Theorem 12.1 holds, according to which the statistics and costs of high-priority types affect the relative ordering of lower priority types.

Theorem 12.2. *In the undiscounted limit $\alpha\downarrow 0$, customer priority increases with increasing h_i/μ_i (if there are no ties in this criterion) where μ_i is the expected service time for an i-customer. If there are no holding costs then customer priority increases with increasing r_i/μ_i, if there are no ties in this criterion.*

Proof. For small α we have

$$\psi_i(\alpha) = 1 - \mu_i\alpha + o(\alpha)$$

so that

$$\psi_i(\alpha_j) = 1 - \alpha\zeta_{ij} + o(\alpha),$$
$$\alpha_j = a_j\alpha + o(\alpha),$$

say. Then from equation (7) we readily find that

$$\zeta_{ij} = \mu_i a_j,$$
$$a_j = \frac{1}{1 - \sum_1^j \lambda_k\mu_k}.$$

If we set $M_i = v_i\alpha^{-1} + o(\alpha^{-1})$ then relation (8) becomes

$$v_{j+1} = \frac{1}{a_j}\max_{i\geq j}\left[\frac{R_i}{\mu_i} + \sum_{k=1}^{j} v_k(a_k - a_{k-1})\right]$$

as long as $a_j > 0$. Since the final sum is independent of i and

$$R_i = r_i + \frac{h_i}{\alpha} + o(1),$$

the assertions of the theorem follow. ■

The optimality of this priority rule in the undiscounted case was proved by Cox and Smith (1961). Customers who tie in the value of the criterion function can actually be given equal priority, unless one aims for a more refined level of optimality (i.e. Blackwell optimality rather than average optimality; see Section 31.6) when the tie must be broken by calculation of a more refined criterion. Harrison (1976) has examined this point.

The condition $a_j > 0$ amounts to

$$\sum_1^j \lambda_k\mu_k < 1,$$

which has a clear interpretation. It states that the traffic intensity of the queue is less than unity (i.e. the server is not saturated) if one restricts attention to the j customer types of highest priority. If

$$\sum_{1}^{j+1} \lambda_k \mu_k > 1$$

then one can serve $(j + 1)$-customers in the intervals when there are no customers of higher priority, but the number of $(j + 1)$-customers waiting in the queue will become infinite. The relative ordering of customer types of even lower priority is then academic.

CHAPTER 15

Structure theorems

The study of structure is potentially a large and rich one, to which this chapter will do scant justice. The aim is to reduce the problem by appeal to its structure. Solution may still be incomplete, but one has made an advance both conceptually and computationally if one can recognize that certain regularities of the problem must imply regularities of the solution. This implication can take a number of forms.

1. PROPERTIES CONSERVED UNDER $\mathscr{L}$

We have repeatedly used the fact that some property of F_0 can be established inductively for all F_s. For example, F_s is linear (Section 3.3), quadratic (Chapter 5), increasing (Exercise 13.5.1) or b-convex (Section 13.4). Such statements have implications also for the form of the optimal policy at various horizons.

These arguments amount to establishing a closure property of the operator $\mathscr{L}$. Let $\mathscr{C}$ be some class of the functions $\phi:\mathscr{X}\to R$ and suppose that $\mathscr{L}$ has the property that $\phi\in\mathscr{C}$ implies that $\mathscr{L}\phi\in\mathscr{C}$. That is, the class $\mathscr{C}$ is closed under the operation $\mathscr{L}$; we might write this $\mathscr{L}\mathscr{C}\subset\mathscr{C}$.

In such a case, if F_0 belongs to $\mathscr{C}$ then so does F_s for all s. At least for some $\mathscr{C}$ it may then also be true that $F_\infty\in\mathscr{C}$, and so that $F\in\mathscr{C}$ if $F_\infty = F$. Apart from saying something about minimal cost (at finite or infinite horizons) this may well also say something about optimal policy.

Note, however, that there are properties which are intrinsically infinite-horizon properties and which could consequently never be established by this argument. Time-invariance of F and stationarity of the optimal policy are the obvious examples.

2. INVARIANCE

A particular example of the $\mathscr{L}$-closure property is that in which the problem is invariant or semiinvariant to a group of transformations of state.

Suppose there exists a group Γ of transformations γ such that for every t, x of $\mathscr{X}$, u of $\mathscr{U}(x,t)$ and γ of Γ:

(a) $\gamma x\in\mathscr{X}$.

(b) There exists an associated transformation γ' such that $\gamma' u\in\mathscr{U}(\gamma x, t)$.

(c) $P(x_{t+1} = \gamma x' | x_t = \gamma x, u_t = \gamma' u; t) = P(x_{t+1} = x' | x_t = x, u_t = u; t)$.
(d) There exists an associated scalar $\nu(\gamma)$ such that

$$K(\gamma x) = \nu(\gamma) K(x), \quad g(\gamma x, \gamma' u, t) = \nu(\gamma) g(x, u, t).$$

Then one easily establishes inductively that $F(\gamma x, t) = \nu(\gamma) F(x, t)$ for all finite horizons T. Furthermore, if $u(x_t, t)$ is the optimal value of u_t at a prescribed finite horizon, then $u(\gamma x, t) = \gamma' u(x, t)$.

The investment and portfolio problems of Sections 3.1, 12.7 and 12.8 gave examples of seminvariance under simple scale transformation of the state and action variables. We encounter a more substantial example in Section 38.6, when time t may also be involved in the transformations.

3. ONE-STEP LOOK-AHEAD RULES

The 'one-step look-ahead' (OSLA) rule was seen to be optimal in the special contexts of Sections 12.4, 12.5 and 12.8. This is a rule applicable for processes for which termination is an option available in all states. It recommends the decision which would be optimal if one compared the alternatives of immediate stopping and of stopping after one further step; hence the name. It is thus a rule optimal at unit horizon; under some circumstances it is optimal at all horizons.

We take the time-homogeneous state formulation; for present purposes this can be regarded as a convenient normalization rather than an essential constraint. The cost functions F_s then obey the relations

$$F_0 = K,$$

$$F_s = \min [K, \mathscr{L}_c F_{s-1}] \qquad (s > 0), \tag{1}$$

where K is the termination cost and $\mathscr{L}_c$ the $\mathscr{L}$-operator restricted to continuation actions.

Let D_s be the set of x such that

$$K \leq \mathscr{L}_c F_{s-1}. \tag{2}$$

This is the optimal stopping set at horizon s if one always terminates in cases of indifference, the usual convention. Then, in particular, D_1 is the set of x such that

$$K \leq \mathscr{L}_c K. \tag{3}$$

Increase of s implies a relaxation in constraints on the policy, so that the sequence $\{F_s\}$ is non-increasing. So then is the sequence $\{D_s\}$, by (2).

Theorem 3.1 (Optimality of the one-step look-ahead rule). *Suppose that the set D_1 is absorbing. Then $D_s = D_1$ for all s.*

The hypothesis is that, under any continuation policy, the process remains in

D_1 once this set is entered. The statement $D_s = D_1$ can be said to imply optimality of the one-step look-ahead rule for any horizon s, and so for any finite horizon.

Proof. The hypothesis implies that if $\phi = K$ in D_1, then also $\mathscr{L}_c\phi = \mathscr{L}_c K$ in D_1. Suppose we have proved that $F_r = K$ in D_1 for $r < s$; the assertion certainly holds for $r = 0$. Then, by (1), for x in D_1,

$$F_s = \min[K, \mathscr{L}_c K] = K. \tag{4}$$

Thus $F_s = K$ in D_1, whence $D_s \supset D_1$, and so $D_s = D_1$ for all s. ■

Extension of Theorem 3.1 to the infinite horizon requires an additional regularity hypothesis (see Section 29.8).

Exercises

1. Suppose a hunter receives a unit bounty for each member of a population captured, but hunting costs him c per unit time. The number r remaining uncaptured is known and will not change by natural causes on the relevant time-scale. The probability of a capture in the next time-unit is $\lambda(r)$, a known increasing function. The probability of more than one capture per unit time is negligible. Show that it is optimal to cease hunting when $\lambda(r) \le c$.
2. Show by example that the hypotheses of Theorem 3.1 do not imply that the optimal continuation action to be taken, in those cases when continuation is optimal, is independent of horizon.

4. TWO-STEP LOOK-AHEAD RULES; INTERCHANGE ARGUMENTS

In Section 3.6 we considered the effect of interchanging the actions proposed at the next two steps. By this comparison one could under certain circumstances deduce the optimal policy. One is then effectively invoking a 'two-step look-ahead' (TSLA) rule.

These results have an immediate stochastic analogue. In fact, Theorem 3.6.1 is valid as it stands, with the substitution of

$$L_u\phi(x,t) = E[g(x_t, x_{t+1}, u_t, t) + \beta(x_t, x_{t+1}, u_t, t)\phi(x_{t+1}, t+1) | x_t = x, u_t = u; t]$$

for the deterministic definition (3.6.2).

Likewise, Theorem 3.6.2, giving easily verifiable forms for condition (d) of Theorem 3.6.1, remains valid if L_u has the time-homogeneous form

$$\begin{aligned} L_u\phi(x,t) &= g(x,u) + \beta E[\phi(x_{t+1}, t+1) | x_t = x, u_t = u] \\ &= g(x,u) + \beta T_u\phi(x,t), \end{aligned}$$

say, and this implicit definition of T_u replaces the previous implicit definition (3.6.4).

The TSLA rule, when optimal, often yields a criterion of index form, in that it chooses u to maximize some simple function of what one might call the state of the uth alternative. The following exercises are examples in point: Exercise 2 is the standard simple search problem. However, as we saw from Section 3.7, the TSLA rule may be optimal without being an index rule.

Exercises

1. A quiz contestant must answer questions from a list of n, presented in any order he chooses. He has probability p_i of answering question i correctly, when he receives reward r_i. He is eliminated from the contest on his first wrong answer, retaining previous winnings. Show that, in order to maximize expected winnings, he should request the questions in order of non-increasing $rp/(1-p)$.
2. A lost object is being sought among n sites. To examine site i (for unit time) costs c_i and reveals the object (if it is there) with probability α_i. If P_i is the probability that the object is in site i, conditional on current information, then show that the policy which minimizes expected cost incurred until the object is found is to examine a site i for which $\alpha_i P_i/c_i$ is greatest.
3. (F. P. Kelly; Cambridge Mathematical Tripos, 1980). The Greek adventurer Theseus is trapped in a room from which lead n passages. Theseus knows that if he enters passage i ($i = 1, 2, \ldots, n$) one of three fates will befall him: he will escape with probability p_i, he will be killed with probability q_i, and with probability r_i ($= 1 - p_i - q_i$) he will find the passage to be a dead end and be forced to return to the room one day after leaving it. The fates associated with different passages are independent. In what order should Theseus attempt the passages if he wishes to maximize his probability of eventual escape?

   If $p_i > p_j$ implies that $p_i/q_i \geq p_j/q_j$ show that there exists an order which simultaneously maximizes the probability of escape within the first t days for each value of $t \in \{1, 2, \ldots, n]$.

5. SUFFICIENCY OF TERMINAL MISS-DISTANCE

In Section 3.5 we showed that, if dynamics are linear in state and instantaneous costs are independent of state, then 'terminal miss-distance' is sufficient for determination of the optimal policy. This is the quantity $\xi_t = \phi_t x_t$ defined in (3.5.2)—the predicted terminal miss-distance if the trajectory were to suffer no further perturbation.

Again, this result has an immediate stochastic analogue, to which we shall appeal in Section 38.6. Theorem 3.5.1 remains valid if for the plant equation (3.5.1) we substitute the stochastic relation

$$x_{t+1} = A_t x_t + \varepsilon_t \tag{1}$$

where the random disturbance ε_t is conditioned only by u_t, t:

$$P(\varepsilon_t | W_t, u_t) = P(\varepsilon_t | u_t, t). \tag{2}$$

The term $B(u_t, t)$ of (3.5.1) has thus been absorbed into ε_t. We leave the reader to verify the generalization that $F(x_t, t) = H(\xi_t, t)$ and obeys

$$H(\xi_t, t) = \inf_{u_t} [g(u_t, t) + E[H(\xi_t + \phi_{t+1}\varepsilon_t, t+1) | u_t, t]]$$

with the corollary that ξ_t, t are sufficient for optimal control.

Exercise

1. Suppose that the pair of scalars (x_t, v_t) constitute the state variable and that these satisfy the plant equations

$$x_{t+1} = x_t + v_t,$$
$$v_{t+1} = v_t + u_t + \varepsilon_t.$$

   Thus, if x_t represents position of a body, then v_t represents its rate of change of position, or velocity, and u_t, ε_t are respectively the intended and white noise components of force applied to it. Suppose the cost function is $\sum^{T-1} u_t^2 + P_0 x_T^2$. Show that the optimal value of u at (x, v, s) is

$$u = -(s-1)P_s(x + vs)$$

   where

$$P_s^{-1} = P_0^{-1} + s(s-1)(2s-1)/6.$$

   Comment on the form of this control for small and large s.

CHAPTER 16

Duality concepts

The simple view of optimization is: the choice of certain conditional distributions $P(u_t | W_t)$ to minimize an expectation $E(C)$. The dynamic programming approach is one that develops only after some thought and that appears indeed as a somewhat indirect attack on the general problem.

In fact, the two approaches are mutually dual, in the technical sense of linear programming. Exploration of this fact yields a second extremal characterization of the minimal cost functions $F(x, t)$ and a linear programming algorithm for the determination of F and the optimal policy. It also reveals a path towards optimization: policy improvement. Finally, it leads one to understanding of the form a 'stochastic maximum principle' must take.

All these ideas are developed for a finite-state, finite-action, finite-horizon Markov decision process. However, this still suggests results of the most general application (see Section 11.4). In particular, the hope is that results may hold in a time-invariant form in the time-homogeneous infinite-horizon case. This is a hope justified frequently, although not invariably, in Part IV.

1. DUAL CHARACTERISATIONS OF THE OPTIMAL POLICY

A feature which will recur in the variants of the basic problem which we consider in the sequel is that F has a second extremal characterization besides that inherent in its definition, as being the *maximum* of a certain class of functions rather than the minimum. A somewhat pedestrian consequence of this second extremal characterization is that F can in principle be determined by linear programming methods. In this section these results emerge as clearly valid for the case of finite-state space, finite-action space and finite horizon. To show whether they then hold under any relaxation of these assumptions requires more advanced argument and some additional assumptions.

We shall assume a state formulation which is time-homogeneous, undiscounted and does not explicitly allow the option of termination. All these assumptions, apparently strong, in fact imply either no constraint at all or else the regularity constraint of finiteness. State formulation can always be achieved, at the cost of losing finiteness and recurrence (see Section 11.4). Time-homogeneity can be achieved formally by incorporation of t into x. The discount factor can be normalized to unity, at least under assumption (11.1.11). Finally, to terminate in state x at cost $K(x)$ can be regarded as the adoption of a decision u^0 such that $g(x, u^0) = K(x)$ and $P(x_{t+1} = x^0 | x_t = x, u_t = u^0) = 1$, where x^0 is the absorbing costless state defined in Section 11.1. If termination at x is obligatory then u^0 is the only action available at x. We shall regard x^0 as being 'out of the system' and shall not include it in $\mathscr{X}$, so that P, the $\mathscr{X} \to \mathscr{X}$ transition matrix, is in general sub-stochastic.

Termination at $t = T$ can be regarded as certainty of leaving the system when x has a value implying that $t = T$. 'Horizon' is thus concealed in the state transition structure. The assumptions of finiteness of horizon and of physical state space are translated into the statement that every x belongs to a set $\mathscr{X}_s$ such that termination is bound to occur within $s + 1$ steps, and by at most v^s possible trajectories, for some finite s and v.

The argument that justified assumption of a time-homogeneous state structure also justifies specification of initial conditions in terms of the initial state and assumption of a time-homogeneous Markov policy. These are all formally achievable if one defines state sufficiently generally.

The given quantities are then the instantaneous cost function $g(x, u)$ and the transition probability

$$p(y | x, u) = P(x_{t+1} = y | x_t = x, u_t = u).$$

The general form of initial conditions consists in the prescription of an initial distribution over state:

$$a(x) = P(x_0 = x).$$

A policy corresponds to a choice of the action distribution

$$\alpha(u | x) = p(u_t = u | x_t = x)$$

on $\mathscr{U}(x)$ for all x of $\mathscr{X}$.

The direct formulation of the optimization problem would now be: choose the distributions $\alpha(u|x)$ to minimize

$$C_p = \sum_x \sum_u \rho(x) \alpha(u|x) g(x, u)$$

where $\rho(\cdot)$ is the unique finite solution of

$$\rho(y) = a(y) + \sum_x \sum_u \rho(x) \alpha(u|x) p(y|x, u). \tag{1}$$

('Finite' in the sense that $0 \le \rho \le s$ if $a(\cdot)$ has its support totally in $\mathscr{X}_s$.) The quantity $\rho(x)$ is to be interpreted as the expected number of visits to state x. If the state description in fact incorporates t, then there can be at most one such visit and $\rho(x)$ is then the probability of a visit.

If P is the matrix with the (xy)th entry $\sum_u \alpha(u|x) p(y|x, u)$ then (1) will have a unique finite solution if and only if $aP^j \to 0$ as $j \to \infty$. That is, if, under the chosen control law $\alpha(\cdot|\cdot)$ and the prescribed initial distribution $a(\cdot)$, the process ultimately leaves the system.

Define

$$\mu(x, u) = \rho(x)\alpha(u|x),$$

the expected number of occurrences of the pair (x, u) of state/action values. In terms of this quantity the optimization problem can be rephrased.

Primal problem. Choose $\mu(\cdot,\cdot)$ to minimize $C_p = \sum_x \sum_u \mu(x, u)g(x, u)$ subject to the constraints $\mu(x, u) \geq 0$ and

$$\sum_u \mu(y, u) = a(y) + \sum_x \sum_u \mu(x, u)p(y|x, u). \tag{2}$$

This is a linear programme, whose dual is immediately interpretable.

Theorem 1.1 (The dual problem and its interpretation).
(a) The linear programme dual to the primal problem above is: choose $\phi(x)$ to maximize $C_d = \sum_x a(x)\phi(x)$ subject to

$$\phi(x) \leq g(x, u) + \sum_y p(y|x, u)\phi(y) \qquad \begin{pmatrix} u \in \mathscr{U}(x) \\ x \in \mathscr{X} \end{pmatrix}. \tag{3}$$

(b) A value of u gives equality for some x in (3) if and only if it is optimal at x.
(c) The $\phi(x)$ solving the dual is independent of $a(x)$ and equal to $F(x)$, at least for x values which have positive probability of occurrence under the initial distribution $a(\cdot)$.

Proof. The proof of statements (a) and (b) follows immediately from the standard theory of linear programming (see, for example, Whittle, 1971). This standard theory is applicable if x takes a bounded set of values. This will effectively be the case, however, if $a(\cdot)$ has its support confined to $\mathscr{X}_s$ for some bounded s. Assertion (b) would usually be expressed: if $\mu(x, u) > 0$ for a particular (x, u) in the solution of the primal, then equality holds in (3) at that (x, u) value for the solution of the dual.

Let $\bar{\phi}(x)$ be the solution of the dual. To obtain the identification $\bar{\phi}(x) = F(x)$ asserted in (c), note that $\bar{\phi}(x)$ is the Lagrangian multiplier associated with the constraint (2) and that $\bar{\phi}(x)$ can then be identified as $\partial \bar{C}_p / \partial a(x)$, where $\bar{C}_p$ is the minimal cost (more strictly, $\bar{\phi}$ is a sub-gradient to $\bar{C}_p$ as a function of $a(\cdot)$). This is enough to identify $\bar{\phi}(x)$ as the expected cost from x if the expected cost has been minimized for an initial distribution $a(\cdot)$. However, the theorem goes further, and asserts that $\bar{\phi}$ is independent of $a(\cdot)$, so that $\bar{C}_p$ actually has the linear expression in $a(\cdot)$: $\bar{C}_p = \sum_x a(x)\bar{\phi}(x) = \sum_x a(x)F(x)$. To see this result we iterate on the horizon. Relations (3) imply that

$$\phi \leq K \qquad (x \in \mathscr{X}_0),$$
$$\phi \leq \mathscr{L}\phi \qquad (x \in \mathscr{X}_s, s > 0),$$

if ϕ is dual-feasible. Equality evidently holds in these relations for the maximizing value $\bar{\phi}$. But F is the unique solution of the equality relations; hence we can identify $\bar{\phi}$ with F. ■

Assertion (c) can be alternatively expressed as the promised second characterization of F:

Corollary 1.1. *F is the greatest solution of*

$$\phi \leq \mathscr{L}\phi. \tag{4}$$

This assertion can be given various special forms in special cases. As indicated in the latter part of the proof of the theorem, it implies that F is the greatest solution of

$$\begin{aligned} \phi &\leq K \qquad (x \in \mathscr{X}_0), \\ \phi &\leq \mathscr{L}\phi \qquad (x \notin \mathscr{X}_0). \end{aligned}$$

For an optimal stopping problem, for which retirement at cost $K(x)$ is an option available in any state, this can be modified to: F is the greatest solution of

$$\begin{aligned} \phi &\leq K \qquad (x \in \mathscr{X}), \\ \phi &\leq \mathscr{L}_c\phi \qquad (x \notin \mathscr{X}_0), \end{aligned}$$

where $\mathscr{L}_c$ is the $\mathscr{L}$-operator with the action choice restricted to the continuation options.

It must again be emphasized that all these results are asserted only for the finite-horizon case, although this is suppressed in our notation. It is often the infinite-horizon case which is of interest, however.

Theorem 1.2. *The assertions of Theorem* 1.1 *and Corollary* 1.1 *remain true in the time-homogeneous infinite-horizon case with the state variable x no longer including specification of t, provided that* (*a*) *state and action spaces are finite and* (*b*) *there exists an optimal infinite-horizon policy which is stationary Markov.*

Proof. Because of condition (b) one need consider only stationary Markov policies, so the choice of policy reduces to the choice of $\alpha(u|x)$. (Recall that x is now assumed to be a physical state variable, which does not include specification of t.) Optimization can thus be characterized by the primal problem as posed above. Its dual is indeed then as specified by Theorem 1.1, if $\mathscr{X}, \mathscr{U}$ are finite. ■

There are indeed cases where an optimal policy which is stationary does not exist and cases where non-finiteness of $\mathscr{X}, \mathscr{U}$ modifies conclusions (see Chapters 26 and 27).

A function ϕ satisfying (4) is said to be $\mathscr{L}$-deficient, and Corollary 1.1 then asserts that F is the greatest $\mathscr{L}$-deficient function. One can interpret (4) as follows. Suppose that in state x one could either terminate at cost $\phi(x)$ or continue for one step and then terminate, taking the optimal continuation action in doing so. Relation (4) then states that the second option is never preferred. That is, the $\mathscr{L}$-deficient functions are the class of terminal cost functions (if termination were permitted) for which immediate termination is optimal.

This characterization of F has a number of uses. The most mundane of these is to see the characterization of infinite-horizon minimal loss as a linear programme, soluble by standard algorithms (Chapters 24 and 32).

2. DUALITY OFF THE OPTIMUM; POLICY IMPROVEMENT

We assume the Markov time-invariant formulation of the last section, but consider a particular policy, not necessarily optimal. The decision variable u can then be suppressed altogether, and the instantaneous cost and transition probability written simply as $g(x)$ and $p(y|x)$ respectively. In this context we shall find it desirable once more to make the distinction between continuation and stopping explicit. We shall thus divide $\mathscr{X}$ into C and D, the continuation and stopping regions respectively under the policy considered. If $V(x)$ is the expected cost incurred from x under this policy, and $K(x)$ the termination cost function, then

$$V(x) = g(x) + \sum_y p(y|x)V(y) \qquad (x \in C)$$

and

$$V(x) = K(x) \qquad (x \in D).$$

We shall suppress arguments and write this last pair of relations in the compressed form

$$V = \begin{cases} g + PV & (C), \\ K & (D). \end{cases} \tag{1}$$

We regard this as a functional relationship with suppressed arguments rather than as a vector relationship. Thus, V is $V(x)$ with the argument suppressed rather than a vector (which would be the array of all such $V(x)$) and P is a linear operator rather than a matrix. We shall sometimes write the first relation of (1) as

$$V = LV.$$

Correspondingly, relation (1.1) now reduces to

$$\rho(x) = \begin{cases} a(x) + \sum_y p(x|y)\rho(y) & (x \in C), \\ 0 & (x \in D) \end{cases}$$

or

$$\rho = \begin{cases} a + P'\rho & (C), \\ 0 & (D), \end{cases} \tag{2}$$

where P' is the operator adjoint to P.

Systems (2) and (1) are Kolmogorov forward and backward equations for the process determined by the policy. They correspond respectively to relations (1.1) and (1.3) for the primal and dual extremal problems of the last section, characterizing the optimal policy in different ways. Note, however, that whereas the ρ of (2) is primal-feasible (rather: $\mu(x, u) = \rho(x)\alpha(u|x)$ is primal-feasible where $\alpha(u|x)$ has the value determined by the policy) the V of (1) is not, in general, dual-feasible. This is because (1) implies that $V \geq \mathscr{L}V$, and V can be dual-feasible only if equality holds.

Let A be an arbitrary set in $\mathscr{X}$ with complement $\bar{A}$, and let $\phi(x), \theta(x)$ be arbitrary functions of state. We shall write an expression such as $\sum_{x \in A} \phi(x) \sum_y \theta(y)p(y|x)$ simply as $\sum_A \phi P\theta$. One then readily verifies the identity

$$\sum_A (\phi P\theta - \theta P'\phi) = \left(\sum_{x \in A} \sum_{y \in \bar{A}} - \sum_{x \in \bar{A}} \sum_{y \in A} \right) \phi(x)p(y|x)\theta(y). \tag{3}$$

This simple but important identity is the version for a discrete variable x and general operator P of Green's theorem relating volume and surface integrals (see, for example, Margenau and Murphy, 1943, p. 156). The sum in the left-hand member of (3) is a 'volume integral'—a sum over the set A. The sum in the right-hand member of (3) is a 'surface integral'. It is a sum over the 'probabilistic boundary' of A, in that it is a sum over possible transitions *into* A and *out of* A. Various choices of A, ϕ and θ lead to various results of interest.

Let us make the choices $A = C$, $\phi = \rho$ and $\theta = V$. We find then, from (1) and (2), that (3) reduces to

$$\sum_C aV = \sum_C \rho g + \sum_{x \in C} \sum_{y \in D} \rho(x)p(y|x)K(y). \tag{4}$$

The left-hand member is the expected total cost, if one starts with the initial distribution a. If $a(x')$ were equal to $\delta(x, x')$ then this quantity would be exactly $V(x)$. The right-hand member of (4) is an expression of this cost in terms of the stochastic development of the process as summarized in ρ. The first term gives the expected cost incurred during continuation; the second term the expected cost incurred at termination.

Relation (4) constitutes no more than a verification of what one would have expected, but these are less obvious consequences of the general identity (3). As a second special case, let $H(x)$ be any function satisfying

$$H = P'H \qquad (C). \tag{5}$$

Since the behaviour of H outside C is not specified, there are many such functions.

Choosing $A = C$, $\phi = H$ and $\theta = V$ in (3) we deduce that

$$\left(\sum_{x\in C}\sum_{y\in D} - \sum_{x\in D}\sum_{y\in C}\right)H(x)p(y|x)V(y) + \sum_{C} H(x)g(x) = 0. \tag{6}$$

We can certainly set $V = K$ in the first double summation, but not in the second. However, if the stopping set D is chosen so as to minimize expected cost, and if the transitions $x \to y$ are 'local', then $V \sim K$ for x 'close' to D. (See Whittle, 1975. The more specific statement is that, if $PK - K$ is small and D is chosen optimally, then $V - K$ is small on the probabilistic boundary of C.) One then has

$$\left(\sum_{x\in C}\sum_{y\in D} - \sum_{x\in D}\sum_{y\in C}\right)H(x)p(y|x)K(y) + \sum_{C} H(x)g(x) \sim 0 \tag{7}$$

and this relation (for varying H) can be used to characterize approximately the stopping set D.

As a final application of Green's identity, choose $A = C$, $\phi = \rho$ and $\theta = G - V$ where $G(x)$ is any function such that

$$G = K \qquad (D). \tag{8}$$

The identity then reduces to

$$\sum_{C} a(G - V) = \sum_{C} \rho(G - LG). \tag{9}$$

The point of relation (9) is seen if G is itself the expected cost for some policy. In fact, let V_i, C_i, L_i, ρ_i be the values of the quantities indicated for policy π_i $(i = 1, 2)$ and let us make the identifications $G = V_1, V = V_2, A = C_2$. Relation (8) then implies that

$$D_1 \subset D_2 \tag{10}$$

and identity (9) becomes

$$\sum_{C_2} a(V_1 - V_2) = \sum_{C_2} \rho_2(V_1 - L_2 V_1). \tag{11}$$

This relation implies the Howard improvement lemma (Theorem 23.2.4, Corollary 26.3.2): $L_2 V_1 \leq V_1$ with inequality somewhere implies that $V_2 \leq V_1$ with inequality somewhere. As sometimes expressed: if π_2 improves π_1, then π_2 dominates π_1.

It is known (see Chapter 27) that the Howard lemma is in general invalid for the case of positive programming, for which there is an incentive to defer termination as long as possible. For such cases, it is often true that, if π_2 improves π_1, then π_2 recommends continuation for some state for which π_1 recommends termination. Condition (10) is thus violated.

Relations (3) to (7) were derived in the optimal stopping context in Whittle (1975). Relation (11) was derived in the case of $P - I$, a diffusion operator by Reid and Teo (1980).

3. THE STOCHASTIC MAXIMUM PRINCIPLE: DISTRIBUTIONAL

There have been many attempts to give a version of the Pontryagin maximum principle in the stochastic case. The true generalization replaces variables by their distributions, and the maximum principle then turns out to be nothing but the dynamic programming principle itself. Uncer certain circumstances it is possible to give a characterization in terms of the actual variables, both primal and dual variables then evolving randomly. However, it has never been demonstrated that this characterization has any computational or conceptual point. In this and the next section the state variable x_t evolves randomly, but we can regard its distribution

$$\rho_t(x) = P(x_t = x) \tag{1}$$

as a 'deterministic state variable'. It satisfies the deterministic plant equation

$$\rho_{t+1}(y) = \sum_x \sum_u \rho_t(x)\alpha_t(u|x)p(y|x, u) \tag{2}$$

where we have specified a time-dependent Markov policy by specification of $\alpha_t(u|x)$. The cost function to be minimized with respect to $\alpha_t(u|x)$ is

$$\tilde{C} = \sum_{t=0}^{T-1} \tilde{g}(\rho_t, \alpha_t, t) + \tilde{K}(\rho_T) \tag{3}$$

where

$$\tilde{g}(\rho_t, \alpha_t, t) = \sum_x \sum_u \rho_t(x)\alpha_t(u|x)g(x, u, t), \tag{4}$$

$$\tilde{K}(\rho_T) = \sum_x \rho_T(x)K(x). \tag{5}$$

This is then a deterministic problem with the state variable ρ_t, decision variable α_t and plant equation (2). The formulation has all the properties of linearity and convexity needed to justify appeal to the maximum principle.

We leave detailed verification of the following assertions to the reader. The conjugate variable

$$\tilde{p}_t = \frac{\partial \min \tilde{C}}{\partial \rho_t(x)}$$

is just $F(x, t)$. The recursion (6.1.5) for this variable is just

$$F(x, t) = \sum_u \alpha_t(u|x)\left[g(x, u, t) + \sum_y p(y|x, u)F(y, t+1) \right]. \tag{6}$$

The instantaneous optimization condition Theorem 6.1.2 (a) is just the statement that $\alpha_t(u|x)$ should be such as to minimize the right-hand member of (6). Thus, the maximum principle yields $F(x, t)$ as the conjugate variable, and the dynamic programming equation as the combined recursion for F and action-optimality condition. In other words, it yields just the dynamic programming principle.

These ideas are, of course, only a slight variant on those of Section 1.

4. THE STOCHASTIC MAXIMUM PRINCIPLE: IN VARIABLES

If derivatives exist it is of course possible to define the random variable.

$$p_t = \frac{\partial F(x_t, t)}{\partial x_t}.$$

Here x_t is the actual value on the random trajectory at time t, and the distribution of p_{t+1} conditional on all variables up to time t depends at most on x_t, u_t and t. Suppose that the plant equation is

$$x_{t+1} = a(x_t, u_t, t) + \varepsilon_{t+1} = a_t + \varepsilon_{t+1}, \tag{1}$$

say, where ε_{t+1} is independent of x_t, u_t, and that costs are undiscounted. Granted differentiability, one will then have, corresponding to (6.1.5),

$$p_t = \frac{\partial g_t}{\partial x_t} + E\left(p_{t+1} \frac{\partial a}{\partial x_t} \middle| x_t, u_t \right) \tag{2}$$

by differentiation of the optimality equation (11.1.9). If the optimal u is located at a stationarity point of the right-hand member of (11.1.9) then one will have

$$\frac{\partial g_t}{\partial u_t} + E\left(p_{t+1} \frac{\partial a_t}{\partial u_t} \middle| x_t, \mathrm{u}_t \right) = 0 \tag{3}$$

at this optimal value. Relation (3) can be converted into a statement that u_t minimizes the 'Hamiltonian' $g(x_t, u_t, t) + E(p_{t+1} a(x_t, u_t, t) | x_t, u_t)$ if g and a are convex and linear in u respectively. In this case a stimulating price (see Section 6.2) also exists.

However, all these statements represent an attempt to force an analogue of the deterministic maximum principle which requires strong conditions (such as (1)) to even begin to work. Even if the attempt were completely successful, the joint random process $\{x_t, u_t, p_t\}$ does not seem constructible in any useful sense.

Part III

Linear/quadratic/Gaussian models

The linear/quadratic/Gaussian model (LQG henceforth) is the LQ model of Chapter 5, modified by the possible addition of Gaussian process and observation noise. Such models have the most extensive and explicit theory of all stochastic decision processes and are realistic for a wide range of applications. Some of their properties survive considerable weakenings of the full LQG hypothesis.

CHAPTER 17

LQG control with incomplete state observation

1. INCOMPLETE STATE OBSERVATION

For Markov models we have hitherto made the observational assumption $W_t = (X_t, U_{t-1}, t)$, implying that the state is observable up to the current instant. Suppose this is modified to $W_t = (Y_t, U_{t-1}, t)$ where Y_t conveys less than complete knowledge of X_t—we can regard Y_t as observations on X_t which are partial, delayed or noise-corrupted. All these features can occur in practice: one may be able to observe only certain aspects of the state (e.g. one can observe the strains on a structure only where sensors are placed), and current observations on the state may in fact refer to the situation as it was some time ago, or be overlaid by observational noise (e.g. guidance of a distant space-craft or of an economy). Such a generalization also formally covers the case when process noise (or any other relevant exogenous variable) falls between the two extremes of being known (i.e. totally predictable) or white (i.e. totally impredictable).

This is the situation of *incomplete state observation*. We shall continue to speak of x_t as the 'underlying' or 'process' state variable, if it possesses all the properties postulated in Section 11.1, except perhaps the observability condition (a). However, if incompletely observable, it cannot be the state variable of the decision process. As we shall see in Chapter 39 the role of the state variable x_t in control optimization is then taken over, subject to some mild additional conditions, by the conditional distribution $P(x_t \mid W_t)$. The variable $P_t(\cdot \mid W_t)$ has all the properties of a state variable, although with the disadvantage that it takes values in a space of much higher dimensionality (the space of all distributions on $\mathscr{X}$) than the structural state space $\mathscr{X}$.

However, for the LQG case results are especially simple, strong and explicit, which is why we treat this case individually and in advance.

For the LQG case it turns out that evaluation of $P(x_t \mid W_t)$ can be reduced to evaluation of the state estimate

$$\hat{x}_t = E(x_t \mid W_t) \tag{1}$$

and of its purely time-dependent covariance matrix. Moreover, questions of state estimation and control optimization are to a large extent decoupled. The importance, properties (including optimality) and method of calculation of the estimate (1) are to a considerable extent independent of control policy, and so of cost function. On the other hand, state estimation has a minimal effect upon policy optimization, in that the optimal control is obtained simply by substituting $\hat{x}_t$ for x_t in the optimal deterministic control. This is the principle of *certainty equivalence*.

In this chapter and the next we separate the two aspects by proving first optimality of a certainty-equivalent policy and then properties of the particular estimate (1). We attempt to isolate the conditions required for each of the two aspects. In particular, many of the characteristic properties of the LQG model do not depend upon the existence of an underlying state structure, and are best understood if no such structure is assumed (see Section 4).

2. CONVENTIONS

If ξ is a vector random variable then we shall write

$$\xi^{(t)} = E_\pi(\xi \mid W_t), \tag{1}$$

and regard this as the estimate of ξ based upon W_t—in some sense, the best estimate one can form at time t. In the particular case of the current value of the state variable x_t we shall denote this quantity simply as

$$\hat{x}_t = x_t^{(t)} = E_\pi(x_t \mid W_t). \tag{2}$$

Note then that also

$$\hat{x}_t = E_\pi(x_t \mid W_t, u_t), \tag{3}$$

since u_t does not affect x_t causally, and, being itself W_t-measurable, does not sharpen the W_t-conditioning.

Let us say that a quantity a is *policy-independent* if $E_\pi(a \mid W_0)$ is independent of π. We shall indicate a policy-independent term simply by $+\cdots$, so that

$$b = c + \cdots$$

implies that $b - c$ is policy-independent.

Note that this concept of policy independence is stronger than that referred to in the theorems of Section 10.2. There we were asking that an expectation of the form $E_\pi[a \mid W_t, u_t]$ should be policy-independent for a given t, meaning that the function $\psi(W_t, u_t)$ thus determined should be independent of π. Now we are asking that $E_\pi[a \mid W_0]$ should be independent of π, which considerably constrains the possible form of $\psi(W_t, u_t)$.

3. CERTAINTY EQUIVALENCE FOR THE MARKOV CASE

We present a relaxed form of Theorem 5.2.1. The quantity

$$\Delta_t = \hat{x}_t - x_t \tag{1}$$

is the error in estimate of the current state at time t. By (1.3), $E_\pi(\Delta_t \mid W_t, u_t) = 0$.

Theorem 3.1. *Suppose that:*

(*a*) *The plant equation has the disturbed linear form*

$$x_{t+1} = A_t x_t + B_t u_t + \varepsilon_{t+1}. \tag{2}$$

(*b*) *The cost structure is identical with that of Theorem* 5.2.1.

(*c*) $E_\pi(\varepsilon_{t+1} \mid W_t, u_t) = 0$, *and the joint covariance matrix of the vector* $(\varepsilon_{t+1}, \Delta_t)$ *is policy-independent.*

Then (*i*)

$$F(W_t) = \hat{x}_t P_t \hat{x}_t + \cdots \tag{3}$$

and (*ii*) *the optimal control is*

$$u_t = K_t \hat{x}_t, \tag{4}$$

where P_t, K_t *are the matrices determined in Theorem* 5.2.1.

Relation (4) is the expression of *certainty equivalence*: the optimal control is obtained just by substituting the estimate of the state $\hat{x}_t$ for the actual state x_t in the expression for the complete-information optimal control. The same feature manifests itself in (3); $F(W_t)$ is just what it was in the complete information case, except for the substitution of $\hat{x}_t$ for x_t and the addition of a term reflecting the increase in costs caused by process and observation noise, but policy-independent.

Proof. The proof is inductive, as ever. The argument for a step in the induction also shows how the induction should be started, at $t = T$. Suppose that, for a given t, relation (3) holds at $t + 1$. Then, with $g(x_t, u_t, t) = (x'Ax + 2u'Sx + u'Qu)_t$ we have

$$\begin{aligned} F(W_t) &= \min_{u_t} E[g(x_t, u_t, t) + \hat{x}_{t+1} P_{t+1} \hat{x}_{t+1} \mid W_t, u_t] + \cdots \\ &= \min_{u_t} E[g(x_t, u_t, t) + x'_{t+1} P_{t+1} x_{t+1} \mid W_t, u_t] + \cdots \\ &= \min_{u_t} E[g(\hat{x}_t, u_t, t) + (A\hat{x} + Bu)'_t P_{t+1} (A\hat{x} + Bu)_t] + \cdots, \end{aligned} \tag{5}$$

whence the assertions follow. Here we have consecutively substituted $\hat{x}_{t+1} = x_{t-1} + \Delta_{t+1}$, $x_{t+1} = (Ax + Bu)_t + \varepsilon_{t+1}$ and $x_t = \hat{x}_t - \Delta_t$, and expanded in powers of $\Delta_{t+1}, \varepsilon_{t+1}$ and Δ_t, taking account of the policy independence of conditional first- and second-order moments of Δ, ε. ■

Note that an assumption of Gaussian distribution for quantities such as Δ and ε has not yet been made or invoked. This will come in Chapter 18, when we look for simpler hypotheses which ensure the validity of conditions (c), and when we consider how the estimates $\hat{x}_t$ are in fact to be calculated.

The hypotheses of the theorem are sufficient to imply that

$$\hat{x}_{t+1} = A_t \hat{x}_t + B_t u_t + e_t \tag{6}$$

where

$$e_t = E[\varepsilon_{t+1} - A_t \Delta_t \,|\, W_{t+1}].$$

However, relation (6) does not supply an effective recursion for $\{\hat{x}_t\}$ unless e_t can be conveniently calculated.

4. CERTAINTY EQUIVALANCE IN THE GENERAL CASE

If one assumes only the structure needed to imply certainty equivalence, then the assertion and proof are strikingly simple.

Consider a T-horizon problem with the total cost function

$$\begin{aligned} C &= Q(u_0, u_1, \ldots, u_{T-1}; \xi) \\ &= Q(U_{T-1}, \xi) \end{aligned} \tag{1}$$

where ξ is a vector random variable. We imagine that ξ subsumes all random exogenous variables such as process and observation noise over $(0, T)$, unknown parameters and the course of the path that one will attempt to follow. These are all quantities which may become observable to an increasing degree in the course of time, but, insofar as they are initially unobservable, will be regarded as random.

We shall assume that ξ takes values in a vector space, as do also the u_t. The estimate $\xi^{(t)} = E_\pi(\xi \,|\, W_t)$ is thus defined. Since all decisions have been taken by time $T - 1$ we can make the convenient convention that, whatever the policy,

$$\xi^{(T)} = \xi. \tag{2}$$

That is, that if one had continued observation for one more step, then the exact value of ξ would have been revealed.

Define also the 'forecast innovation'

$$\zeta_{t+1} = \xi^{(t+1)} - \zeta^{(t)}, \tag{3}$$

the change in estimate of ξ as the new observables at time $t + 1$ become available.

Theorem 4.1 (Certainty equivalence). *Assume that:*

(a) Q is a quadratic form in the joint variable (U_{T-1}, ξ); positive-definite in U_{T-1}.

(b) $E_\pi(\xi \,|\, W_t, u_t) = E_\pi(\xi \,|\, W_t)$.

(c) $cov_\pi(\zeta_{t+1})$ is policy-independent.

Then the optimal value of u_t is that obtained by minimizing $Q(U_{T-1}, \xi^{(t)})$ with respect to $u_t, u_{t+1}, \ldots, u_{T-1}$.

Proof. Define the freely-minimized expression

$$Q_t(U_{t-1}, \xi) = \min_{u_t, u_{t+1}, \ldots, u_{T-1}} Q(U_{T-1}, \xi). \tag{4}$$

This will then also be a quadratic form in its arguments and positive definite in U_{t-1}. Recall the partially optimized cost function

$$G(W_t) = \inf_{\pi} E_\pi(C \mid W_t) \tag{5}$$

of (10.4.1). We shall show inductively that

$$G(W_t) = Q_t(U_{t-1}, \xi^{(t)}) + \cdots; \tag{6}$$

the assertion of the theorem will emerge as valid simultaneously.

Relation (6) is true for $t = T$, by the normalizing convention (2). For a given value of t, suppose the relation established for $t+1$, $t+2, \ldots, T$. Then the optimality equation yields

$$G(W_t) = \min_{u_t} E[Q_{t+1}(U_t, \xi^{(t+1)}) \mid W_t, u_t] + \cdots$$

where the minimizing value of u_t will be optimal. Now, hypothesis (b) implies that

$$E_\pi(\zeta_{t+1} \mid W_t, u_t) = E_\pi(\xi^{(t+1)} - \xi \mid W_t, u_t) = 0,$$

and so hypotheses (a) to (c) imply that

$$\begin{aligned} E_\pi[Q_{t+1}(U_t, \xi^{(t+1)}) \mid W_t, u_t] &= E_\pi[Q_{t+1}(U_t, \xi^{(t)} + \zeta_{t+1}) \mid W_t, u_t] \\ &= E_\pi[Q_{t+1}(U_t, \xi^{(t)}) \mid W_t, u_t] + \cdots \\ &= Q_{t+1}(U_t, \xi^{(t)}) + \cdots \end{aligned}$$

where $+\cdots$ denotes, as ever, policy-independent terms. We thus deduce that

$$G(W_t) = \min_{u_t} Q_{t+1}(U_t, \xi^{(t)}) + \cdots \tag{7}$$

which completes the induction, establishing (6). It also establishes the assertion of the theorem, because the minimizing value in (7) is exactly that defined in the theorem. ■

The theorem is said to express certainty equivalence, because it states that the optimal action at time t is that which would have been optimal if ξ had actually been known to be equal to $\xi^{(t)}$—its current estimate.

Let $u_s^{(t)}$ be the value of u_s obtained when $Q(U_{T-1}, \xi^{(t)})$ is minimized with respect to $u_t, u_{t+1}, \ldots, u_{T-1}$. Then, for $s < t$, $u_s^{(t)}$ is just the decision that already *has* been taken at time s, optimal or not. For $s = t$, $u_t^{(t)}$ is the optimal decision to be taken currently, by Theorem 3.1. For $s > t$ we must regard $u_s^{(t)}$ as the best assessment one can make at time t of the optimal decision one will make at time s (see Exercise 1).

Presented this way, one sees a subjectively valuable aspect of optimization that a dynamic programming approach regrettably suppresses: current decisions are related to a provisional forward plan of decisions.

Hypotheses (b) and (c) assert that estimation errors in ξ are to a sufficient degree policy-independent. Otherwise expressed, the information gained from observations is essentially independent of policy, and so of actions. We shall see in the next chapter that, roughly speaking, these hypotheses are ensured if both the plant equation and observations are linear in past state variables, and if injected random variables such as process noise and observation noise are jointly Gaussian and policy-independent.

Exercises and comments

1. Show that, under the hypotheses of Theorem 3.1,

$$u_s^{(t)} = E(u_s^{(s)} | W_t)$$

if all decisions are taken optimally.

2. *Tracking and buffering*. Assume the plant equation (2.2) and the cost structure of Section 5.2 with x observable up to the current instant. Let ξ denote the process noise sequence $\{\varepsilon_t\}$ and suppose that this satisfies the hypotheses of Theorem 3.1. (These hypotheses are certainly valid if both $\{\varepsilon_t\}$ and the information gained on it are policy-independent.) Then, by Corollary 5.4.1 and Theorem 3.1, the optimal control is

$$u_t = K_t x_t + \sum_{j=0}^{T-t-1} K_{tj}^* E(\varepsilon_{t+j} | W_t) \tag{8}$$

where

$$K_{tj}^* = -D_t^{-1} B_t \Gamma'_{t+1} \Gamma'_{t+1} \cdots \Gamma'_{t+j} P_{t+j+1}.$$

That is, future disturbances ε_{t+j} are replaced in (8) by their forecasts, $E(\varepsilon_{t+j} | W_t)$. If process noise is white, and unpredictable, as in Section 12.1, then $E(\varepsilon_{t+j} | W_t) = E(\varepsilon_{t+j}) = 0$.

3. Consider the LQ dam regulation problem of Exercise 5.6.2. Suppose that x_s, α_s, u_{s-1} are observable for $s \leq t$ at t and that inflow $\{\alpha_t\}$ obeys the linear Markov relation

$$(\alpha_{t+1} - \mu) = \beta(\alpha_t - \mu) + \varepsilon_t^*$$

where ε_t^* is an exogenous white noise sequence. Show that then

$$E(\alpha_{t+j} | W_t) = \mu + \beta^j(\alpha_t - \mu) \qquad (j \geq 0)$$

and that the optimal infinite-horizon control rule is

$$u_t = \mu + (1 - \Gamma)\left[x_t - x_t^* + \frac{\beta(\alpha_t - \mu)}{1 - \beta\Gamma}\right].$$

5. RESTRICTED CERTAINTY EQUIVALENCE

A statistician would conjecture that optimal solutions derived on the Gaussian assumption continue to enjoy some kind of restricted optimality even in the non-Gaussian case. This turns out to be the case in the present instance.

Let y_t denote the actual observation made at time $t(t>0)$; we shall regard $\hat{x}_0 = E(x_0 | W_0)$ as summary of the observations made up to time $t=0$. Then the information W_t at time t can be conveniently divided into *process history* $W_t^* = (\hat{x}_0, Y_t, U_{t-1})$ and other quantities which are specified rather than observed, such as t itself and specification of the process dynamics.

The estimate $\xi^{(t)} = E(\xi | W_t)$ could more properly be written $\xi^{(t)} = E(\xi | W_t^*)$, since non-random variables such as t and process specification are parametrizing rather than conditioning (see Section 10.2). This conditioned expectation can be characterized by the fact that, for given t,

$$E[(\xi^{(t)} - \xi)\phi(W_t^*)] = 0 \tag{1}$$

for any function ϕ of W_t^*.

Suppose now there is an estimate $\bar{\xi}^{(t)}$ of ξ, more special in that it is required to be *linear* in W_t^* in its dependence on W_t, and correspondingly more weakly characterized in that, for given t, one requires that

$$E[(\bar{\xi}^{(t)} - \xi)\phi(W_t^*)] = 0 \tag{2}$$

only for *linear* functions ϕ of W_t^*. We shall express this characterization as

$$\bar{\xi}^{(t)} - \xi \perp W_t^*. \tag{3}$$

We shall see in Section 18.7 that $\bar{\xi}^{(t)}$ is uniquely characterized by these properties as being the linear least-square estimate and also that $\bar{\xi}^{(t)} = \xi^{(t)}$ in the Gaussian case. However, for the moment we simply ask whether a version of certainty equivalence (Theorem 3.1) holds if property (1) is weakened to property (2).

Define the innovations

$$\bar{\zeta}_{t+1} = \bar{\xi}^{(t+1)} - \bar{\xi}^{(t)} \tag{4}$$

in the estimate $\bar{\xi}$.

Theorem 5.1 (Restricted certainty equivalence). *Assume that:*

(*a*) *Q is a quadratic form in the joint variable (U_{T-1}, ξ); positive definite in U_{T-1}.*

(*b*) *$\bar{\xi}^{(t)}$ is linear in W_t^* and $\xi - \bar{\xi}^{(t)} \perp (W_t^*, u_t)$ under π.*

(*c*) *$\mathrm{cov}_\pi(\bar{\zeta}_{t+1})$ is policy-independent.*

Then the optimal value of u_t which is linear in W_t^* is that obtained by minimizing $Q(U_{T-1}, \bar{\xi}^{(t)})$ with respect to $u_t, u_{t+1}, \ldots, u_{T-1}$.

That is, use of a certainty equivalence rule with the estimate $\bar{\xi}^{(t)}$ of ξ yields the optimal *linear* control rule.

Proof. Let $\bar{G}_t(\pi)$ be the value of $E(C)$ when policy π is followed up to time $t-1$, but $u_t, u_{t+1}, \ldots, u_{T-1}$ are chosen optimally among linear rules. We shall demonstrate inductively that

$$\bar{G}_t(\pi) = E_\pi Q_t(U_{t-1}, \bar{\xi}^{(t)}) + \cdots \tag{5}$$

with the definition (4.4) of Q_t.

Assume the result true for $t+1$; we have then

$$\bar{G}_t(\pi) = \min_{u_t} E_\pi Q_{t+1}(U_t, \bar{\xi}^{(t+1)}) + \cdots \tag{6}$$

where the minimum is over functions $u_t = u_t(W_t)$ which are linear in W_t^*.

Now, the orthogonality hypotheses (b) will imply that $\bar{\zeta}_{t+1} \perp \bar{\xi}^{(t)}, U_t$. From this and the covariance hypothesis (c) it follows that

$$\begin{aligned} \bar{G}_t &= \min_{u_t} E_\pi Q_{t+1}(U_t, \bar{\xi}^{(t)}) + \cdots \\ &\geq E_\pi \min_{u_t} Q_{t+1}(U_t, \bar{\xi}^{(t)}) + \cdots \\ &= E_\pi Q_t(U_{t-1}, \xi^{(t)}) + \cdots \end{aligned} \tag{7}$$

where the minimizing operation in the second line is a free one. But the freely minimizing u_t will in fact be linear in W_t^*; hence this will be the optimal linear u_t and equality holds in (7). The induction establishes both relation (5) and the assertion of the theorem. ■

CHAPTER 18

State estimation with linear structure

1. STATE ESTIMATION AS A GOAL IN ITSELF

The *Bayesian* or *conditional* state estimate $\hat{x}_t = E(x_t | W_t)$ was introduced in Chapter 17 just because it provided a certainty-equivalent for purposes of optimal control under certain conditions. The same held, in the non-state case, for the estimate $\xi^{(t)} = E(\xi | W_t)$ of the process input vector ξ.

A first task must be to determine sufficient conditions that these estimates indeed satisfy the hypotheses of Theorems 17.3.1 and 17.4.1 (Section 2). Roughly speaking, a sufficient condition is that the observables can be transformed to be representable as linear functions of ξ (or of past and present state plus observation noise) and that all noise be policy-independent Gaussian. A second task must be to consider how these estimates should be calculated. The classic algorithm here is the Kalman filter (see Section 3); if one has state structure and an appropriately related observation structure then the state estimates are generated by a recursion (3.3) closely related to the plant equation itself.

However, these estimates are of interest in themselves, as optimal estimates (in the least-square sense) whatever the policy, and so whatever the cost functions. Moreover, if one preserves the linear/Gaussian assumptions for the plant equation and observation structure, but does not require quadratic costs or all of the policy-independence assumptions of Theorem 2.2, the quantities $\hat{x}_t = E(x_t | W_t)$ and $V_t = \mathrm{cov}(x_t | W_t)$ remain sufficient for purposes of policy-optimization (see Section 41.11). In this more general class of cases the policy may affect the information gained, a feature which the hypotheses required for certainty equivalence in Theorems 17.3.1 and 17.4.1 explicitly excluded.

Finally, we note that some of these results transfer also to the non-Gaussian case, if one uses the linear least-square estimates $\bar{x}_t$ and $\bar{\xi}^{(t)}$ and settles for a control policy among linear policies (see Sections 6 to 8).

2. SUFFICIENT CONDITIONS THAT BAYESIAN ESTIMATES PROVIDE CERTAINTY EQUIVALENTS

For general random vectors x, y we shall use the notation

$$\mu_x = E(x),$$

$$V_{xy} = E((x - \mu_x)(y - \mu_y)')$$

to denote mean vectors and cross-covariance matrices. We shall appeal to the fact that, if x and y are jointly normally distributed with zero means, then the distribution of x conditional on y is normal with mean and covariance matrix

$$E(x|y) = V_{xy}V_{yy}^{-1}y, \tag{1}$$

$$\operatorname{cov}(x|y) = V_{xx} - V_{xy}V_{yy}^{-1}V_{yx} \tag{2}$$

respectively (cf. Theorem 6.3).

We consider first the non-state version of the LQG problem, formulated in Theorem 17.4.1.

Theorem 2.1. *Suppose that the input vector ξ is Gaussian with known policy-independent parameters and that the observation at time t has the form*

$$y_t = c_t'\xi + d(W_{t-1}, u_{t-1}). \tag{3}$$

That is, it is representable as a time-dependent linear function of ξ plus an arbitrary function of (W_{t-1}, u_{t-1}).

Then the hypotheses (b) and (c) of Theorem 17.4.1 are satisfied. Thus, under the additional hypothesis (a) $\xi^{(t)}$ provides a certainty equivalent of ξ for purposes of optimal LQ control.

Proof. The only policy dependence in the specification of ξ and Y_t as joint random variables is the presence of the term $d(W_{t-1}, u_{t-1})$ in (3), which means that the expectation of y_t may depend upon U_{t-1}. However, since this term is calculable before y_t is observed, the situation is equivalent to that in which the observation at time t is $\tilde{y}_t = y_t - d(W_{t-1}, u_{t-1})$, known to have the form

$$\tilde{y}_t = c_t'\xi.$$

The joint distribution of $\xi, \tilde{Y}_T$ is then Gaussian with known policy-independent parameters. Thus $E_\pi(\xi|W_t, u_t) = E(\xi|\tilde{Y}_t)$ and hypothesis (b) of Theorem 17.4.1 is satisfied. By (1), $\xi^{(t)}$ is a policy-independent linear function of ξ. So then is ζ_{t+1}. The distribution of ζ_{t+1} conditional on (W_t, u_t) is the distribution conditional on $\tilde{Y}_t$; this is Gaussian and policy-independent. Hypothesis (c) of Theorem 32.3.1 is thus satisfied. ■

Consider now the state-structured problem envisaged in Theorem 17.3.1.

Theorem 2.2. *Assume the plant equation (17.3.2). Suppose also that W_0 specifies that $x_0 \sim N(\hat{x}_0, V_0)$, that the observation at time t has the form*

$$y_t = \sum_{s=0}^{t} c_{st}x_t + d(W_{t-1}, u_{t-1}) + \eta_t, \tag{4}$$

that $\Delta_0 = \hat{x}_0 - x_0$, $\{\varepsilon_t\}$ and $\{\eta_t\}$ are jointly Gaussian with known policy-independent parameters and that $E[\varepsilon_{t+1}|\Delta_0, \varepsilon_s, \eta_s;\ 0 < s \le t] = 0$.

Then hypothesis (c) of Theorem 17.3.1 is satisfied. Thus, under the additional hypothesis (b), $\hat{x}_t$ provides a certainty equivalent of x_t for purposes of optimal LQ control.

Proof. One can use the plant equation to express x_t in terms of x_0 and u_{s-1}, ε_s $(0 < s \leq t)$, and so write (4) as

$$y_t = c_{ot}^* \Delta_0 + \sum_{s=1}^{t} c_{st}^* \varepsilon_s + d^*(W_{t-1}, u_{t-1}) + \eta_t \tag{5}$$

for some c^*, d^*. As in the previous theorem we can redefine the observations as

$$\tilde{y}_t = c_{ot}^* \Delta_0 + \sum_{s=1}^{t} c_{st}^* \varepsilon_t + \eta_t.$$

Solution of the plant equation for x_t will represent it in a similar form to (5):

$$x_t = x_t^* + \tilde{x}_t$$

where x_t^* is a function of $\tilde{x}_0$ and U_{t-1}, calculable at time t without error, and $\tilde{x}_t$ is a linear function of Δ_0 and $\varepsilon_s (1 \leq s \leq t)$ with policy-independent coefficients. The estimation error Δ_t is the error in the estimate of $\tilde{x}_t$ from $\tilde{y}_s$ $(1 \leq s \leq t)$, and is a linear function of Δ_0, ε_s and η_s $(1 \leq s \leq t)$ with policy-independent coefficients.

The joint distribution of Δ_t, Δ_{t+1}, ε_{t+1} and $\tilde{y}_s$ $(1 \leq s \leq t)$ is thus Gaussian with policy-independent parameters. The distribution of Δ_t, Δ_{t+1} and ε_{t+1} conditional on (W_t, u_t) is then the distribution conditional on $\tilde{Y}_t$ and is policy-independent. The conditional means of Δ_t and ε_{t+1} are zero, by definition and hypothesis respectively. The hypotheses on Δ and ε of Theorem 17.3.1 are thus satisfied. ■

Although the estimate $\hat{x}_t$ provides a certainty equivalent for x_t under these conditions, it must also be calculable recursively if it is to be sufficient. That is, $\hat{x}_{t+1}$ must be calculable from $\hat{x}_t$, u_t, y_{t+1} and t. The Kalman filter provides such a recursion, deducible under more specific hypotheses on the form (4) of the observation.

3. RECURSIVE ESTIMATION; THE KALMAN FILTER

We shall make assumptions which ensure that:

(a) The distribution of x_t conditional on W_t is normal; $N(\hat{x}_t, V_t)$, say.

(b) The posterior mean $\hat{x}_t$ and covariance V_t can be calculated recursively. This will imply (Section 41.11) that $\hat{x}_t$ and V_t are sufficient for the Markov decision problem.

Hypotheses will in one sense be stronger than those of Theorem 2.2, in order to secure recursivity. However, in another sense they can be more general. If costs and termination rules are varied from those of the LQG case then $(\hat{x}_t, V_t)$ may well still be sufficient, but $\hat{x}_t$ will generally not provide a certainty equivalent and there is no point in postulating conditions which make it so. Essentially, one may allow the information yielded by the observations to be policy-dependent.

Theorem 3.1. *Suppose that*

(*a*) *Conditional on* W_0, *the initial state* x_0 *is distributed* $N(\hat{x}_0, V_0)$.

(*b*) $W_t = (W_0; y_1, y_2, \ldots, y_t; u_0, u_1, \ldots, u_{t-1}; t) = (W_0, Y_t, U_{t-1}, t)$.

(*c*) *The state* x_t *and observation* y_t *obey the recursions linear in state*

$$x_{t+1} = a_t + A_t x_t + \varepsilon_{t+1}, \tag{1}$$

$$y_{t+1} = c_t + C_t x_t + \eta_{t+1}, \tag{2}$$

where, conditional on (x_t, W_t, u_t), *the pair* $(\varepsilon_{t+1}, \eta_{t+1})$ *are jointly normally distributed with zero mean and covariance matrix* $\begin{bmatrix} N_t & L_t \\ L_t' & M_t \end{bmatrix}$. *Here* a_t, A_t, c_t, C_t, L_t, M_t *and* N_t *may all be functions of* (W_t, u_t).

Then, conditional on W_t, *the current state* x_t *is distributed* $N(\hat{x}_t, V_t)$. *The conditional mean and variance obey the updating recursions*

$$\hat{x}_{t+1} = a_t + A_t \hat{x}_t + H_t \zeta_{t+1}, \tag{3}$$

$$V_{t+1} = [N + AVA' - (L + AVC')(M + CVC')^{-1}(L + AVC')']_t, \tag{4}$$

where

$$\zeta_{t+1} = y_{t+1} - c_t - C_t \hat{x}_t, \tag{5}$$

$$H_t = [(L + AVC')(M + CVC')^{-1}]_t. \tag{6}$$

Conditional on W_t, u_t *the random variable* ζ_{t+1} *is distributed with zero mean and covariance matrix* $(M + CVC')_t$.

Proof. This is scarcely longer than the statement. Suppose that indeed $x_t \equiv N(\hat{x}_t, V_t)$ conditional on W_t; hypothesis (a) provides a starting point for the induction. The distribution will be the same conditional on (W_t, u_t), since x_t is causally independent of u_t and u_t is W_t-measurable. Let us for simplicity omit the subscript t on all quantities which are constant conditional on (W_t, u_t), except perhaps $\hat{x}_t$. It follows then from the assumptions (c) that, conditional on W_t, u_t, the quantities $(x_t - \hat{x}_t,\ x_{t+1} - a - Ax_t,\ y_{t+1} - c - Cx_t)$ are jointly normally distributed with zero mean and the covariance matrix

$$\begin{bmatrix} V & \cdot & \cdot \\ \cdot & N & L \\ \cdot & L' & M \end{bmatrix}$$

We can eliminate x_t to deduce that, conditional still on (W_t, u_t), the quantities (x_{t+1}, y_{t+1}) are jointly normally distributed with mean $(a + A\hat{x}_t, c + C\hat{x}_t)$ and covariance matrix

$$\begin{bmatrix} N + AVA' & L + AVC' \\ L' + CVA' & M + CVC' \end{bmatrix}.$$

It follows from formulae (2.1), (2.2) that, conditional on (W_t, u_t, y_{t+1}), i.e. conditional on W_{t+1}, the state value x_{t+1} is normally distributed with the mean $\hat{x}_{t+1}$ given by (3) and the covariance matrix V_{t+1} given by (4). ■

Note that the theorem does not imply that the unconditional distribution of x_t is normal. The facts that process history may affect the mean of x_t non-linearly and affect the distribution of ε_t preclude this in general. But it is the conditional statement which is important.

Note also that the theorem is true independently of control policy, and so of costs. Controls are a part of observational history, and so are regarded as fixed quantities when one conditions on that history. It follows, then, that Theorem 3.1 is quite dissociated from the question of control optimization.

The fact that (under LQ assumptions) control policy does not affect estimation procedures and that (under LQG assumptions) optimal controls are certainty-equivalent versions of optimal deterministic controls is sometimes referred to as the *separation principle*. It implies that optimization of estimate and control can to this extent be decoupled.

Recursion (3) is what is normally referred to as the *Kalman filter*. It has the appealing interpretation that the state estimate $\hat{x}_t$ obeys exactly the same dynamic equation as does the state variable x_t itself, except that the process noise ε_{t+1} is replaced by a correction term $H_t\zeta_{t+1} = H_t(y_{t+1} - c_t - C_t\hat{x}_t)$, allowing for the deviation of the observation y_{t+1} from the value $c_t + C_t\hat{x}_t$ predicted for it on the basis of W_t. This deviation is what will later be known as the *innovation* in the $\{y_t\}$ process (see (6.11)).

The Kalman recursion is more usually stated for the particular case $a_t = B_t u_t$, $c_t = 0$, with the quantities A, B, C, L, M and N functions of t at most. We state this standard form for reference:

$$x_{t+1} = A_t\hat{x}_t + B_t u_t + H_t(y_{t+1} - C_t\hat{x}_t). \tag{7}$$

The final assertion of Theorem 3.1 then implies also that $\{\zeta_t\}$ is a white noise Gaussian process.

More general forms of the filter can be asserted, best proved by projection methods (see Section 8). However, one cannot significantly generalize without losing recursivity. Of course, relations (3), (4) cannot be regarded as recursions unless they allow $\hat{x}_{t+1}$, V_{t+1} to be calculated in terms of $\hat{x}_t$, V_t, u_t, y_{t+1} and t. This implies that the quantities listed as (W_t, u_t)-dependent in (c) of the theorem should in fact be $(\hat{x}_t, V_t, u_t, t)$-dependent at most. Under these conditions $(\hat{x}_t, V_t)$ are sufficient for a general Markov cost structure (see Section 41.11).

In relation (4) we recognize a Riccati equation of the form (5.2.4) obeyed by P_t. Indeed, under Markov LQG assumptions there is a duality between optimization of control and of estimation. The separation principle is one manifestation of this; the occurrence of the separate but analogous forms of the backward Riccati recursion for P_t and the forward Riccati recursion for V_t is another.

The analogue to Γ_t is the matrix

$$\Omega_t = (A - HC)_t$$

because we see from (1), (3) that

$$\Delta_{t+1} = \Omega_t \Delta_t + (H_t \eta_{t+1} - \varepsilon_{t+1}).$$

So, if Ω_t has a limit Ω, and $\mathrm{cov}(H_t \eta_{t+1} - \varepsilon_{t+1})$ is non-singular, then stability of estimation error requires that Ω be a stability matrix, just as stability of control required the same of Γ.

Some of the most interesting special cases of incomplete state observation are degenerate, and better handled by the alternative forms of the next section.

4. ALTERNATIVE FORMS FOR THE RECURSIONS

As in Section 5.7, the Riccati recursion for V_t and associated matrix evaluations can be given dual forms, which are advantageous for certain purposes (see the degenerate cases of the exercises or the risk-sensitive version of LQG control in Chapter (19).

Theorem 4.1. *The Riccati recursion* (3.4) *has the alternative expression*

$$V_{t+1} = [N - LM^{-1}L' + (A - LM^{-1}C)(V^{-1} + C'M^{-1}C)^{-1}(A - LM^{-1}C)']_t \tag{1}$$

Correspondingly,

$$H_t = [LM^{-1} + (A - LM^{-1}C)(V^{-1} + C'M^{-1}C)^{-1}C'M^{-1}]_t, \tag{2}$$

$$\Omega_t = [(A - LM^{-1}C)(V^{-1} + C'M^{-1}C)^{-1}V^{-1}]_t. \tag{3}$$

We defer proof to Chapter 19, when the significance of the alternative forms emerges in a more general setting.

Exercises

1. Consider the case of zero process noise, so that N, L are zero. The problem is then essentially that of estimating the value of the initial state x_0. Note that, at least if A_t is non-singular, relation (1) reduces to the linear recursion for V_t^{-1}:

$$A_t' V_{t+1}^{-1} A_t = (V^{-1} + C'M^{-1}C)_t.$$

2. Another special case would be that in which the current state was observable, but the plant equation was

$$x_{t+1} = \mu + A_t x_t + B_t u_t + \varepsilon_{t+1}$$

where the vector μ is unknown. By augmenting the state variable or otherwise, deduce recursive equations for the estimation of μ.

5. CONVERGENCE TO STATIONARITY

One expects that, if the model is time-homogeneous and effective estimation of all aspects of state is possible, the V_t (and so H_t) will converge to a limt as $t \to \infty$. That is, the covariance matrix of estimation errors will stabilize at a limit value V, and the Kalman filter, (3.3) or (3.7), will become time-invariant in form. Indeed, one will expect this to be true even if y_t is generated from process history by a more general relation than (3.2), provided only that the relation is linear and time-invariant. However, we shall consider the case (3.2) alone, because $\{V_t\}$ is then generated by the recursion (3.4), which is a complete forward analogue of the recursion (5.2.4) for $\{P_t\}$. We need then only state the analogue of Theorems 5.5.1 and 5.5.2, whose proofs transcribe faithfully.

Theorem 4.1. *Suppose that:*
(a) $M > 0$, *(b)* $N = ZZ'$, *where* $[A, Z, \cdot]$ *is controllable,* (c) *the system* $[A, \cdot, C]$ *is either observable or detectable.*

Then: (i) *the equation*

$$V = N + AVA' - (L + AVC')(M + CVC')^{-1}(L + AVC')' \tag{1}$$

has a unique non-negative definite solution V.

(ii) $V_t \to V$ *for any finite non-negative definite* V_0.

(iii) The matrix $\Omega = A - HC = \lim(A - H_t C)$ *is a stability matrix.*

Hypothesis (c) is the important one, stating essentially that the observables do indeed yield information on all components of state. Conditions (a) and (b) are convenient as they exclude certain degeneracies; they state that no component of state is free from observation noise and that process noise ultimately affects all components of state. However, the degeneracies excluded are ones that are of prime interest; it may well be that many components of state can be observed with zero error.

The matrix Ω is of interest for the reason explained at the end of Section 3. The pair of relations

$$x_{t+1} = Ax_t + Bu_t + \varepsilon_{t+1},$$

$$y_{t+1} = Cx_t + \eta_{t+1}$$

describe the system in that they relate 'input' u and 'output' y, with state x appearing only as an intermediary. We noted in Exercise 4.1.4 that this representation of the system is not unique; the state variable can be transformed without altering the relation of y to u. The equation pair

$$\hat{x}_{t+1} = A\hat{x}_t + Bu_t + H\zeta_{t+1},$$

$$y_{t+1} = C\hat{x}_t + \zeta_{t+1}$$

can be regarded as just such a transformed representation, satisfying the normalization that the white noise input $\{\varepsilon_t, \eta_t\}$ is now such that the first component is linearly predictable from the second.

Just as in Chapter 5, one will rarely if ever use the Riccati equation (1) as the actual algorithm for calculation on V_t. The square-root algorithm (Section 5.9) or the Kailath–Chandrasakhar algorithm (Exercise 5.7.2) are computationally superior.

6. LINEAR LEAST-SQUARE ESTIMATES

For the purposes of this section x, y, etc., will be arbitrary random vectors, not necessarily bearing the same connotation as in the control context. We shall assume these to be measured about their mean values, so that $E(x) = 0$, etc., and

$$V_{xy} = E(xy'). \tag{1}$$

If $V_{xy} = 0$ we shall say that x and y are *uncorrelated*, expressed in symbols as $x \perp y$. This is indeed an orthogonality relation. We generalize it to processes by saying that $\{x_t\} \perp \{y_t\}$ if $x_s \perp y_t$ for all s, t.

Let us use $\mathscr{E}(x|y)$ to denote the linear function of y that approximates x best in mean square. That is,

$$\mathscr{E}(x|y) = a + by$$

where the vector a and matrix b are chosen to minimize

$$D(a, b) = E[x - a - by)'M(x - \alpha - by)]$$

for M a positive definite matrix. The estimate thus characterized turns out to be independent of M.

Theorem 6.1. *Define the linear least-square estimation error,* $\delta = x - \mathscr{E}(x|y)$. *Then* $\mathscr{E}(x|y)$ *is characterized by*

$$E(\delta) = 0, \tag{2}$$

$$\delta \perp y. \tag{3}$$

These conditions can alternatively be written as

$$a = 0, \tag{4}$$

$$bV_{yy} = V_{xy}. \tag{5}$$

The determinations

$$\mathscr{E}(x|y) = V_{xy}V_{yy}^{-1}y, \tag{6}$$

$$V_{\delta\delta} = V_{xx} - V_{xy}V_{yy}^{-1}V_{yx} \tag{7}$$

hold and are unique, with V_{yy}^{-1} defined as a generalized inverse if necessary.

Proof. The convex function $D(a, b)$ is minimal and only at a stationary point and it has stationary values for finite a, b. The stationarity conditions can be written

$$ME(\delta) = 0,$$

$$ME(\delta y') = 0.$$

These relations are equivalent to (2), (3), which are in turn plainly equivalent to (4), (5). Since D has finite stationary points, equation (5) has a solution. Its general solution is

$$b = V_{xy}V_{yy}^{-1} + c$$

where the columns of c are arbitrary elements of the null space of V_{yy}. With $c = 0$ one readily verifies expressions (6), (7). But $c'V_{yy}c = 0$ implies that $c'y = 0$ in mean square, and $V_{xy}c = 0$. Expressions (6) and (7) thus hold for all solutions of (5). ■

The operator $\mathscr{E}(x|y)$ is plainly a projection operator, projecting x on to y, one may say. The following properties, immediate consequences of Theorem 6.1, will be recognized as properties of projection operators.

Theorem 6.2. *(a) $\mathscr{E}(\cdot|y)$ is linear and $\mathscr{E}(1|y) = 1$.*
(b) If $y \perp z$ then

$$\mathscr{E}(x|y,z) = \mathscr{E}(x|y) + \mathscr{E}(x|z). \tag{8}$$

Up to now we have made hypotheses only on first and second moments. The following theorem is then striking and important.

Theorem 6.3. *Suppose the joint distribution of x and y is Gaussian. Then the distribution of x conditional on y is Gaussian, with*

$$E(x|y) = \mathscr{E}(x|y), \tag{9}$$

$$\text{cov}(x|y) = V_{\delta\delta}. \tag{10}$$

That is, the expectation of x conditional on y is just the linear least-square estimate of x derived from y, and the conditional covariance matrix is just the covariance matrix of linear least-square estimation errors.

Proof. The assertions follow from two properties of the Gaussian (normal) distribution: (a) linear functions of jointly Gaussian variables are jointly Gaussian and (b) jointly Gaussian variables which are uncorrelated are statistically independent.

So $\delta = \mathscr{E}(x|y) - x$ and y, linear functions of x and y, are jointly Gaussian. They are independent, by (3). Hence the distribution of $x - \mathscr{E}(x|y)$ conditional on y is identical with its unconditional distribution, which is Gaussian with zero mean and the covariance matrix $V_{\delta\delta}$. ■

Finally, we introduce the important concept of an innovation. Consider a stochastic process $\{y_t\} = \{y_0, y_1, y_2, \ldots\}$. Suppose that y_t is the new observable that becomes available at time t. Define the *innovation* ζ_t at time t:

$$\zeta_t = y_t - \mathscr{E}(y_t | Y_{t-1}) \tag{11}$$

with $\zeta_0 = y_0$. This can be regarded as the new information which becomes available at time t, as it is the part of the observation y_t which was not linearly predictable from previous observations. One might also have considered the definition

$$\zeta_t^* = y_t - E(y_t | Y_{t-1}),$$

but there will be reasons for wanting to retain linearity of relationships. Moreover, in the Gaussian case the two definitions are identical.

Theorem 6.4 *The innovations process is uncorrelated.*

That is, $\zeta_s \perp \zeta_t$ for $s \neq t$.

Proof. It follows from (3) that $\zeta_t \perp Y_{t-1}$. Since ζ_s is a linear function of $y_0, y_1, y_2, \ldots, y_s$ it follows then that $\zeta_t \perp \zeta_s$ for $s < t$. ■

In fact, the $\{\zeta_t\}$ sequence is just that obtained from the $\{y_t\}$ sequence by Gram–Schmidt orthogonalization, and one sequence is obtained from the other by an invertible linear transformation with a triangular matrix.

If the $\{y_t\}$ were stationary and orthogonalization had commenced at $t = -\infty$ rather than $t = 0$, then $\{\zeta_t\}$ would also be stationary and would be second-order white noise.

Exercise

1. Note that in general

$$\mathscr{E}(x|y) = E(x) + b(y - E(y))$$

with b satisfying (4).

7. RESTRICTED CERTAINTY EQUIVALENCE

We return to the material of Section 17.5 and shall show that, under appropriate conditions, the linear least-squares estimate $\mathscr{E}(\xi|W_t^*)$ has the properties demanded of $\bar{\xi}(t)$ in Theorem 17.5.1. Here W_t^* denotes the actual process history (Y_t, U_{t-1}).

Theorem 7.1. *Suppose that the input vector ξ is such that $E(\xi)$, $cov(\xi)$ are policy-independent and that observations have the form*

$$Y_t = c_t'\xi + d_t(Y_{t-1}, U_{t-1})$$

where d_t is linear in Y_{t-1}, U_{t-1}. Then the linear least-square estimate $\bar{\xi}^{(t)} = \mathscr{E}(\xi|W_t^*)$ and the correspondingly forecast innovations $\bar{\zeta}_{t+1} = \bar{\xi}^{(t+1)} - \bar{\xi}^{(t)}$ have the properties (b), (c) demanded in Theorem 17.5.1.

Thus $\bar{\xi}^{(t)}$ provides a certainty equivalent for the purposes of optimal linear control of an LQ process.

Proof. This follows very much the course of the proof of Theorem 2.1. That is, ξ and the effective observables $\tilde{y}_t = c_t'\xi$ have first and second moments which are policy-independent. The same will then be true of ξ, the estimates $\bar{\xi}^{(t)}$ and the innovations $\bar{\zeta}_t$. From this policy independence and the relation $\xi - \xi^{(t)} \perp \tilde{Y}_t$ we deduce the validity of conditions (b), (c) of Theorem 17.5.1. ■

8. THE KALMAN FILTER BY PROJECTION METHODS

Define the linear least-square estimate of state

$$\bar{x}_t = \mathscr{E}(x_t|W_t^*) \tag{1}$$

where W_t^* is the process history $(\bar{x}_0, Y_t, U_{t-1})$. Then this can be generated by the Kalman filter, even if process and observation noise are not supposed Gaussian. Suppose that the assumptions of Theorem 3.1 are modified in that the quantities $\delta_0 = \bar{x}_0 - x_0$ and (ε_t, η_t) $(t > 0)$ are assumed to be mutually uncorrelated with zero means (but not necessarily Gaussian), a_t and c_t to be t-dependent linear functions of (W_t^*, u_t) and A_t, C_t, L_t, M_t and N_t to be t-dependent at most.

One verifies that $\bar{\zeta}_{t+1} = y_{t+1} - \mathscr{E}(y_{t+1}|W_t^*, u_t)$ is orthogonal to (W_t^*, u_t), and thus that

$$\begin{aligned}\bar{x}_{t+1} &= \mathscr{E}(x_{t+1}|W_t^*, u_t) + \mathscr{E}(x_{t+1}|\bar{\zeta}_{t+1})\\ &= a_t + A_t\bar{x}_t + H_t\bar{\zeta}_{t+1}\end{aligned}$$

for some H_t, dependent only on t. This is just the Kalman filter (3.3), stated for the linear estimate $\bar{x}_t$. The value of H_t must be such as to minimize the mean-square value of the deviation

$$\bar{x}_{t+1} - x_{t+1} = (A_t - H_tC_t)\delta_t + (H_t\eta_{t+1} - \varepsilon_{t+1}).$$

This characterization yields evaluation (3.6) again; the matrix V_t, still obeying the recursion (3.4), is now identifiable as $\text{cov}(\delta_t)$.

CHAPTER 19

Risk-sensitive LQG control

1. RISK-SENSITIVE CRITERIA

Apart from occasional specializations, the cost function over horizon T for the LQG problem has been taken as (5.2.2), for both deterministic and stochastic versions of the problem. Let us denote this by $\mathbb{C}$, to avoid confusion with the matrix C of the observation relation (18.3.2). In the stochastic case the actual criterion has been the expectation $E(\mathbb{C})$, in that this has been the quantity minimized. We shall now adopt the more general criterion

$$\gamma(\theta) = -\frac{2}{\theta}\log E(e^{-\theta\mathbb{C}/2}) \tag{1}$$

where θ is a real scalar. The quantity $-\theta$ can be regarded as a measure of risk-aversion on the part of the optimizer, in a sense to be explained. Since $E(\mathbb{C})$ is the limit of $\gamma(\theta)$ as θ converges to zero from either side, then $\gamma(0)$ is effectively the criterion hitherto employed.

In fact, if $\theta \operatorname{var}(\mathbb{C})$ is small then

$$\gamma(\theta) \sim E(\mathbb{C}) - \frac{\theta}{4}\operatorname{var}(\mathbb{C}),$$

which illustrates in a local kind of way that the cases $\theta = 0$, $\theta > 0$ and $\theta < 0$ correspond respectively to what an economist would term *risk-neutral*, *risk-preferring* and *risk-averse* attitudes on the part of the optimizer (see DeGroot, 1970, p. 99, or Bertsekas, 1976, p. 17).

In the case $\theta > 0$ one is maximizing the expectation of a convex decreasing function of $\mathbb{C}$. For a given value of $E(\mathbb{C})$ the introduction of statistical variation in $\mathbb{C}$ improves the value of the criterion, because the occurrence of some values of $\mathbb{C}$ smaller than $E(\mathbb{C})$ is regarded as more than outweighing the necessary occurrence of some values greater than $E(\mathbb{C})$. Indeed, the mean-square criterion $E(\mathbb{C})$ is often criticized as giving excessive weight to large deviations, and so to large $\mathbb{C}$ values; criterion (1) with θ positive will then be an improvement in this respect.

In the case $\theta < 0$ one is minimizing the expectation of a convex increasing function of $\mathbb{C}$, and takes the opposite evaluation: the penalty inherent in the occurrence of values of $\mathbb{C}$ greater than $E(\mathbb{C})$ outweighs the alleviation in penalty caused by the necessary occurrence of some values less than $E(\mathbb{C})$. That is, the optimizer is particularly sensitive to the occasional large deviation.

We shall refer to the cases $\theta > 0$ and $\theta < 0$ as the *optimistic* and *pessimistic* cases respectively, as it will turn out that the form of the criterion leads one to act as though all uncertainties were likely to turn out to one's advantage or disadvantage according to whether $\theta > 0$ or $\theta < 0$. It will be a matter of interest to determine how far optimism or pessimism can be pushed before a crisis of some kind occurs (see Section 6).

We shall show that all desirable features of the risk-neutral treatment persist in the risk-sensitive case, with appropriate definition. In the case of complete state observation, optimal control is linear Markov (Section 5). In other cases a certainty equivalence property holds (Section 3), the problems of control optimization and state estimation can be decoupled (Section 4), solved separately (Sections 5 and 7) and then coupled by appeal to the certainty equivalence principle (Section 8). All the familiar results survive, including generation of state estimates by a first-order linear recursion: the Kalman filter. Indeed, the approach provides a treatment which is remarkably concise and inclusive by the standards of conventional treatments for the risk-neutral case.

We find it convenient to work with the modified versions of the Riccati recursions, which show transparently the effect of risk-sensitivity: to modify the control-power matrix $BQ^{-1}B'$ and the observation information matrix $C'M^{-1}C$ to $BQ^{-1}B' + \theta N$ and $C'M^{-1}C + \theta R$ respectively.

Repeated appeal is made to certain standard results concerning the maximization and integration of Gaussian densities. For convenience these are summarized.

Lemma 1.1. *Let $S(u, v, w)$ be a positive definite quadratic form in the components of vectors u and v, where w is a fixed parameter. Suppose this attains its minimum at $u = u^*$, $v = v^*$. Then*

$$\max_u \int \mathrm{e}^{-S(u,v,w)}\,\mathrm{d}v \propto \mathrm{e}^{-S(u^*,v^*,w)}$$

and the maximum is attained at $u = u^*$. More explicitly,

$$\int \mathrm{e}^{-S(u,v,w)}\,\mathrm{d}v = (2\pi)^{n/2}|S_{vv}|^{-1/2}\mathrm{e}^{-S(u^*,v^*,w)}\mathrm{e}^{-T/2}$$

where $n = \dim(v)$, $T = (u - u^*)'[S_{uu} - S_{uv}S_{vv}^{-1}S_{vu}](u - u^*)$ and $\begin{bmatrix} S_{uu} & S_{uv} \\ S_{vu} & S_{vv} \end{bmatrix}$ is the partitioned matrix of the quadratic form.

The material of Sections 5 and 6 is essentially due to Jacobson (1973, 1977), although he did not cast the Riccati equation in the alternative form (5.3) which reveals the transformation (5.7). The remaining material of the chapter is due to Whittle (1981b). Partial solutions of the problem for the cases $R = 0$ and $N = 0$

have been given by Speyer, Deyst and Jacobson (1974), Speyer (1976) and Kumar and van Schuppen (1981).

2. FORMULATION

For simplicity we shall assume a time-homogeneous formulation, without cross-product terms $u'Sx$ in the cost function or cross-covariance L between process and observation noise. All results remain valid (with appropriate modification) without those assumptions, which are made purely for convenience.

Thus, we assume the plant equation

$$x_{t+1} = Ax_t + Bu_t + \alpha_{t+1} + \varepsilon_{t+1} \tag{1}$$

where the sequence $\{\alpha_t\}$ is known and $\{\varepsilon_t\}$ is a white noise process with the covariance matrix N. The cost function over horizon T is

$$\mathbb{C} = \sum_{t=0}^{T-1} (x_t'Rx_t + u_t'Qu_t) + x_T'\Pi_T x_T. \tag{2}$$

The terminal cost is written $x'\Pi_T x$, rather than $x'P_T x$ or $x'P(0)x$ as previously, for reasons which will transpire in Section 4.

If state observation is not complete (i.e. if x_t is not observable at time t) then it is supposed that, conditional on W_0, x_0 is $N(\hat{x}_0, V_0)$, and that the new observable at time t is y_t, where

$$y_{t+1} = Cx_t + \eta_{t+1}. \tag{3}$$

Here $\{\eta_t\}$ is a white noise process with the covariance matrix M, and the two noise processes are supposed to be jointly Gaussian and mutually independent.

The observable W_t at time t thus consists of the process history $(\hat{x}_0, Y_t, U_{t-1})$, plus, of course, the knowledge of process structure implied by specification of relations (1) to (3) and the distributional statements on $x_0, \{\varepsilon_t, \eta_t\}$.

3. STRESS; CERTAINTY EQUIVALENCE

For later convenience let us define the quadratic forms

$$r_t = x_t'Rx_t,$$

$$q_t = u_t'Qu_t,$$

$$n_t = (x_{t+1} - Ax_t - Bu_t - \alpha_{t+1})'N^{-1}(x_{t+1} - Ax_t - Bu_t - \alpha_{t+1}),$$

$$m_t = (y_{t+1} - Cx_t)'M^{-1}(y_{t+1} - Cx_t),$$

so that the cost function $\mathbb{C}$ can be written

$$\mathbb{C} = \sum_0^{T-1} (r_t + q_t) + x_T'\Pi_T x_T$$

and the residual sum of squares (the quadratic function in the exponent of the joint density of X_T, Y_T) can be written

$$\mathbb{D} = (x_0 - \hat{x}_0)' V_0^{-1} (x_0 - \hat{x}_0) + \sum_0^{T-1} (n_t + m_t).$$

Define the *total stress* as

$$\mathbb{S} = \mathbb{C} + \frac{1}{\theta}\mathbb{D}.$$

We can regard $\mathbb{C}$ as a component of stress due to deviation of x and u from desired values; $\mathbb{D}$ as a component of stress due to discrepancy between values of the various variables x, y and u. These components are combined positively in the case $\theta > 0$ and differentially in the case $\theta < 0$.

Let us say that a variable renders $\mathbb{S}$ *extreme* if it minimizes $\mathbb{S}$ in the case $\theta > 0$, maximizes $\mathbb{S}$ in the case $\theta < 0$.

Theorem 3.1 (Certainty equivalence). *Suppose the stress $\mathbb{S}$ freely minimized with respect to $u_t, u_{t+1}, \ldots, u_{T-1}$ and freely extremized with respect to $x_0, x_1, \ldots, x_T$ and $y_{t+1}, y_{t+2}, \ldots, y_T$ for the given value of W_t. The order in which these operations are applied is irrelevant, and the value of u_t thus determined is the optimal value of control at time t.*

In the case $\theta < 0$ these calculations have a meaning only if $\mathbb{S}$ is negative definite in the extremizing variables.

This theorem generalizes the statements of certainty equivalence in the risk-neutral case, expressed by Theorems 17.3.1 and 17.4.1. As expressed by Theorem 17.4.1 the statement was that, if $\mathbb{C}$ is written as a quadratic function $Q(U_{T-1}, \xi)$ of the control sequence U_{T-1} and the process input vector ξ, then the optimal value of u_t is that obtained by minimizing $Q(U_{T-1}, E(\xi | W_t))$ with respect to $u_t, u_{t+1}, \ldots, u_{T-1}$. Note that $E(\xi | W_t)$ is the value of ξ minimizing $\mathbb{D}$ for prescribed W_t.

In the present case the $\mathbb{D}$-minimizing values of unobservables at time t are replaced by $\mathbb{S}$-extremizing values. These values are to be regarded as the estimates of current unobservables which extremize total stress. The values of $u_{t+1} \cdots u_{T-1}$ determined in the theorem, chosen to minimize total stress, are likewise current estimates of optimal future decisions.

Proof. Take the argument initially in the case $\theta > 0$. Using f to denote probability density, define

$$G_t(W_t) = f(W_t | W_0) E(e^{-\theta\mathbb{C}/2} | W_t)$$

where the expectation is calculated on the assumption that $u_t, u_{t+1}, \ldots, u_{T-1}$ have been optimally determined. Recall that 'optimality' implies the choice of π to maximize $E_\pi(\mathrm{e}^{-\theta\mathbb{C}/2}|W_0)$. The aim is then to evaluate $G_0(W_0)$, and with it the optimal control for all t. Note that

$$G_T(W_T) \propto \int \mathrm{e}^{-\theta\mathbb{S}/2}\,\mathrm{d}x_T$$

$$\propto \exp\left(-\frac{\theta}{2}\min_{x_0, x_1, \ldots, x_T} \mathbb{S}\right), \tag{1}$$

the second relation following from Lemma 1.1. It follows as a special case of the general optimality equation (10.4.3) that

$$G_t(W_t) = \max_{u_t} \int G_{t+1}(W_{t+1})\,\mathrm{d}y_{t+1}, \tag{2}$$

where the maximization is free, and that the value of u_t thus determined is the optimal control at time t.

If we set

$$G_t(W_t) = \exp\left(-\frac{\theta}{2}\phi_t(W_t)\right)$$

then it follows from (1) and Lemma 1.1 that $\phi_T(W_T)$ is a positive definite quadratic form in Y_T, U_{T-1}. Appeal to (2) for $t = T-1, T-2, \ldots$ and Lemma 1.1 then establishes recursively that $\phi_t(W_t)$ is a positive definite quadratic form in Y_t, U_{t-1} and that the recursion (2) amounts to

$$\phi_t(W_t) = \text{constant} + \min_{u_t, y_{t+1}} \phi_{t+1}(W_{t+1}), \tag{3}$$

the 'constant' being a function of t alone. From this and the effective specification of ϕ_T in (1) we deduce the theorem in the case $\theta > 0$.

In the case $\theta < 0$ the same argument holds except that u_t is minimized out in (2), y_{t+1} is maximized out in (3) and $x_0, x_1, \ldots, x_T$ is maximized out in (1). However, the integrals with respect to unobservables in (1), (2) will be convergent, and evaluated by extremization, as in (1), (3), if and only if $\mathbb{S}$ is negative definite in these unobservables.

If $\theta > 0$ then the order of the minimizations can certainly be interchanged. If $\theta < 0$ and the negative-definiteness condition is fulfilled, then we are dealing with quadratic forms possessing a saddle-point. The operations of min and max commute under these conditions. ■

4. PAST AND FUTURE STRESSES

Define the *past stress* (at time t) as

$$P_t(x_t, W_t) = \operatorname*{ext}_{x_0 \cdots x_{t-1}} \left[\theta^{-1}(x_0 - \hat{x}_0)' V^{-1}(x_0 - \hat{x}_0) + \sum_{s=0}^{t-1} (r_s + q_s + \theta^{-1}(n_s + m_s)) \right] \tag{1}$$

and the *future stress* as

$$\begin{aligned} F_t(x_t) &= \min_{u_t \cdots u_{T-1}} \operatorname*{ext}_{x_{t+1} \cdots x_T} \operatorname*{ext}_{y_{t+1} \cdots y_T} \left[\sum_{s=t}^{T-1} (r_s + q_s + \theta^{-1}(n_s + m_s)) + x_T' \Pi_T x_T \right] \\ &= \min_{u_t \cdots u_{T-1}} \operatorname*{ext}_{x_{t+1} \cdots x_T} \left[\sum_{s=t}^{T-1} (r_s + q_s + \theta^{-1} n_s) + x_T' \Pi_T x_T \right] \end{aligned} \tag{2}$$

Here 'ext' stands for 'extremum' and so must be read as 'min' or 'max' according to whether $\theta > 0$ or $\theta < 0$. These quantities can be regarded as the components of total stress attributable to past and future if the current state value x_t is specified, future actions are optimized out and variables currently unobservable are extremized out. Note that the u-minimizations in (2) are free.

The notations P and F are appropriate, in view of the association with past and future respectively. If $\theta = 0$, then the quantity $F_t(x_t)$ will indeed be identical with the minimal future cost $F(x_t, t)$ for the deterministic case with complete state observation. The reason we use Π_t to denote the matrix of $F_t(x_t)$ (seen to be a quadratic form in (5.1)) is to release the symbol P_t so that it can be used to denote past stress.

Expressions (1) and (2) obviously obey the recursions

$$P_{t+1}(x_{t+1}, W_{t+1}) = \operatorname*{ext}_{x_t} [P_t(x_t, W_t) + r_t + q_t + \theta^{-1}(m_t + n_t)], \tag{3}$$

$$F_t(x_t) = \min_{u_t} \operatorname*{ext}_{x_{t+1}} [r_t + q_t + \theta^{-1} n_t + F_{t+1}(x_{t+1})], \tag{4}$$

with the end conditions

$$P_0(x_0, W_0) = \theta^{-1}(x_0 - \hat{x}_0)' V_0^{-1}(x_0 - \hat{x}_0), \tag{5}$$

$$F_T(x_T) = x_T' \Pi_T x_T. \tag{6}$$

The forward recursion (3) revises the current state information as a new observation is incorporated; the backward recursion (4) revises the current action criteria as the horizon recedes by a single step.

In terms of these quantities Theorem 3.1 can be rewritten:

Theorem 4.1 (Alternative expression of certainty equivalence). *Let $u(x_t, t)$ be the minimizing value of u_t in (4) and let $\bar{x}_t$ be the value of x_t extremizing $P_t(x_t, W_t) + F_t(x_t)$, a function of W_t. Then the optimal value of u_t is $u(\bar{x}_t, t)$.*

This is indeed no more than a rephrasing of Theorem 3.1, as the reader can verify. However, it does bring out certainty equivalence in another way. If the state is completely observed, then the optimal control is just $u_t = u(x_t, t)$. The optimal control in the case of imperfect state observation is obtained simply by replacing x_t by its minimum-stress estimate $\bar{x}_t$, which is a version of certainty equivalence.

5. COMPLETE STATE OBSERVATION

In the case of complete state observation the optimal control rule is fully determined by solving the recursion (4.4) for future stress. For simplicity we shall consider the homogeneous case $\alpha = 0$ for the moment, and return to the deterministically disturbed case in Section 9.

We shall define the matrix $J = BQ^{-1}B'$ and shall use the convention of writing $U > V$ or $U \geq V$ if $U - V$ is positive definite or non-negative definite respectively, where U and V are symmetric matrices of equal order.

Theorem 5.1. *The future stress is quadratic in the state variable*

$$F_t(x_t) = x_t' \Pi_t x_t, \tag{1}$$

and the optimal control is linear:

$$u(x_t, t) = K_t x_t. \tag{2}$$

The matrix Π_t obeys the Riccati recursion

$$\Pi_t = R + A'(J + \theta N + \Pi_{t+1}^{-1})^{-1} A \tag{3}$$

and K_t has the evaluation

$$K_t = -Q^{-1}B'(J + \theta N + \Pi_{t+1}^{-1})^{-1} A. \tag{4}$$

For these assertions to be true in the case $\theta < 0$ it is necessary that $\Pi_s < -(\theta N)^{-1}$ $(s > t)$.

Proof. The argument is the familiar inductive one, but there are a few points that need to be brought out. Relation (1) is certainly true at T. Assuming it true at $t+1$, we have then

$$F_t(x) = \min_u \operatorname*{ext}_\varepsilon \left[x'Rx + u'Qu + \frac{1}{\theta}\varepsilon' N^{-1}\varepsilon + (Ax + Bu + \varepsilon)'\Pi_{t+1}(Ax + Bu + \varepsilon) \right]. \tag{5}$$

Consider first the case $\theta > 0$. Then one sees from (5) that it is as though one were in the conventional (i.e. deterministic, state-observed and risk-neutral) case of Chapter 5, with ε acting as an extra component of control. One can recover the treatment of this case from that of Chapter 5 by the substitutions

$$u \to \begin{bmatrix} u \\ \varepsilon \end{bmatrix}, \qquad B \to [B \; I], \qquad Q \to \begin{bmatrix} Q & \cdot \\ \cdot & (\theta N)^{-1} \end{bmatrix}.$$

With these substitutions in the treatment of Theorem 5.2.1 we obtain a recursion for Π_t and a determination of K_t. These are not the determinations (3), (4) given, however, which are generalizations rather of the alternative forms of Section 5.7. To establish these forms we write (5) rather as

$$F_t(x_t) = \min_u \operatorname*{ext}_\varepsilon \max_z \left[x'Rx + u'Qu + \frac{1}{\theta}\varepsilon' N^{-1}\varepsilon + 2z'(Ax + Bu + \varepsilon) - z'\Pi_{t+1}^{-1} z \right] \tag{6}$$

where z can in fact be identified as the conjugate variable $\frac{1}{2}\partial F_{t+1}/\partial x_{t+1}$. Writing down conditions for stationarity with respect to u, ε and z, solving these for u, ε in terms of z and then solving for z, one obtains evaluations (3), (4). These are also valid in the case $\theta < 0$, provided only that the bracketed expression in (5) is indeed then negative definite in ε, i.e. that $\Pi_{t+1} + (\theta N)^{-1} < 0$. ■

As noted from (5), in the case of $\theta > 0$ it is as though ε were available to take over some of the role of control, so that actual control u could be chosen more moderately than in the case $\theta = 0$. This is the optimistic view: process noise will be such as to direct the state somewhat in the desired direction. This fact is illustrated by formulae (3), (4), which differ from the corresponding formulae for the risk-neutral case by the substitution

$$BQ^{-1}B' \to BQ^{-1}B' + \theta N. \tag{7}$$

The matrix $J = BQ^{-1}B'$ can be regarded as a measure of 'control power'; this is now effectively increased by a multiple of the process noise power matrix.

In the case $\theta < 0$ one has the contrary situation: the extremizing ε in (5) will be such as to increase the last term, and one is effectively making the pessimistic assumption that process noise will be such as to direct the state in an undesired direction. Correspondingly, formula (7) now reflects a decrease in effective control power.

6. INSTABILITY AND BREAKDOWN

As $\theta \to \pm\infty$ the optimizer's attitude moves to an extreme of optimism or pessimism, and it is a question of how far it can move before control breaks down. In the one case, the attitude that uncertainties work to one's advantage may lead to a control which is too lax for stability. In the other, the attitude that uncertainties work to one's disadvantage may lead to a conviction of inability to control, and paralysis. These ideas are best considered in the stationary infinite-horizon limit, obtained either by holding t fixed as $T \to \infty$ or by holding $T = 0$ as $t \to -\infty$. It follows by the arguments of Section 5.5 that infinite-horizon limits exist if, for instance, $R > 0$, $J + \theta N \geq 0$ and the system is controllable. That is, Π_t has a unique-finite, positive definite limit for any finite non-negative definite Π_0.

Furthermore, consider the matrix Γ_t defined by

$$Ax_t + B\hat{u} + \hat{\varepsilon} = \Gamma_t x_t \tag{1}$$

where $\hat{u}, \hat{\varepsilon}$ are the extremal values obtaining in (5.5). Then Γ_t also has a limit, verified as

$$\Gamma = (I + (J + \theta N)\Pi)^{-1} A,$$

which is a stability matrix. That is, all its eigenvalues lie strictly inside the unit circle.

However, the actual dynamic equation that the controlled system follows in the limiting case is

$$x_{t+1} = \Gamma^* x_t + \varepsilon_{t+1}$$

where $\Gamma^* x_t = Ax_t + B\hat{u}_t = (A + BK)x_t$ or

$$\Gamma^* = (I + \theta N \Pi)\Gamma. \tag{2}$$

For stability of the controlled process one would require that Γ^* be stable. In the case $\theta > 0$ one might suspect from (2) that it need not be if θ were large enough, and this in fact turns out to be the case.

Equation (5.3) suggests that $\Pi \to R$ as $\theta \to +\infty$, and hence that $K \to 0$, $\Gamma^* \to A$. That is, that the controlled system approaches the uncontrolled system arbitrarily closely as $\theta \to +\infty$, with the certainty of ultimate instability if the uncontrolled system is unstable.

Consider now the case $\theta < 0$. Then $I + \theta N^{1/2} \Pi N^{1/2} < I$, and one certainly has stability as long as $I + \theta N^{1/2} \Pi N^{1/2} \geq 0$, but the situation is unclear if $I + \theta N^{1/2} \Pi N^{1/2} \not\geq 0$. In fact, the solution is unacceptable on other grounds in this latter case.

We know from Theorem 5.1 that it is not sufficient that $\{\Pi_t\}$ should converge, but it is also necessary that $\Pi_t < -(\theta N)^{-1}$ at all stages. If the inequality is violated for some t then the criterion function for total costs incurred from any earlier time has infinite expectation. One may say that the optimizer's pessimistic assessment of the value of ε so outweighs his knowledge of its statistical distribution that his expected cost criterion becomes infinite. Optimization then becomes pointless; the optimizer suffers a type of neurotic breakdown. One convinces oneself easily that such a breakdown will always occur if θ is sufficiently large and negative (see Whittle, 1981b).

7. INCOMPLETELY OBSERVED STATE

In the case of incompletely observed state we must also solve recursion (4.3).

Theorem 7.1. *$P_t(x_t, W_t)$ has the form*

$$P_t(x_t, W_t) = \theta^{-1}(x_t - \hat{x}_t)' V_t^{-1}(x_t - \hat{x}_t) + \cdots \tag{1}$$

where $+\cdots$ indicates terms independent of x_t. The matrix V_t and vector $\hat{x}_t$ satisfy the recursions

$$V_{t+1} = N + A(V_t^{-1} + C'M^{-1}C + \theta R)^{-1}A', \tag{2}$$

$$\hat{x}_{t+1} = A\hat{x}_t + Bu_t + \alpha_{t+1} + A(V_t^{-1} + C'M^{-1}C + \theta R)^{-1}(C'M^{-1}(y_{t+1} - C\hat{x}_t) - \theta R\hat{x}_t). \tag{3}$$

For these assertions to be true in the case $\theta < 0$ it is necessary that

$$V_s < -(\theta R)^{-1} \qquad (s \leq t).$$

In the risk-neutral case $\hat{x}_t$ would be the Bayesian estimate of state, $E(x_t \mid W_t)$, and V_t its covariance matrix. In the present case $\hat{x}_t$ is the value of state which extremizes past stress; relative to $E(x_t \mid W_t)$ it will be biassed optimistically or pessimistically according by whether $\theta > 0$ or $\theta < 0$. The matrix V_t^{-1} is a measure of the sharpness of the extreme.

In (2) we recognize one form of the familiar Riccati equation, and in (3) we recognize a version of the Kalman filter, updating the state estimate $\hat{x}_t$. This differs from the risk-neutral filter, partly in that the innovation $y_{t+1} - C\hat{x}_t$ and the matrix multiplying it have θ-dependent values, but also in the occurrence of the biassing term $-\theta R\hat{x}_t$. This is the risk-induced bias mentioned.

The similarity of structure in Theorems 5.1 and 7.1 is apparent, with, for example, the 'single-observation information matrix' $C'M^{-1}C$ replacing the 'control-power matrix' $J = BQ^{-1}B'$.

The effect of risk is to modify this matrix to $C'M^{-1}C + \theta R$, just as J was modified to $J + \theta N$. That is, it is as though the pressures exerted by cost-stress supplied information (positive or negative) additional to that supplied by observations.

Proof. Relation (1) holds at $t = 0$ by hypothesis; assume it to hold at t. Inserting expression (1) into (4.3) we see that the quadratic form holds also for $P_{t+1}(x_{t+1}, W_{t+1})$. The centring value $\hat{x}_{t+1}$ will be the value of x_{t+1} extremizing $P_{t+1}(x_{t+1}, W_{t+1})$: writing down stationary conditions for the form $P_t + r_t +$

$q_t + \theta^{-1}(m_t + n_t)$ with respect to x_{t+1} and x_t we deduce immediately the risk-modified Kalman filter relation (3).

This form has the matrix

$$\begin{bmatrix} \beta & \gamma \\ \gamma' & \delta \end{bmatrix} = \begin{bmatrix} N^{-1} & -N^{-1}A \\ -A'N^{-1} & V_t^{-1} + C'M^{-1}C + \theta R + A'NA \end{bmatrix} \tag{4}$$

when regarded as a quadratic form in x_{t+1}, x_t. We shall then have, by Lemma 1.1,

$$V_{t+1}^{-1} = \beta - \gamma\delta^{-1}\gamma'.$$

Appealing to the identity

$$(\beta - \gamma\delta^{-1}\gamma')^{-1} = \beta^{-1} + \beta^{-1}\gamma(\delta - \gamma'\alpha^{-1}\gamma)^{-1}(\gamma')^{-1}\beta^{-1},$$

obtained by comparing the alternative pivotal inversions of the partitioned matrix (4), we deduce recursion (2) for V_t. ■

8. THE COUPLING OF STATE ESTIMATION AND OPTIMAL CONTROL

A feature of the risk-neutral case is the separation theorem: in the case of complete state observation the optimal control law (5.2) is independent of process noise characteristics and in the case of incomplete state observation the state estimate $\hat{x}_t$ and its covariance are unaffected by the control policy. The two are coupled by the certainty equivalence assertion: the optimal control in the case of incomplete observation is $u_t = K_t\hat{x}_t$.

In the risk-sensitive case we do not have this complete separation. The occurrence of the term θN in the formulae of Theorem 5.1 means that process noise affects the optimal control rule; the occurrence of the term θR in the formulae of Theorem 7.1 means that state penalties affect state estimation. Nevertheless, there is still a separation, in that the problems of determining control rule and state estimate can be decoupled by separate solution of the backward recursion (4.4) and the forward recursion (4.3). The two problems can finally be coupled by application of the risk-sensitive certainty equivalence principle: Theorem 4.1.

We can now appeal to Theorems 4.1, 5.1 and 7.1 to obtain the following specific evaluation.

Theorem 8.1. *The optimal control rule is*

$$u_t = K_t\bar{x}_t$$

where

$$\bar{x}_t = (I + \theta V_t\Pi_t)^{-1}\hat{x}_t \tag{1}$$

and $\Pi_t, K_t, V_t, \hat{x}_t$ *have the evaluations asserted in Theorems* 5.1 *and* 7.1.

The evaluation (1) of $\bar{x}_t$ will be modified in the case $\alpha \neq 0$ (see equation (9.2)).

9. TRACKING AND BUFFERING

The results of Section 5 were obtained under the supposition $\alpha = 0$. We shall now restore this term, so that the system is effectively buffering an injected deterministic disturbance α_{t+1}. The tracking of a desired $\{x_t, u_t\}$ profile can be reduced to this case (see Section 5.4).

The assertions of Theorem 5.1 will now hold with

$$F_t(x_t) = (x_t - x_t^*)'\Pi_t(x_t - x_t^*) + \cdots,$$

where Π_t is as before and $+\cdots$ indicates terms independent of x_t. The term x_t^* represents the value of x_t extremizing future stress, and so the ideal value of x_t, and has to be calculated. It is a quantity dual to the corresponding quantity $\hat{x}_t$ in past stress, the state estimate. One might think that one could obtain a recursion for it dual to that obeyed by $\hat{x}_t$, the Kalman filter (7.3), and just as easily. This is not quite true, because relation (4.4) involves two extremizing operations, whereas relation (4.3) involves but one. What one finds is the following.

Theorem 9.1. *Define* $\sigma_t = -\Pi_t x_t^*$. *Then*

$$\sigma_t = \Gamma_t'(\Pi_{t+1}\alpha_{t+1} + \sigma_{t+1}) \tag{1}$$

where Γ_t *is the matrix determined in* (6.1):

$$\Gamma_t = [I + (J + \theta N)\Pi_{t+1}]^{-1}A.$$

Relation (1) generalizes (5.4.7). We now see this relation as the backward dual of the Kalman filter, even though it is not immediately recognizable as such.

Proof. The analogue of (5.6) will now be

$$F_t(x_t) = \min_u \operatorname*{ext}_\varepsilon \max_z \left[x'Rx + u'Qu + \frac{1}{\theta}\varepsilon' N^{-1}\varepsilon \right.$$
$$\left. + 2z'(Ax + Bu + \alpha_{t+1} + \varepsilon - x_{t+1}^*) - z'\Pi_{t+1}^{-1}z \right].$$

The value of x extremizing this is x_t^*. If one writes down the stationarity conditions of the bracket with respect to x, u, ε and z, uses those for u and ε to eliminate these variables and then solves the consequent pair of equations for x, one deduces that

$$Rx_t^* + A'(J + \theta N + \Pi_{t+1}^{-1})^{-1}(Ax_t^* + \alpha_{t+1} - x_{t+1}^*) = 0.$$

This reduces immediately to (1). ■

The solution (8.1) for $\bar{x}_t$ now becomes

$$\begin{aligned} \bar{x}_t &= (I + \theta V_t\Pi_t)^{-1}(\hat{x}_t + \theta V_t\Pi_t x_t^*) \\ &= (I + \theta V_t\Pi_t)^{-1}(\hat{x}_t - \theta V_t\sigma_t). \end{aligned} \tag{2}$$

CHAPTER 20

LQG homing

One of the items in the specification of the LQG model (Sections 5.2 and 12.1) was that termination occurred at a prescribed time T, with a terminal cost quadratic in x. It turns out that, for an interesting class of cases in continuous time, this demand can be relaxed, and one can allow termination on first entry into a general (x, t) set D, with a general terminal loss function $K(x, t)$. The problem then becomes one of reaching D at minimal cost, with linear/Gaussian dynamics and quadratic costs on the way—hence the term 'LQG homing'.

The problem is also interesting as one in which state deviations are not costed explicitly except at termination. Nevertheless, these terminal costs propagate themselves back into the free-flight region, in that the optimal guidance rules are such as if a cost were indeed incurred when the trajectory in state space deviated from some ideal path. However, this ideal path is derived from the postulates of the model rather than being itself postulated.

Most of the material of this chapter is taken from Whittle and Gait (1969). It might have been included in Chapter 38 (control of diffusion processes) but is justified here as an LQG variant of independent interest.

The solution derived in the stochastic case turns out to be an interesting analogue of the solution derived via the Pontryagin maximum principle in the deterministic case. We thus begin with a treatment of continuous time LQ problems by the maximum principle.

1. CONTINUOUS TIME LQ PROCESSES AND THE MAXIMUM PRINCIPLE

The material of Chapter 5, for deterministic LQ processes, has an obvious analogue in the case of continuous time. For simplicity, we give the analogue only of Theorem 5.2.1, and in the case where the cross-term $u'Sx$ is absent from the cost function.

So, suppose $x \in R^n$, $u \in R^n$, the plant equation is

$$\dot{x} = Ax + Bu \tag{1}$$

and the cost function is

$$\frac{1}{2}\int^{T} (x'Rx + u'Qu)\mathrm{d}t + K(x(T)) \tag{2}$$

where R and Q are positive definite and A, B, R, Q are conceivably t-dependent.

Theorem 1.1. *Suppose that, in addition to the above assumptions, the stopping time T is prescribed and the terminal cost function is*

$$K(x) = \tfrac{1}{2}x'P(T)x. \tag{3}$$

Then the minimal cost incurred from time t with $x(t) = x$ is

$$F(x, t) = \tfrac{1}{2}x'P(t)x \tag{4}$$

where P satisfies the matrix Riccati equation

$$\dot{P} + R + A'P + PA - PBQ^{-1}B'P = 0. \tag{5}$$

The optimal control is

$$u = -Q^{-1}B'Px, \tag{6}$$

all quantities being evaluated at time t.

An analogue of the recursive proof of Theorem 5.2.1 is evident (see Exercise 1). We shall rather use the maximum principle, as in Section 6.5, because this treatment also transfers largely to the case of more general terminal conditions. The introduction of the factor $\frac{1}{2}$ into the quadratic costs is then convenient.

Proof. The system has the Hamiltonian

$$H = \tfrac{1}{2}(x'Rx + u'Qu) + p(Ax + Bu). \tag{7}$$

The equation for the dual variables is

$$\dot{p} = -x'R - pA \tag{8}$$

and, since u minimizes H,

$$u = Q^{-1}B'p'. \tag{9}$$

The plant equation (1) can thus be written

$$\dot{x} = Ax - BQ^{-1}B'p'. \tag{10}$$

It follows from the assumed terminal conditions that

$$p = F_x = x'P \tag{11}$$

at T. From solution of the pair of linear differential equations (8), (10) subject to condition (11) at T it follows that (11) holds for $t \leq T$ for some $P(t)$ (cf. Section 6.5). Differentiating relation (11) with respect to time, and substituting for $\dot{p}$, $\dot{x}$ and p from (8), (10) and (11) respectively, one deduces that $x\,[\;]x = 0$ where $[\;]$ represents the left-hand member of (5). Since x is arbitrary, we deduce that P satisfies (5).

Relations (9), (11) imply (6). Relation (11) implies (4), since $F(0, t) = 0$ and the P generated by (5) is symmetric. ■

The point of using this derivation is that relations (7) to (10) remain true whatever the terminal conditions. That is, under assumptions (1) and (2), relations (7) to (10) hold for any terminal conditions, provided only that the maximum principle is valid. This conclusion yields valuable results for problems which are LQG except at termination (see Section 2); these results have stochastic analogues (see Section 4).

Exercise

1. Show that under assumptions (1), (2) the optimality equation takes the form

$$\min_u \left[\tfrac{1}{2}(x'Rx + u'Qu) + F_x(Ax + Bu) + F_t\right] = 0$$

where F_x is the row vector $\partial F/\partial x$, etc. Show that this implies the solution (4) to (6) for F and u under the terminal conditions of Theorem 1.1.

2. PROBLEMS WHICH ARE LQ EXCEPT AT TERMINATION

It is convenient to denote the pair (x, t) jointly by ξ or by $\xi(t)$ if the interest is in a particular trajectory $(x(t), t)$. We shall consider the general termination rule: termination occurs on the first entry to a prescribed ξ-set D, with the terminal cost $K(\bar{\xi})$, where $\bar{\xi}$ is the termination coordinate. In general, terminal values will be distinguished by a bar (e.g. $\bar{\xi}$, $\bar{x}$, $\bar{t}$, $\bar{p}$) and we shall set $s = \bar{t} - t$.

We shall assume that the problem has the LQ form (1.1), (1.2) up to termination, and make the further specializations that $R = 0$ and all matrices are constant in time. The first additional assumption, that state costs are incurred first at termination, implies a fairly crucial simplification. The second assumption, of time-homogeneity before termination, is made only for convenience.

Theorem 2.1. *Under the above assumptions, the minimal cost incurred from a position (x, t) is*

$$F(x, t) = \inf_{\xi \in D} \left[K(\bar{\xi}) + \tfrac{1}{2}(\bar{x} - e^{As}x)'\Phi(s)^{-1}(\bar{x} - e^{As}x)\right] \tag{1}$$

and the optimal control is

$$u = Q^{-1}B'e^{A's}\Phi(s)^{-1}(\bar{x} - e^{As}x) \tag{2}$$

provided that the orbit thus generated does not enter D before the time $\bar{t}$ determined by the ξ-minimization in (1). Here $s = \bar{t} - t$ and $\Phi(s)$ is the matrix

$$\Phi(s) = \int_0^s e^{A\tau}BQ^{-1}B'e^{A'\tau}\,d\tau. \tag{3}$$

Proof. Relation (1.8) now reduces to

$$\dot{p} = -pA \tag{4}$$

whence

$$p = \bar{p}e^{As} \tag{5}$$

and, by (1.9),

$$u = -Q^{-1}B'e^{A's}\bar{p}'. \tag{6}$$

From (6) and the plant equation (1.1) we deduce that

$$\bar{x} - e^{As}x = -\Phi(s)\bar{p}', \tag{7}$$

$$\tfrac{1}{2}\int_t^{\bar{t}} u'Qu\,dt = \tfrac{1}{2}\bar{p}\Phi(s)\bar{p}'. \tag{8}$$

Substituting for $\bar{p}$ from (7) into (6) and (8) we deduce respectively expression (2) for the optimal control at time t and the bracketed expression in (1) for the total cost incurred from time t. The terminal value $\bar{\xi}$ is to be chosen in D to minimize this total cost, whence one deduces expression (1). ■

Should the orbit determined by Theorem 2.1 not be such that $\bar{t}$ is in fact the time of first entry to D then one is in the situation encountered previously in Sections 3.4 and 7.3. In such a case the optimal trajectory breaks into a number of sections of free LQ flight, all but the last of which terminate in a 'grazing encounter' with D, at which the value of p is reset (see the example at the end of the section and Exercise 1).

As a very simple example, consider the scalar case with the plant equation $\dot{x} = u$, termination region $x \le 0$ and termination cost $K(t)$. One could imagine this to be a grossly oversimplified representation of the landing-path of an aeroplane at a height x above the ground, moving with constant (unit) horizontal velocity, so that $K(t)$ is the cost if one lands at time t (= horizontal coordinate t). The function $K(t)$ will be minimal at the ideal landing point. The model is less than rudimentary, in that it treats the plane as a point mass, incorporates neither linear nor angular momenta and neglects variations in horizontal velocity and all aerodynamic characteristics of the plane, etc. Nevertheless, it has some interest.

For this model we have $A = 0$, $\Phi = Q^{-1}s$, $\bar{x} = 0$, so that the control rule (2) becomes simply

$$u = -\frac{x}{s} = -\frac{x}{\bar{t} - t}. \tag{9}$$

Here $\bar{t}$ must be constant for a given orbit and is the minimizing value defined in

$$F(\xi) = \inf_{\bar{t} > t}\left[K(\bar{t}) + \frac{Qx^2}{2(\bar{t} - t)}\right]. \tag{10}$$

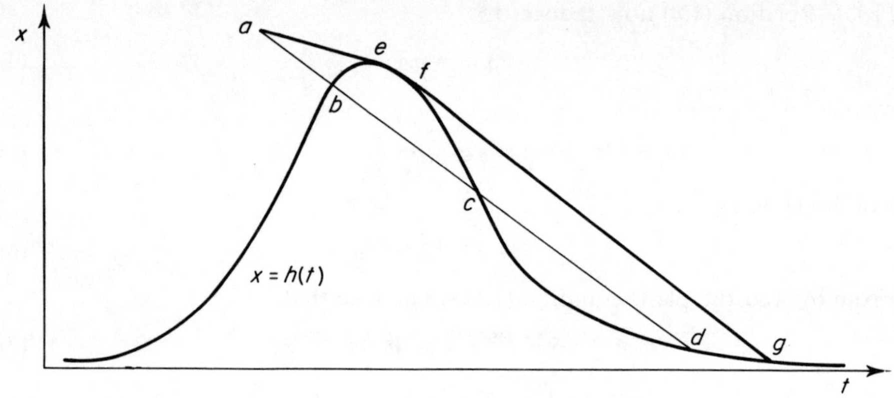

Figure **2.1** Landing with a hill in the way

It follows from (9) and the plant equation that

$$\frac{dx}{x} = -\frac{dt}{\bar{t} - t},$$

so the optimal trajectory is a straight line path from (x, t) to $(0, \bar{t})$. Indeed, since the value $\bar{t}$ minimizing (10) is constant along an orbit, we deduce that $x/(\bar{t} - t)$ is constant along an orbit. With a straight line path, $(0, \bar{t})$ is indeed the coordinate of first entry into the terminal region $x \leq 0$.

Suppose, however, that the topography of the ground as it is traversed is represented by a function $x = h(t)$, so that D is now characterized by $x \leq h(t)$. Suppose that the plane is in position a in Figure 2.1, that the topography h includes the hill indicated and that the ideal landing point is on the other side of the hill. A straight line path such as *abcd* might be optimal, except that it is impossible in that one is asking the aircraft to bore through the hill; b rather than d is the coordinate of first entry to D. The optimal feasible path will be rather of the form *aefg*: two straight line segments *ac* and *fg* separated by a terrain-hugging section *ef*.

Exercises

1. Modify the last example to allow for vertical momentum. More specifically, assume the state variable to consist of vertical coordinates x and vertical velocity v, obeying the plant equation $\dot{x} = v, \dot{v} = u$. Assume the instantaneous cost rate $\frac{1}{2}Qu^2$ and the terminal cost $K(v, t)$ in the termination set $x \leq 0$.

   Show that the optimal control is

$$u = -\frac{6}{s^2}(x + vs) + \frac{2}{s}(v - \bar{v})$$

on an optimal trajectory and that x is cubic in s on such a trajectory. Show that

$$F(\xi) = \inf_{\xi \in D} \left\{ K(v,t) + 6Q\left[\frac{v^2 + v\bar{v} + \bar{v}^2}{3s} + \frac{x(v+\bar{v})}{s^2} + \frac{x^2}{s^3}\right]\right\}$$

if the trajectory consists of a single free-flight section; show that it may not do so.

2. Develop the theory of this section without the assumption of time-homogeneity.

3. PROBLEMS WHICH ARE LQG EXCEPT AT TERMINATION; A SIMPLE EXAMPLE

The results of the last section have analogues in the stochastic case, under appropriate further assumptions. We shall assume process noise only, and assume this to be white noise (abbreviated WN). More specifically, we shall say that a process $\{\varepsilon(t)\}$ is WN(0, N) if it is Gaussian, $E(\varepsilon) = 0$ and

$$E[\textstyle\int \alpha(t)'\varepsilon(t)\,dt]^2 = \int \alpha(t)'N\alpha(t)\,dt$$

for appropriately regular vector functions α.

In this section we consider the simplest interesting example: the scalar case with the plant equation

$$\dot{x} = u + \varepsilon$$

and instantaneous cost rate

$$g = \tfrac{1}{2}(L + Qu^2).$$

One is thus effectively costing time and control effort to termination. Assume termination conditions to be time-homogeneous. Since motion is on the real line and continuous, one effectively specifies the general case by assuming the continuation region to be an interval (b_1, b_2) containing the starting point x and by prescribing the cost of terminating at $x = b_i$ as K_i $(i = 1, 2)$.

The optimality equation is

$$\inf_u [\tfrac{1}{2}(L + Qu^2) + uF_x + \tfrac{1}{2}NF_{xx}] = 0 \tag{1}$$

(see Exercise 1) with the boundary conditions

$$F(b_i) = K_i \qquad (i = 1, 2). \tag{2}$$

It follows from (1) that F and the optimal u satisfy

$$u = -Q^{-1}F_x, \tag{3}$$

$$L - Q^{-1}F_x^2 + NF_{xx} = 0. \tag{4}$$

In the deterministic case $N = 0$ one then has

$$F_x = \pm \sqrt{(QL)} \tag{5}$$

and

$$u = \pm \sqrt{\left(\frac{L}{Q}\right)}. \tag{6}$$

Which of these options prevails depends upon which boundary point one seeks. If one aims for b_2 one travels for a distance $b_2 - x$ at velocity $\sqrt{(L/Q)}$, and so for a time $(b_2 - x)\sqrt{(Q/L)}$, and incurs a terminal cost K_2. This makes a total cost of $(b_2 - x)\sqrt{(QL)} + K_2$. There will be a corresponding cost for termination at b_1, and, from a given starting point x, one will seek the terminal point which offers smaller loss. Thus

$$F(x) = \min\left[(x - b_1)\sqrt{(QL)} + K_1, (b_2 - x)\sqrt{(QL)} + K_2\right] \tag{7}$$

There are two cases, according to whether there is or is not a point in (b_1, b_2) for which the two courses offer equal terminal loss. In the first case, there is a change in form of F at the break-point and a discontinuity in F_x at that point. In the

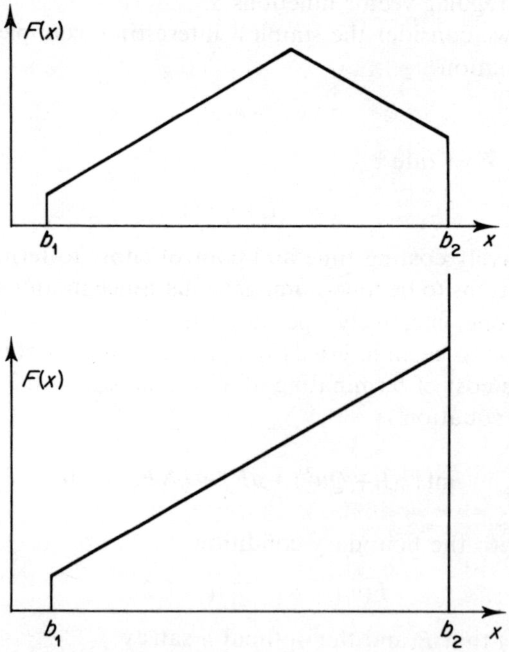

Figure 3.1 The minimal cost function in cases where the optimal target point changes in the interior of the interval and where it does not

second, one terminal point (say b_1) is more favourable from any starting point in (b_1, b_2), and one will always aim for it. This means, however, that $F(b_2 -)$ will be smaller than $F(b_2) = K_2$, and there will be a discontinuity in F itself at the boundary point b_2.

Both types of discontinuity are innocuous in that in both cases F and F_x are continuous along an optimal trajectory. The discontinuity in F_x in the first case marks the break between two different optimal orbits. The discontinuity in F in the second case marks the break between an orbit which is optimal, and is available if $x < b_2$ and the region $x \geq b_2$ in which it is no longer available.

Note also that these two types of behaviour resolve the problem of matching a differential equation (4), which is first order in the case $N = 0$, to boundary conditions at two points (2). In effect, the boundary which is inaccessible under an optimal policy supplies no boundary condition.

In the stochastic case $N > 0$ equation (4) is of second order, and one should be able to match the solution to both boundary values without discontinuity. This is related to the fact that one may aim for a particular boundary point, but process noise may carry one to the other.

To solve (4) in the stochastic case, transform from F to

$$\phi(x) = e^{-F(x)/QN}, \tag{8}$$

in terms of which equation (4) takes the linear form

$$QN^2 \phi_{xx} - L\phi = 0. \tag{9}$$

This has the general solution

$$\phi = c_1 \exp\left[\frac{x}{N}\sqrt{\left(\frac{L}{Q}\right)}\right] + c_2 \exp\left[-\frac{x}{N}\sqrt{\left(\frac{L}{Q}\right)}\right]. \tag{10}$$

The constants c can be chosen so that both boundary conditions (2) are met, and F is thus determined by a single analytic expression in $[b_1, b_2]$. This evaluation of F will be greater and smoother than that for the deterministic case; the graphs corresponding to Figure 3.1 will have the appearance indicated in Figure 3.2.

In the first case, the change in target boundary point will occur at the x value for which $F_x = 0$; see (3). However, F is smooth, because random variation may well drive one to either of the two possible ultimate terminations.

In the second case, F_x increases as one approaches the unwelcome termination point $x = b_2$, reflecting one's increasing efforts to avoid it. Avoidance is never certain, however, as is reflected in the fact that F is now continuous also at this boundary. This might be termed the 'fly-paper' effect, if one regards the moving point as a fly trying to avoid the fate of becoming stuck on the fly-paper at $x = b_2$. A deterministic fly, whose path is fully under its own control, can approach arbitrarily closely to the fly-paper with impunity, knowing he can avoid entrapment. Hence the discontinuity in F at b_2, reflecting the difference between being close and being stuck. A stochastic fly cannot guarantee his escape; the

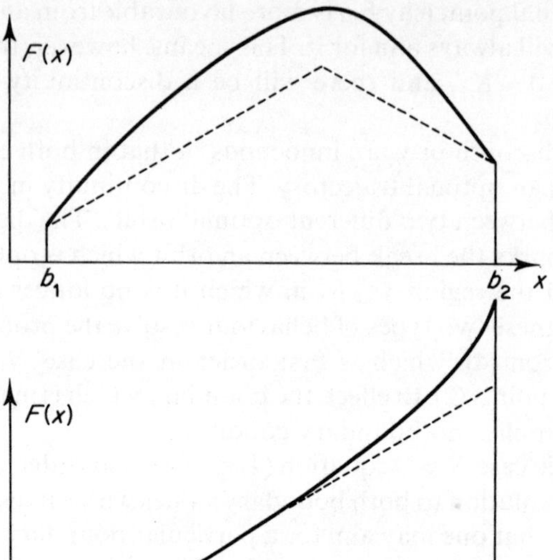

Figure 3.2 The analogue of Figure 3.1 in the stochastic case. The lower diagram represents the 'fly-paper effect'

nearer he is to the paper, the more certain it is that he will be carried on to it. Hence the continuity of F at b_2. This also explains why the fly tries so much harder in the stochastic case than in the deterministic case to escape the neighbourhood of the fly-paper (as evidenced by the greater magnitude in the stochastic case of F_x, and so of $-u$, for x near b_2). One may say that the penalty of ending on the fly-paper 'propagates' into the free-flight region in the stochastic case, causing the fly to take avoiding action while still at a distance from the paper.

Exercises

1. See any text as diffusion processes, e.g. Cox and Miller (1965, pp. 208, 246); Prabhu (1965, p. 87) or Wong (1971, p. 180). The heuristic argument is that the assumptions on ε imply that, if $\mathrm{d}\zeta$ is the integral of ε over time $\mathrm{d}t$, then $E(\mathrm{d}\zeta) = 0$, $E(\mathrm{d}\zeta\,\mathrm{d}\zeta') = N\,\mathrm{d}t$ and moments of higher order than the second are $o(\mathrm{d}t)$.
2. Consider the problem of this section in the case where the terminal set consists of a single point. Show that both u and the optimal F are independent of N, at least if L and Q are not both zero.

4. PROBLEMS WHICH ARE LQG EXCEPT AT TERMINATION; A GENERAL CLASS OF CASES

We shall show that Theorem 2.1, concerning processes which are LQ except at termination and involve state only at termination, can be extended to some stochastic versions of the problem. The example of the last section was such a case.

The assumptions of Section 2 can in fact be relaxed in some ways for the result we shall prove. We shall assume that the plant equation is

$$\dot{x} = a + Bu + \varepsilon \tag{1}$$

where a is an arbitrary n-vector function of $\xi = (x, t)$ and ε is WN(0, N). We also allow the matrices B and N to be ξ-dependent. The aim is to minimize the expected value of the cost function

$$C = \tfrac{1}{2}\int^{T} u'Qu\,\mathrm{d}t + K(\xi(T)) \tag{2}$$

where the matrix Q may also be ξ-dependent and T is the moment of first entry to a prescribed ξ-set D.

Theorem 4.1. *Suppose that in addition to the above conditions the relation*

$$N = \alpha B Q^{-1} B' \tag{3}$$

holds for some scalar α. Suppose also that the uncontrolled process

$$\dot{x} = a + \varepsilon \tag{4}$$

starting from a point $\xi = (x, t)$ is certain to terminate in D. Then

$$\mathrm{e}^{-F(\xi)/\alpha} = E_{\xi}(\mathrm{e}^{-K(\bar{\xi})/\alpha}) \tag{5}$$

where E_{ξ} is an expectation over the coordinate $\bar{\xi}$ of first entry of the uncontrolled process (4) *into D, conditional on starting from ξ.*

Proof. The optimality equation for the decision process is

$$\inf_u[\tfrac{1}{2}u'Qu + F_t + F_x(a + Bu) + \tfrac{1}{2}\mathrm{tr}(NF_{xx})] = 0 \tag{6}$$

(see Exercise 3.1) with the boundary condition

$$F(\xi) = K(\xi) \qquad (\xi \in D). \tag{7}$$

It follows from (4) that

$$u = -Q^{-1}B'F'_x \tag{8}$$

and

$$F_t + F_x a - \tfrac{1}{2}F_x BQ^{-1}B'F'_x + \tfrac{1}{2}\mathrm{tr}(NF_{xx}) = 0. \tag{9}$$

Now, if (3) holds, then the transformation

$$\phi(\xi) = e^{-F(\xi)/\alpha} \tag{10}$$

linearizes (9) to

$$\phi_t + \phi_x a + \tfrac{1}{2}\mathrm{tr}(N\phi_{xx}) = 0. \tag{11}$$

This is subject to the boundary condition

$$\phi(\xi) = e^{-K(\xi)/\alpha} \qquad (\xi \in D). \tag{12}$$

Now, if the right-hand member of (5) were denoted $\phi(\xi)$, then (11), (12) are just the equation and boundary condition that the characterization of the theorem would require of it. The solution of (11), (12) is unique, however, in view of the condition of certain absorption. We can thus identify the two characterizations of ϕ, whence (5) follows. ■

The linearizing transformation (10) is just that employed for the example of the last section. The need for a relation (3) is interesting, in view of the connection between $BQ^{-1}B'$ and N that emerged in Section 19.5. The condition may be said to require that control-power and noise-power matrices be proportional.

Relation (5) is not obviously an analogue of relation (2.1). Indeed, it constitutes a solution for $F(\xi)$ only in that it relates F to the expected terminal cost ϕ for the uncontrolled process (4). One must essentially determine the distribution of the coordinate $\bar{\xi}$ of first passage into D for the process (4). This first-passage problem is itself usually difficult to solve. However, in cases where it can be solved, and the problem is analogous to that of Section 2 in that $a(\xi) = Ax$ and all matrices are constant, one recognizes expression (5) for $F(\xi)$ as a stochastic analogue of (2.1).

For example, consider the stochastic analogue of the scalar landing problem treated in the latter part of Section 2, with the plant equation $\dot{x} = u + \varepsilon$. Ths first-passage problem for the uncontrolled process $\dot{x} = \varepsilon$ has a well-known solution in this case: if s is the time of first passage from coordinate x to $x = 0$, then this has the probability density

$$f(s) = \sqrt{\left(\frac{x^2}{2\pi N}\right)} e^{-x^2/2Ns} s^{-3/2} \qquad (s \geq 0) \tag{13}$$

(see Cox and Miller, 1965, p. 221). Solution (5) becomes

$$e^{-F/NQ} = \sqrt{\left(\frac{x^2}{2\pi N}\right)} \int_0^\infty \exp\left[-\frac{K(t+s) + Qx^2/2s}{NQ}\right] \frac{ds}{s^{3/2}}. \tag{14}$$

The numerator in the exponent of (14) is simply the bracketed expression in (2.10), since $s = \bar{t} - t$.

If N is decreased indefinitely, so that one approaches the deterministic case, then solution (14) will converge exactly to expression (2.10), the operations of integration and exponentiation in (14) yielding ultimately the minimizing operation of (2.10).

This example and ideas associated with it have a good deal of independent interest, and are discussed further in the next two sections.

Exercises

1. Suppose that the instantaneous cost rate has the form

$$g = \tfrac{1}{2}(L + u'Qu).$$

Show that relation (11) then becomes

$$\frac{L}{2\alpha}\phi = \phi_t + \phi_x a + \tfrac{1}{2}\mathrm{tr}(N\phi_{xx}). \tag{15}$$

2. Consider an LQG version of the path-optimization problem of Section 3.4, with the plant equation $\dot{x} = u + \varepsilon$, where ε is $\mathrm{WN}(0, \alpha I)$. The control variable u is unconstrained, but there is an in-flight cost rate $\tfrac{1}{2}(L + |u|^2)$, and terminal conditions are time-invariant. The whole problem is then time-invariant, and F is a function of x.

   Show that, if $\alpha = 0$, then F obeys the eikonal equation

$$|\nabla F|^2 = L.$$

   Show that, if $\alpha > 0$, then $\phi = \mathrm{e}^{-F/\alpha}$ obeys the wave equation

$$\nabla^2 \phi = \frac{L}{\alpha^2}\phi.$$

5. IDEAL PATHS

The approach of a craft to a landing or rendezvous can be viewed as a homing problem, susceptible to optimization. There are treatments of approach-control optimization in which an ideal approach path is *prescribed*. However, the ideal approach path should emerge from the model itself, taking into account dynamics, disturbances, control costs and costs due to 'landings' which are unintended and unfortunate. The costs defined on the termination region should propagate into the free-flight region, and so define the desirable path in flight.

In fact, suppose one has determined the minimal cost function $F(x, t)$ for a perfectly general optimization problem. One can define the 'ideal trajectory' in the state space as the path $\tilde{x}(t)$ such that

$$F(\tilde{x}(t), t) = \min_x F(x, t) \tag{1}$$

with an obvious ε-version if the minimum does not exist. If the problem is time-invariant, so that F is a function of x alone, then some component of state can be taken as the independent variable, in terms of which the points on a trajectory can be parametrized (e.g. absolute distance to the target point).

Under certain circumstances optimal control can appear as an attempt to converge on to the ideal path thus defined. If F_{xx} exists in the neighbourhood of the path, then, for small deviations $x(t) - \tilde{x}(t)$ one has

$$F(x(t), t) \sim F(\tilde{x}(t), t) + \tfrac{1}{2}(x(t) - \tilde{x}(t)' H(t)(x(t) - \tilde{x}(t)) \tag{2}$$

where

$$H(t) = F_{xx}(\tilde{x}(t), t)$$

Suppose that the problem is LQ insofar as dynamics are linear in u and in-flight costs are quadratic in u. That is, the plant equation is

$$\dot{x} = Bu + \cdots$$

and the optimality equation is

$$\min_u [\tfrac{1}{2}u'Qu + F_x Bu] + \cdots = 0, \tag{3}$$

where $+\cdots$ indicates terms independent of u, although the matrices B and Q may be (x, t)-dependent. The optimal control is then

$$u = -Q^{-1}B'F_x' \tag{4}$$

and, by (2), this is

$$u \sim -(Q^{-1}B'H)(x(t) - \tilde{x}(t)), \tag{5}$$

which corresponds exactly to an attempt to draw x to the ideal trajectory. Essentially, we have approximated the original problem by a time-dependent LQG problem, but with the ideal approach path and effective costs of deviation from it *deduced* from the non-LQG aspects of the original problem.

Note that, although control u is zero on the ideal path (at least under the LQ assumptions on u), the representative point may well not automatically stay on the path (i.e. the path may not be a 'ballistic orbit'). Even in the absence of stochastic disturbances, the uncontrolled plant dynamics will in general shift x off the path, and a correcting control is needed if the path is to be held.

6. THE SCALAR LANDING PROBLEM

We shall illustrate the ideas of Section 5 by completion of the analysis of the stochastic scalar landing problem, formally solved by (4.14). Define the integral

$$I(\psi(s)) = \int_0^\infty \psi(s) \exp\left[-\frac{M(s)}{NQ}\right] s^{-3/2} ds. \tag{1}$$

Here $M(s)$ is the numerator in the exponent of the stochastic solution (4.14); this occurs also in the deterministic solution (2.10). In the limit of small N, when the problem is almost deterministic, the integral will have the approximate evaluation

$$I(\psi(s)) \sim \sqrt{\left(\frac{2\pi NQ}{\bar{s}^3 M''(\bar{s})}\right)} \mathrm{e}^{-M(\bar{s})/NQ}\psi(\bar{s}) \tag{2}$$

where $\bar{s}$ is the positive value of s minimizing $M(s)$. This is the usual Laplace evaluation (see de Bruijn, 1958, p. 60), obtained by approximating $M(s)$ quadratically in the neighbourhood of $\bar{s}$. The quantity $\bar{s}$ should be regarded as an approximate evaluation of time-to-go.

It is exactly true that the optimal control is

$$u = -Q^{-1}F_x = \frac{N}{x} - x\frac{I[s^{-1}]}{I[1]} \tag{3}$$

(see (4.8), (4.14)). Inserting the approximation (2) we deduce that

$$u \sim \frac{N}{x} - \frac{x}{\bar{s}} \tag{4}$$

and that the ideal approach path approximately satisfies

$$x^2 = N\bar{s} \tag{5}$$

Control (4) is just the deterministic control (2.9) with the addition of an *upward* correction N/x which comes into operation at low heights. This correction is another manifestation of the 'fly-paper' effect of Section 3. In the deterministic case one made a linear approach to the landing point, whatever the initial height. In the stochastic case, one will gain height if one is low and distant from the target, in an effort to avoid being carried by random disturbances to premature landing. One would expect that this height-gaining correction would cease to operate as one approached (or if one had overshot) the ideal landing point. In such a case approximation (2) in fact ceases to be valid, and an appropriate approximation yields

$$u \sim -\sqrt{\left[\frac{2K'(t)}{Q}\right]} \tag{6}$$

instead of (4). The calculation is sketched in Exercise 1.

To complete the specification (5) of the ideal approach path, we must evaluate $\bar{s}$ in terms of t. The determining relation is

$$K'(t + \bar{s}) = \frac{Qx^2}{2\bar{s}^2}, \tag{7}$$

so, from (5) and (7),

$$\bar{s}K'(t+\bar{s}) \sim \frac{NQ}{2} \tag{8}$$

for the termination time $t + \bar{s}$ of an ideal approach path. If K is minimal at t_0 (so that t_0 is the ideal landing point) then (8) has the approximate solution

$$\bar{s} \sim t_0 - t, \tag{9}$$

the term neglected being of order $N/(t_0 - t)$ (so that either $t_0 - t$ should be large or $t_0 - t$ should be moderate, and positive, but N small).

From (5), (9) we deduce that

$$x \sim \surd[N(t_0 - t)] \tag{10}$$

on an optimal approach path. That is, the ideal landing point is also the termination point of an ideal trajectory, along which one descends on a $\surd[N(t_0 - t)]$ curve. This again illustrates the fly-paper effect. If $N = 0$ one will come in at zero height—the most economical trajectory, and not dangerous if one has perfect control. If $N > 0$ then one does *not* have perfect control, and so approaches cautiously.

With a more realistic model, incorporating vertical momentum, the approach path (10) will of course not be ideal—the ideal path will be almost tangential to the ground at landing.

Exercise

1. Suppose $K(t)$ is minimal at t_0, and increasing thereafter, so that t_0 would be the ideal landing point if control costs were no consideration. If this point has been overshot, so that $t > t_0$, then it is solely the presence of the term x^2/s in $M(s)$ that gives $M(s)$ a stationary minimum in $s \geq 0$. If x is itself of order N, then approximation (2) will be invalid, and a better one is to set

$$\exp\left[-\frac{M(s)}{NQ}\right] \sim \exp\left[-\frac{M(0) + sM'(0)}{NQ}\right]$$

whence

$$F(\xi) \sim K(t) + x\surd[2QK'(t)].$$

Show that this formula is exact if $K(t)$ is linear for $t > t_0$ and that it leads to the approximate control (6).

CHAPTER 21

LQG control by spectral methods

1. RECURSIVE AND STATIONARY FORMULATIONS

It was realized quite early that the problems of optimizing the control of LQG systems are relatively tractable, but the realization came about in two different ways.

The dynamic programming approach of Chapter 5 and Section 12.1 is one. This depends entirely upon the notions of state variable and of simple recursion. The approach does not demand stationarity of the relevant processes, even if the stationary case is one of special importance. The principal early contributor in this area was Kalman (1960; and with Bucy, 1961).

Another approach is that associated with the optimal filtration of stationary processes. That is, the determination of linear functions of observables which will best approximate some unobservable of interest in mean square. This was the approach particularly developed by Wiener (1949) for signal prediction, signal extraction, etc. However, it was realized independently by a number of authors (Newton, 1952; Newton, Gould and Kaiser, 1957; Holt, Modigliani, Muth and Simon, 1960; Whittle, 1963) that the same techniques could be extended to the optimal linear control of linear systems, if the optimality criterion were quadratic. To postulate stationarity is essential if one is to be able to use the powerful Fourier techniques associated with the stationary case. However, only the variables actually occurring in the criterion function need be stationary, usually control variables and deviations (see Sections 5 and 6).

We shall refer to these two approaches as the 'recursive' and 'stationary' approaches respectively. The 'stationary' approach is more special in that it does demand stationarity, less special in that it does not demand Markov behaviour, but can in principle deal with very general processes and operations. The virtues of recursivity do emerge also in the stationary approach, however, in that cases where the processes have spectral densities which are rational of low degree will yield optimal filters whose frequency transfer functions are also rational of low degree. That is, processes generated by low-order stochastic recursions have optimal controls which are also generated by low-order recursions. A somewhat related feature is the usefulness of an 'innovations' representation of the process of observables.

It is interesting that a basic mathematical feature of the optimization finds somewhat different expression in the two approaches. The problem of solving a matrix Riccati equation in the recursive approach has, as analogue, the problem of canonical factorization of a complex function, possibly matrix-valued, in the stationary approach. We examine the connection between the two problems in Sections 7 and 8.

The approach of this chapter gives optimal control rules (in the stationary régime) under the LQG hypotheses postulated. If the injected variables are not Gaussian, the approach still gives the *linear* control which is optimal for the prescribed quadratic cost function (cf. Sections 17.5 and 18.7).

2. NOTATION

We shall work wholly in discrete time. The observed process will be denoted $\{y_t\}$, and we shall usually suppose that this has been observed up to the current instant. That is, at time t one has observed $y_t, y_{t-1}, y_{t-2}, \ldots$. The series we wish to control will be denoted $\{x_t\}$ (no state variable connotation being implicit in the notation). This may or may not be the same as $\{y_t\}$. We shall assume $x \in R^n$, $y \in R^r$, although until Section 6 we consider only the scalar case $n = r = 1$.

Autocovariances of a stationary process $\{y_t\}$ will be denoted

$$v_s = E(y_t y'_{t-s}),$$

with an additional label y if one has to distinguish between several processes. The autocovariance generating function (AGF, henceforth) will be denoted

$$g(z) = \sum_{-\infty}^{\infty} v_s z^s.$$

In the more regular of the applications to be considered, v_s will decay exponentially with increasing $|s|$, and $g(z)$ will be analytic in an annulus $\rho < |z| < \rho^{-1}$ containing the unit circle.

The AGF is related to the *spectral density function* (SDF)

$$f(\omega) = \sum_s v_s e^{i\omega s}$$

(ω being frequency) by the simple transformation of the argument

$$z = e^{-i\omega}.$$

We shall use T to denote the backward translation operator

$$T y_t = y_{t-1}. \tag{1}$$

More complicated operators (translation-invariant linear filters) can then be constructed:

$$\alpha(T) = \sum_s \alpha_s T^s, \tag{2}$$

such that

$$\alpha(T)y_t = \sum_s \alpha_s y_{t-s}. \tag{3}$$

We shall often denote both the operator $\alpha(T)$ and the corresponding generating function

$$\alpha(z) = \sum \alpha_s z^s \tag{4}$$

(itself an $r \times r$ matrix, if y is an r-vector) simply by α. There will in general be no ambiguity; the first interpretation holds when one is working in the time-domain, the second when one is working in the z-transform or frequency domain. If the s-summation in (2) is infinite, then of course conditions are required if expressions (3) and (4) are to be meaningful.

When working in the transform domain, we shall use $\bar{\alpha}$ to denote $\alpha(z^{-1})' = \sum \alpha_s' z^{-s}$ and $|\alpha|^2$ to denote $\alpha(z)\alpha(z^{-1})'$. The operator $\mathscr{A}$ denotes the 'coefficient of z^0 in the power expansion on $|z| = 1$'. So

$$\mathscr{A}(\alpha) = \alpha_0,$$

$$\mathscr{A}(|\alpha|^2) = \sum_s \alpha_s \alpha_s'$$

if the expansion $\sum \alpha_s z^s$ is valid on $|z| = 1$.

We shall also define restricted expansions:

$$[\alpha(z)]_m^n = \sum_m^n \alpha_s z^s,$$

$$\{\alpha(z)\}_m^n = \sum_m^n \alpha_s^* z^s$$

where $\sum \alpha_s z^s$ is the expansion of $\alpha(z)$ valid on the unit circle and $\sum \alpha_s^* z^s$ the expansion valid at $z = 0$, i.e., the formal expansion in non-negative powers.

More particularly,

$$[\alpha(z)]_+ = \sum_0^\infty \alpha_s z^s,$$

$$[\alpha(z)]_- = \sum_{-\infty}^0 \alpha_s z^s.$$

An expression such as $\sum_m^n$ (where m or n may be $\pm\infty$) denotes a partial power series $\sum_m^n c_s z^s$ where the coefficients c_s may be unknown, but for which the series is supposed convergent at least on the unit circle.

A generating function $\psi(z)$ representing an operator $\psi(T)$ is said to be *realizable* if $\psi_s = 0$ for $s < 0$. It is said to be *stable* if $\psi(z)$ is analytic in $|z| \leq 1$, which implies that it is realizable and $\sum_0^\infty \psi_s z^s$ is convergent in $|z| \leq 1$. It is said to be *invertible* if $\psi(z)^{-1}$ is stable. It is said to be *canonical* if it is both stable and invertible.

The canonical factorization of a scalar-valued function $g(z)$ is a factorization

$$g(z) = \psi_1(z)\psi_2(z^{-1}) \tag{5}$$

or $g = \psi_1\bar{\psi}_2$, where ψ_1 and ψ_2 are canonical. If $g(z)$ is symmetric in that $g = \bar{g}$ then the canonical factorization will also be symmetric:

$$g(z) = \psi(z)\psi(z^{-1}), \tag{5'}$$

sometimes written simply as $g = |\psi|^2$. We shall postulate regularity conditions ensuring that such a factorization exists, if required. In the case of g a matrix-valued function there will be two factorizations (5), depending upon the order in which the factors are taken (see Sections 7 and 8).

If $g(z)$ is the AGF of a stationary process $\{y_t\}$, then the factorization (5') corresponds to the existence of a *moving-average representation:*

$$y_t = \sum_0^\infty \psi_s \varepsilon_{t-s} \tag{6}$$

where the $\{\varepsilon_t\}$ are uncorrelated, with unit variance, and are determined in terms of the y_t by an inverse *autoregressive* representation:

$$\sum_0^\infty \theta_s y_{t-s} = \varepsilon_t \tag{7}$$

of the process $\{y_t\}$. The generating functions $\psi(z) = \sum_0^\infty \psi_s z^s$ and $\theta(z)$ of (6) and (7) are related by $\theta(z) = \psi(z)^{-1}$. Formulae (6) and (7) can indeed be written simply as

$$y_t = \psi(T)\varepsilon_t,$$
$$\varepsilon_t = \psi(T)^{-1} y_t,$$

and these formulae are sometimes further condensed to $y = \psi\varepsilon$ and $\varepsilon = \psi^{-1}y$. Operators will in general be interpreted in their realizable form (e.g. $\psi(T)^{-1}$ is expanded in non-negative powers of T in order to define $\theta(T)$), and this understanding will be correct if the operators are canonical.

If ψ_0 is non-singular then one can rescale the ε's to write (6) rather as

$$\begin{aligned} y_t &= \sum_0^\infty (\psi_s\psi_0^{-1})(\psi_0\varepsilon_{t-s}) \\ &= \sum_0^\infty \hat{\psi}_s \zeta_{t-1}, \end{aligned} \tag{8}$$

say. Thus $\hat{\psi}_0 = I$ and $\{\zeta_t\}$ is, like $\{\varepsilon_t\}$, a white noise sequence, but with covariance matrix $\psi_0\psi_0'$. The process $\{\zeta_t\}$ is often termed the *innovation process of* $\{y_t\}$; ζ_t is the part of y_t not linearly predictable from $y_{t-1}, y_{t-2}, \ldots$ (see Section 18.6). Relation (8) represents y_t as a function of innovations up to time t.

3. REGULATION IN THE SCALAR CASE

'Stationarity' implies stationarity of all variables occuring in the criterion function, and so of the policy. This implies restriction to infinite-horizon problems, and to problems such as regulation to a set point or the tracking of a variable which can be followed with stationary error, at least. We begin by solving a scalar regulation problem.

Suppose that the variable x_t to be controlled is scalar, obeying the plant equation

$$\alpha x_{t+1} = u_t + \varepsilon_{t+1}, \tag{1}$$

where $\{\varepsilon_t\}$ is a stationary process with the AGF g_ε. Here the control variable u_t is itself a scalar and α is an operator into the past:

$$\alpha x_t = \sum_0^\infty \alpha_s x_{t-s} \tag{2}$$

with $\alpha_0 \neq 0$. We shall suppose that the observables upon which u_t can be based are just $x_t, x_{t-1}, \ldots$, and that the control is constructed linearly:

$$u_t = \beta x_t = \sum_0^\infty \beta_s x_{t-s}. \tag{3}$$

The operator β is to be constructed so as to minimize

$$C = E[x_t^2 + \lambda u_t^2], \tag{4}$$

the expectation being calculated on the supposition that (x_t, u_t) are jointly stationary. The coefficient λ thus weights control effort relative to state deviation.

Substituting for u from (3) into (1) and solving formally we then have

$$x = (\alpha - \beta T)^{-1} \varepsilon \tag{5}$$

so that

$$u = \beta(\alpha - \beta T)^{-1} \varepsilon. \tag{6}$$

These relations give x and u in terms of the disturbing series ε. If we write (5) as

$$x_t = \theta \varepsilon_t = \sum_0^\infty \theta_s \varepsilon_{t-s} \tag{7}$$

then (6) can be written

$$u = T^{-1}(\alpha\theta - 1)\varepsilon. \tag{8}$$

The value of θ_0 is known, and equal to α_0^{-1}, by (5). Choice of $\beta_0, \beta_1, \beta_2, \ldots$ can be regarded as choice of $\theta_1, \theta_2, \ldots$. The optimization is more conveniently posed in this way, as θ occurs linearly in (7), (8), whereas β occurs non-linearly in (5), (6). One operator is, of course, easily determined in terms of the other by

$$\beta = T^{-1}(\alpha - \theta^{-1}). \tag{9}$$

Theorem 3.1. *Assume the canonical factorizations*

$$g_\varepsilon = |\psi|^2, \tag{10}$$

$$1 + \lambda|\alpha|^2 = |D|^2. \tag{11}$$

Then the generating function θ *corresponding to the optimal control* (3) *is*

$$\theta = \frac{1}{D\psi}\left\{\left[\frac{\lambda\psi\bar{\alpha}}{\bar{D}}\right]_1^\infty + \frac{D_0\psi_0}{\alpha_0}\right\} \tag{12}$$

Proof. In view of (7), (8) the cost function (4) to be minimized can be written in terms of θ as

$$C = \mathscr{A}(|\theta|^2 + \lambda|1 - \alpha\theta|^2)g_\varepsilon \tag{13}$$

where α and θ are the generating functions of the corresponding operators. If (13) is written $C = \mathscr{A}[H(z)]$, then the equation of the derivative of C with respect to θ_s to zero gives a relation which can be written

$$\mathscr{A}\left(\frac{\partial H}{\partial \theta}z^s + \frac{\partial H}{\partial \bar{\theta}}z^{-s}\right) = 0$$

if θ and $\bar{\theta}$ are regarded as functionally independent quantities. Since the two terms in the square bracket are mutual complex conjugates, we have, in effect

$$\mathscr{A}\left[\frac{\partial H}{\partial \bar{\theta}}z^{-s}\right] = 0 \qquad (s > 0)$$

or

$$\frac{\partial H}{\partial \theta} = \sum_{-\infty}^{0}.$$

That is,

$$[(1 + \lambda|\alpha|^2)\theta - \lambda\bar{\alpha}]g_\varepsilon = \sum_{-\infty}^{0}, \tag{14}$$

which, in view of the canonical factorizations (10), (11), can be written

$$|D\psi|^2\theta - \lambda\bar{\alpha}|\psi|^2 = \sum_{-\infty}^{0}. \tag{15}$$

Division of (15) by $\bar{D}\bar{\psi}$ does not change the character of the undetermined $\sum_{-\infty}^{0}$ term, and we obtain

$$D\psi\theta = \frac{\lambda\psi\bar{\alpha}}{\bar{D}} + \sum_{-\infty}^{0}. \tag{16}$$

Since the left-hand member has an expansion purely in non-negative powers of z (for $|z| \leq 1$) we must be able to identify it with the corresponding part in the expansion of the right-hand member; i.e.

$$D\psi\theta = \left[\frac{\lambda\psi\bar{\alpha}}{\bar{D}}\right]_1^\infty + \pi_0. \tag{17}$$

Here π_0 is a constant, identifiable as

$$\pi_0 = D_0\psi_0\theta_0 = \frac{D_0\psi_0}{\alpha_0}. \tag{18}$$

These last two relations imply solution (12) for the optimal θ. ■

Solution (12) is explicit if the operations of z-transformation and its inversion, truncations of power series and canonical factorization are all supposed executable. The combination of these operations used to solve the problem is simply the Wiener–Hopf technique. The optimization of linear operators on a half-realized series, as in (3), will typically yield a Wiener–Hopf problem.

We have slurred over questions of convergence. Discussion of these is not difficult in the simpler cases. However, the mere fact that one is optimizing will imply that stability will be achieved if stability can be achieved, and θ is then stable as an operator. Note that the operator α itself need not correspond to a stable system, i.e. the unstabilized system

$$\alpha(T)x_t = \varepsilon_t \tag{19}$$

need not be stable and $\alpha(z)$ need not be invertible. The optimized system

$$[\alpha(T) - T\beta(T)]x_t = \varepsilon_t \tag{20}$$

will necessarily be stable, if stabilization by some control (3) is possible.

Exercises

1. Note that if process noise ε is white then $\theta = D_0/\alpha_0 D$. Note also that if the uncontrolled process is Markov of order p, in that ε is white and $\alpha_s = 0$ for $s > p$, then the same is true of the controlled process (20): $\theta^{-1}x = \varepsilon$.
2. Suppose the process Markov of order p, as defined in (1). Let the roots of $\alpha(z) = 0$ be denoted $\xi_1, \xi_2, \ldots, \xi_p$. Show that, as $\lambda \uparrow +\infty$, and so control becomes more expensive, then the roots of $\theta(z)^{-1} = 0$ tend to $\xi_1^{\pm 1}, \xi_2^{\pm 1}, \ldots, \xi_p^{\pm 1}$, where the $\pm$ option is taken which places the root outside the unit circle.
3. Repeat the analysis of this section for the modified criterion $C = E[(x_t - \bar{x})^2 + \lambda(u_t - \bar{u})^2]$, where $\bar{x}$ and $\bar{u}$ are prescribed constants.

4. A SCALAR TRACKING PROBLEM

Suppose we again assume the plant equation (3.1), but modify the criterion to

$$C = E[(x_t - \xi_t)^2 + \lambda u_t^2] \tag{1}$$

where ξ is a stationary process with the AGF g_ξ, independent of process noise. The aim is thus to make x follow ξ, without an excessively large exertion of the control u. We shall suppose x_t to be observable with lag h and ξ_t with lag k (which may be either positive or negative), so that the control will have the form

$$\begin{aligned} u_t &= \beta x_t + \gamma \xi_t \\ &= \sum_{s=h}^{\infty} \beta_s x_{t-s} + \sum_{s=k}^{\infty} \gamma_s \xi_{t-s}. \end{aligned} \tag{2}$$

Substituting (2) into (3.1) and solving for x we obtain

$$\begin{aligned} x &= (\alpha - \beta T)^{-1}\varepsilon + (\alpha - \beta T)^{-1}\gamma T \xi \\ &= \theta\varepsilon + \phi\xi, \end{aligned} \tag{3}$$

say, where

$$\theta(T) = \sum_0^{\infty} \theta_s T^s,$$

$$\phi(T) = \sum_{k+1}^{\infty} \phi_s T^s.$$

The inverse relations to definitions (3) are (3.9) and

$$\gamma = T^{-1}\theta^{-1}\phi. \tag{4}$$

Determination of β and γ is equivalent to the determination of θ_s $(s > h)$ and ϕ_s $(s > k)$. We see from definition (3) that $\sum_0^h \theta_s z^s$ is to be identified with the sum of the first $h + 1$ terms in the expansion of $\alpha(z)^{-1}$ in non-negative powers of z. The full expansion of $\alpha(z)^{-1}$ will be invalid in $|z| \leq 1$ if the plant is intrinsically unstable, but $[\alpha(z) - z\beta(z)]^{-1}$ will have a valid expansion, whose first $h + 1$ terms must agree with those in the formal expansion of $\alpha(z)^{-1}$.

We find from (2), (3) that

$$u = T^{-1}(\alpha\theta - 1)\varepsilon + T^{-1}\alpha\phi\xi \tag{5}$$

so that the expression for the average cost becomes

$$C = \mathscr{A}[(|\theta|^2 + |1 - \alpha\theta|^2)g_\varepsilon + (|1 - \phi|^2 + \lambda|\alpha\phi|^2)g_\xi]. \tag{6}$$

Theorem 4.1. *Assume the canonical factorizations* (3.11) *and*

$$g_\varepsilon = |\psi_\varepsilon|^2, \qquad g_\xi = |\psi_\xi|^2. \tag{7}$$

Then the generating functions θ, ϕ corresponding to the optimal control (2) *are*

$$\theta = \frac{1}{D\psi_\varepsilon}\left\{\left[\frac{\lambda\psi_\varepsilon\bar{\alpha}}{\bar{D}}\right]_{h+1}^{\infty} + \left\{\frac{D\psi_\varepsilon}{\alpha}\right\}_0^h\right\}, \tag{8}$$

$$\phi = \frac{1}{D\psi_\xi}\left[\frac{\psi_\xi}{\bar{D}}\right]_{k+1}^{\infty}. \tag{9}$$

Proof. Formula (8) is simply a generalization of (3.12) to the case of lag h in observations on x. The proof goes exactly as in Theorem 3.1, except that relation (3.17) now becomes

$$D\psi_\varepsilon\theta = \left[\frac{\lambda\psi_\varepsilon\bar{\alpha}}{\bar{D}}\right]_{h+1} + \pi_h,$$

where π_h is an undetermined polynomial of degree h. Its identification follows from

$$\{\theta - \alpha^{-1}\}_0^h = 0, \tag{10}$$

as indicated earlier, whence (8) follows.

Minimizing C with respect to ϕ we find that

$$[(1+\lambda|\alpha|^2)\phi - 1]g_\xi = \sum_{-\infty}^{k},$$

whence (9) follows by the Wiener–Hopf type of argument employed in the proof of Theorem 3.1. ■

Exercise

1. Suppose ψ_ξ is a polynomial of degree q. Show that $\phi = 0$ if $q \le k$. Interpret.

5. THE TRACKING OF NON-STATIONARY PROCESSES

Consider the tracking problem of the last section. Suppose, however, the process $\{\xi_t\}$ is not stationary but can be regarded as being generated by a solution

$$\chi\xi_t = \varepsilon_t^* \tag{1}$$

where $\{\varepsilon_t^*\}$ is stationary but χ^{-1} is an unstable operator. That is, the sequence $\{\xi_t\}$ calculated recursively from (1) develops non-stationarily.

If we assume that χ has the form

$$\chi = \sum_1^m (1 - \kappa_j^{-1}T) \tag{2}$$

then instability of χ corresponds to the fact that $|\kappa_j| \le 1$ for some j. We may as well suppose this so for all j, for factors for which $|\kappa_j| > 1$ can be divided out in (1) (in a realizable fashion: by multiplying by $\sum_0^\infty \kappa_j^{-s}T^s$) and we then achieve a reduced form of (1), with ε_t^* still stationary and all factors of χ^{-1} unstable.

The tracking problem of Section 4 will be well defined even for non-stationary ξ if the quantities u_t and $x_t - \xi_t$ can be exhibited as obtainable from the stationary series $\{\varepsilon_t\}$ and $\{\varepsilon_t^*\}$ by stable operators. In their dependence upon $\{\varepsilon_t\}$ there is no problem, but we see from (4.3), (4.5) and (1) that it is necessary that $1 - \phi$ and $\alpha\phi$

should both be representable as the product of χ and a stable operator. This will be achieved if

$$1 - \phi = \delta\chi, \tag{3}$$

$$\alpha = \tilde{\alpha}\chi \tag{4}$$

where δ and $\tilde{\alpha}$ are stable operators. We can attempt to determine δ optimally under this condition.

Theorem 5.1. *Suppose the plant equation is $\alpha x = Tu + \varepsilon$ and the cost function is* (4.1), *where ξ is generated by* (1) *with $\{\varepsilon_t^*\}$ stationary and χ realizable but χ^{-1} not necessarily stable. Lags of h and k in the observation of x and ξ are assumed. Suppose that $\alpha = \tilde{\alpha}\chi$ with $\tilde{\alpha}$ stable; alternatively expressed: $\{\alpha\xi_t\}$ is stationary. Then tracking with stationary error and control can be realized by a control of form* (4.2). *The operators ϕ and θ of* (4.3) *have the following optimal determinations: θ as in* (4.8) *and ϕ equal to $1 - \delta\chi$ with*

$$\delta = \frac{1}{D\psi_{\varepsilon^*}}\left[\frac{\lambda|\tilde{\alpha}|^2\bar{\chi}\psi_{\varepsilon^*}}{\bar{D}}\right]_{k+1}^{\infty} + \frac{1}{D\psi_{\varepsilon^*}}\left\{\frac{D\psi_{\varepsilon^*}}{\chi}\right\}_0^k. \tag{5}$$

Here D is the canonical factor defined in (3.11).

Proof. As far as δ is concerned, one has to minimize

$$\{[|\delta|^2 + \lambda|\tilde{\alpha}(1 - \delta\chi)|^2]|\psi_{\varepsilon^*}|^2\}$$

with respect to δ_s $(s > k)$ with $\delta_0, \delta_1, \ldots, \delta_k$ determined by

$$\{\delta\chi\}_0^k = 1.$$

The derivation of (5) then follows the pattern of that of (4.8). ■

The essential assumption is (4). The alternative expression of it, that $\{\alpha\xi_t)$ is stationary, implies that the plant can itself cancel the unstable factors of χ^{-1}.

Condition (4) holds trivially when χ is invertible, i.e. χ^{-1} is stable, and evaluation (5) should then reduce to (4.9). In fact we find easily:

Corollary 5.1. *Define ψ_ξ as*

$$\psi_\xi = \chi^{-1}\psi_{\varepsilon^*} \tag{6}$$

whether χ is invertible or not. Then (5) *implies the optimal evaluation of ϕ:*

$$\phi = \frac{1}{D\psi_\xi}\left[\frac{\psi_\xi}{\bar{D}}\right]_{k+1}^{\infty} + \frac{1}{D\psi_\xi}[D\psi_\xi]_0^k - \frac{1}{D\psi_\xi}\{D\psi_\xi\}_0^k. \tag{7}$$

Expression (7) is easily deducible from (3) and (5), and reduces to (4.9) in the case when χ is invertible and the last two terms of (7) cancel.

6. EXACTLY TRACKED SEQUENCES

A topic closely related to that of Section 5 is the idea of sequences which the controlled plant would track exactly in the absence of process noise. We can characterize these as sequences $\{\bar{\xi}_t\}$ for which

$$\phi\bar{\xi}_t = \bar{\xi}_t. \tag{1}$$

Let us say that $\chi(z)$ is *totally uninvertible* if $\chi(z^{-1})$ is invertible. That is, the expansion of $\chi(z)^{-1}$ on $|z| = 1$ contains only non-positive powers of z.

Theorem 6.1. (*a*) *Suppose that* χ *is stable but totally uninvertible and that the hypotheses of Theorem* 5.1 *hold:* $\chi\xi$ *is stationary and* χ *divides* α, *in that* $\chi^{-1}\alpha$ *is stable. Then any solution* $\bar{\xi}_t$ *of*

$$\chi\bar{\xi}_t = 0 \tag{2}$$

is tracked exactly by the optimally controlled plant.

(*b*) *Suppose that* χ *is stable and invertible, that* $\psi(z) = \chi(z)\psi_\xi(\lambda)$ *has no poles or zeros at the zeros of* $\chi(z)$ *and that* $\chi(z)$ *divides* $|\alpha(z)|^2$, *in that* $\alpha(z)\alpha(z^{-1})/\chi(z)$ *has no poles at the zeros of* $\chi(z)$. *Then any solution of* (2) *is tracked exactly by the optimally controlled plant.*

Proof. For the optimally controlled plant we have

$$\phi = 1 - \delta\chi \tag{3}$$

where δ is stable. To prove that (2) implies (1) we have then to show that $\delta(z)$ has no poles at the zeros of $\chi(z)$. In case (a) this would be impossible, for if $\delta(z)$ had poles inside the unit circle it would not be stable. In case (b) we have, from (5.5),

$$\delta = \frac{1}{D\psi}\left[\frac{\lambda|\alpha^2|\psi}{\chi\bar{D}}\right]_{k+1}^{\infty} + \frac{1}{D\psi}\left\{\frac{D\psi}{\chi}\right\}_0^k.$$

The poles of this expression could only come from zeros of $|D|^2$, poles or zeros of ψ or poles of $|\alpha|^2/\chi$. Hypothesis excludes the second and third of these contingencies. The first is excluded by the fact that $|\alpha|^2 = 0$ and so $|D|^2 = 1$ at a zero of $\chi(z)$. ■

Note the relaxation of hypothesis in the second case: that χ divides $|\alpha|^2$ rather than α.

The principal application of the theorem is a limiting version of case (a), for $\chi = (1 - z)^p$, so that χ only just fails to be invertible. Suppose that $\alpha(z)$ has $(1 - z)^p$ as a factor, for integral p. The plant control can then be designed so that ξ is tracked with stationary error, where ξ is generated by

$$(1 - T)^p\xi_t = \varepsilon_t^*$$

and $\{\varepsilon_t^*\}$ is stationary with $g_\varepsilon^*(1) \neq 0$. Then the plant thus controlled will follow the sequences $\xi_t = t^j$ $(j = 0, 1, 2, \ldots, p-1)$ exactly. That is, its first p dynamic lags are zero.

7. THE VECTOR REGULATION PROBLEM; COMPARISON WITH STATE SPACE METHODS

Consider again the plant equation (3.1), but under the supposition that x and u are vectors of the same dimensionality, n. The operator $\alpha(T)$ will then be an $n \times n$ matrix of operators. If (3.1) is truly a recursion determining x_t, then α must have the realizable form

$$\alpha(T) = \sum_{s=0}^{\infty} \alpha_s T^s \tag{1}$$

with α_0 non-singular. Again, $\alpha(z)$ will denote the corresponding matrix of generating functions.

If the problem is one of regulation (to $x = 0$) then we can take the criterion function as

$$C = E(x_t' R x_t + u_t' Q u_t). \tag{2}$$

Actually, if we are to take the spectral approach, there is no difficulty in giving R the more general form $\sum R_s T^s$, and ditto for Q, so that C is the expectation of a more general quadratic functional of the x, u processes (a non-negative definite functional, presumably). However, for simplicity, we shall keep to the 'instantaneous' version (2).

If we consider the particular case

$$\alpha(T) = I - AT \tag{3}$$

for constant A and suppose ε_t to be a vector white noise process, then we are back in just the state space formulation of Section 12.1, except that we have effectively set $B = I$. We shall examine this special case shortly, to see how the two approaches compare.

To assume $B = I$ is undoubtedly convenient, because the control injected into the plant equation is then unrestricted in form. One can recover the general case by redefinition of u and Q, at least if one is willing to allow Q to go to infinity in various ways, so as to effectively enforce relations between the controls acting on different components of αx_{t+1}.

As explained in Section 2, the AGF g_ε will now be an $n \times n$ matrix of generating functions, and $\bar{\alpha}(z)$ is now to be understood as $\alpha(z^{-1})'$. Relations such as (3.3), (3.7) and (3.9) will persist, with α, β, θ all interpreted as matrices, of operators (see (1)) or generating functions.

Theorem 7.1. *Assume the factorizations*

$$g_\varepsilon = \psi_\varepsilon \bar{\psi}_\varepsilon, \tag{4}$$

$$R + \bar{\alpha} Q \alpha = \bar{D} D \tag{5}$$

with D and ψ_ε canonical. Then the generating function $\theta(z)$ corresponding to the optimal control $u = T^{-1}(\alpha\theta - I)\varepsilon$ is

$$\theta = D^{-1}[\bar{D}^{-1}\bar{\alpha}Q\psi_\varepsilon]_1^\infty \psi_\varepsilon^{-1} + D^{-1}D_0\alpha_0^{-1}\psi_{\varepsilon 0}\psi_\varepsilon^{-1}. \tag{6}$$

Relation (6) is the matrix analogue of (3.12), and its proof is analogous. The only point one has to observe with care is the order of terms; note the differing orders of the canonical factors in (4) and (5). The calculations are indicated in rather more detail for the more general case of Theorem 8.1.

Solution (6) simplifies considerably if process noise is white, when the first term in (6) vanishes.

Corollary 7.1. *Suppose that $g_\varepsilon(z) = N$, where the noise covariance matrix N is non-singular. Then, for the optimal control,*

$$\theta = D^{-1}D_0\alpha_0^{-1}, \tag{7}$$

$$\beta = z^{-1}(\alpha - \alpha_0 D_0^{-1} D). \tag{8}$$

Note that these results are independent of N—a manifestation of certainty equivalence.

If we revert completely to the state space case and assume both white process noise and the simple recursive plant operator (3) then D will be of first order in χ:

$$D(z) = D_0(I - \Gamma z), \tag{9}$$

say. It thus follows from (8) that

$$\beta(z) = \Gamma - A. \tag{10}$$

That is, the optimal control is Markov, and the matrix Γ of (9) is indeed identifiable with the matrix Γ of Section 5.5, characterized by the fact that the optimally controlled process obeys $x_{t+1} = \Gamma x_t + \varepsilon_{t+1}$.

We have now to link evaluation of the canonical factor (9) with evaluation of the limit solution P of the Riccati equation (5.2.4) and consequent evaluation of Γ. We find from (7) that the minimal value of average cost is

$$C = \operatorname{tr} N[\bar{D}_0 D_0 - Q]. \tag{11}$$

This is to be identified with $\operatorname{tr}(NP)$ (see 12.1.7), whence we deduce the identification

$$\bar{D}_0 D_0 = P + Q. \tag{12}$$

From (5), (9), (12) we thus deduce that

$$R + (I - A'z^{-1})Q(I - Az) = (I - \Gamma'z^{-1})(P + Q)(I - \Gamma z), \tag{13}$$

whence we deduce the identities

$$R + A'QA = P + \Gamma'(P + Q)\Gamma, \tag{14}$$

$$QA = (P + Q)\Gamma. \tag{15}$$

Eliminating Γ from (14), (15) we deduce the equilibrium form of the Riccati equation (5.2.4) for P (with $B = I, S = 0$) as one expects. Eliminating P we deduce the equation

$$R + (A' - Q\Gamma^{-1}Q^{-1})Q(A - \Gamma) = 0 \tag{16}$$

for Γ. If $\Gamma = \Gamma_1$ is a root of this equation, then so is

$$\Gamma = \Gamma_2 := Q^{-1}\Gamma_1'Q.$$

One seeks the matrix root of (16) which is a stability matrix. This point is made somewhat differently by the role of Γ in the canonical factorization (13).

8. VECTOR REGULATION WITH OBSERVATION NOISE

We assume the vector plant equation (3.1) and cost criterion (7.2), but assume now that one observes a noise-corrupted version of x:

$$y_t = x_t + \eta_t, \tag{1}$$

observations being available at time t on y_{t-s} for $s \geq h$. The control will then be of the form

$$u_t = \beta y_t = \sum_{s=h}^{\infty} \beta_s y_{t-s}. \tag{2}$$

One could suppose more generally (as in Chapter 18) that one observes

$$y_t = Cx_t + \eta_t \tag{3}$$

where C may be either a constant matrix or a matrix of time-distributed operators, $\sum C_s T^s$. However, division of (3) by C and a redefinition of y and η will restore the case (1). This can be effectively achieved even if C is singular, if one allows certain components of the transformed η to have infinite variance. In such a case the corresponding components of x are unobservable, in that they are observable only with infinite error.

This normalization of C to I corresponds completely to the normalization of B to I in Section 7, achieved by redefinition of Q and u.

We assume $\{\varepsilon_t\}$ and $\{\eta_t\}$ jointly stationary and, for simplicity, mutually independent.

It follows from (3.1), (1) and (2) that

$$x = (\alpha - \beta T)^{-1}[\varepsilon + \beta T\eta], \tag{4}$$

$$u = \beta(\alpha - \beta T)^{-1}\varepsilon + \beta(\alpha - \beta T)^{-1}\alpha\eta. \tag{5}$$

Defining again

$$\theta = (\alpha - \beta T)^{-1} \tag{6}$$

we can write these relations as

$$x = \theta\varepsilon + (\theta\alpha - 1)\eta, \tag{7}$$

$$u = T^{-1}(\alpha\theta - I)\varepsilon + T^{-1}(\alpha\theta\alpha - \alpha)\eta. \tag{8}$$

whence

$$\begin{aligned} C = \operatorname{tr}\{[\bar{\theta}R\theta + (\overline{\alpha\theta - I})Q(\alpha\theta - I)]g_\varepsilon\} \\ + \operatorname{tr}\{[(\overline{\theta\alpha - I})R(\theta\alpha - I) + \bar{\alpha}(\overline{\alpha\theta - I})Q(\alpha\theta - I)\alpha]g_\eta\}. \end{aligned} \tag{9}$$

Theorem 8.1. *Define the canonical factorizations*

$$R + \bar{\alpha}Q\alpha = \bar{D}D, \tag{10}$$

$$g_\varepsilon + \alpha g_\eta \bar{\alpha} = F\bar{F} \tag{11}$$

and the quantity

$$Z = \bar{\alpha}Qg_\varepsilon + (R + \bar{\alpha}Q\alpha)g_\eta\bar{\alpha}. \tag{12}$$

Then the optimal θ is determined by

$$D\theta F = [\bar{D}^{-1}Z\bar{F}^{-1}]_{h+1}^{\infty} + \{D\alpha^{-1}F\}_0^h. \tag{13}$$

Proof. Minimizing expression (9) with respect to θ_s $(s > h)$ we deduce the relations

$$\bar{D}D\theta F\bar{F} - Z = \sum_{-\infty}^{h}. \tag{14}$$

The coefficients θ_s $(0 \le s \le h)$ are determined by (4.10); from this and (14) we deduce (13) by the usual arguments. ■

The separate occurrence of the analogous factorizations (10), (11) is an indication of the analogy and separateness of the two problems: of control of x (in the absence of either process or observation noise) and estimation of x (regardless of the control rule employed). This is even more evident if we specialize to the Markov structure of earlier chapters for plant and observations.

Theorem 8.2. *Suppose that plant is simply recursive, $\alpha(z) = I - Az$, and noise is white, $g_\varepsilon(z) = N$, $g_\eta(z) = M$, so that the factorizations* (10), (11) *have first-order* *factors*

$$R + \bar{\alpha} Q \alpha = (I - \Gamma' z^{-1}) D_0' D_0 (I - \Gamma z), \tag{15}$$

$$N + \alpha M \bar{\alpha} = (I - \Omega z) F_0 F_0' (I - \Omega' z^{-1}). \tag{16}$$

Suppose also that $h = 1$. *Then the solution* (13) *for* θ *reduces to*

$$\theta = (I - \Gamma z)^{-1} [I - (\Gamma + \Omega - A) z] (I - \Omega z)^{-1}. \tag{17}$$

Proof. The first term in the right-hand member of (13) vanishes, since the square bracket contains no terms in z^s for $s > 1$. We thus deduce that

$$(I - \Gamma z)\theta(I - \Omega z) = \{(I - \Gamma z)(I - Az)^{-1}(I - \Omega z)\}_0^1,$$

whence (17) follows. ■

Corollary 8.1. *Relation* (7) *with* θ *evaluated by* (17) *is equivalent to the pair of relationships*

$$\begin{aligned} x_{t+1} &= A x_t + K \hat{x}_t + \varepsilon_t, \\ \hat{x}_{t+1} &= A \hat{x}_t + K x_t + H(x_t - \hat{x}_t + \eta_t), \end{aligned} \tag{18}$$

where

$$\begin{aligned} K &= \Gamma - A, \\ H &= A - \Omega. \end{aligned} \tag{19}$$

Proof. The equivalence is easily confirmed: eliminating $\hat{x}$ from (18) one deduces the relation (7) for x with θ evaluated by (19). ■

Relations (18) of course constitute the expression of the control rule $u_t = K\hat{x}_t$ and of the Kalman filter for the recursive evaluation of $\hat{x}_t$. We proved in the last section that the evaluation of Γ from the canonical factorization (15) was equivalent to the evaluation of P from the static Riccati equation. The analogous results hold for the determinations of Ω and V. The mutual symmetry and separateness of these two sets of calculations is also clear.

Bibliography

GENERAL REFERENCES

Aitken, A. C. (1959) *Determinants and Matrices* (9th edition), Oliver and Boyd.
Åström, K. J. (1970). *Introduction to Stochastic Control Theory*, Academic Press.
Bellman, R. (1957). *Dynamic Programming*, Princeton University Press, New Jersey.
Bertsekas, D. P. (1976). *Dynamic Programming and Stochastic Control*, Academic Press.
Bryson, A. E., and Ho, Y. C. (1969). *Applied Optimal Control*, Ginn.
Davis, M. H. A. (1977). *Linear Estimation and Stochastic Control*, Chapman and Hall.
DeGroot, M. (1970). *Optimal Statistical Decisions*, McGraw-Hill.
Derman, C. (1970). *Finite State Markovian Decision Processes*, Academic Press.
Dynkin, E. B., and Yushkevich, A. A. (1979). *Controlled Markov Processes*, English edition, Springer.
Fleming, W. H., and Rishel, R. W. (1975). *Deterministic and Stochastic Optimal Control*, Springer.
Gihmann, I. I., and Skorohod, A. V. (1979). *Controlled Stochastic Processes*, English edition, Springer.
Hinderer, K. (1970). *Foundations of Non-stationary Dynamic Programming with Discrete Time Parameter*, Springer.
Howard, R. A. (1960). *Dynamic Programming and Markov Processes*, M. I. T. and Wiley.
Jacobs, O. C. R. (1974). *Introduction to Control Theory*, Oxford University Press.
Kailath, T. (1980). *Linear Systems*, Prentice-Hall.
Kushner, H. (1976). *Introduction to Stochastic Control*, Holt, Rinehart and Winston.
Luenberger, D. G. (1979). *Introduction to Dynamic Systems*, Wiley.
Ross, S. M. (1970). *Applied Probability Models with Optimisation Applications*, Holden-Day.
Willems, J. C. (1971). *The Analysis of Feedback Systems*, M.I.T. Press, Cambridge, Mass.
White, D. (1969). *Dynamic Programming*, Oliver & Boyd.

SPECIFIC REFERENCES

Arrow, K. R., Harris, T., and Marschak, J. (1951. Optimal Inventory policy, *Econometrica*, **19**, 250–272.
Arrow, K. J., Karlin, S., and Scarf, H. (1958). *Studies in the Mathematical Theory of Inventory and Production*, Stanford University Press.
Bather, J. A. (1962). Optimal regulation policies for finite dams, *J. Soc. Indust. Appl. Math.*, **10**, 395–423.
Bather, J. A. (1963). The optimal regulation of dams in continuous time, *J. Soc. Indust. Appl. Math.*, **11**, 33–63.
Bather, J. A. (1966). A continuous time inventory model, *J. Appl. Prob.*, **3**, 538–549.
Bather, J. A. (1968). A diffusion model for the control of a dam, *J. Appl. Prob.*, **5**, 55–71.

Berkowitz, L. D. (1974). *Optimal Control Theory*, Springer.

Bierman, G. (1977). *Factorisation Methods for Discrete Estimation*, Academic Press.

Black, W. (1965). Discrete sequential search, *Inf. and Control*, **8**, 159–162.

Brockett, R. W. (1970). *Finite Dimensional Linear Systems*, Wiley.

Canon, M., Cullum, C., and Polak, E. (1966). Constrained minimisation problems in finite-dimensional spaces, *SIAM J. Control*, **4**, 528–547.

Canon, M. D., Cullum, C. D., and Polak, E. (1970). *Theory of Optimal Control and Mathematical Programming*, McGraw-Hill.

Cayley, A. (1874). Mathematical questions and their solutions, Problem No. 4528, *The Educational Times*, **27**, 189.

Cayley, A. (1875). Mathematical questions and their solutions, solution to Problem No. 4528, *The Educational Times*, **27**, 237.

Cox, D. R., and Miller, H. D. (1965). *The Theory of Stochastic Processes*, Methuen.

Cox, D. R., and Smith, W. L. (1961). *Queues*, Methuen.

De Bruijn, N. G. (1958). *Asymptotic Methods in Analysis*, Amsterdam.

Dubins, C., and Savage, C. (1965). *How to Gamble If You Must*, McGraw-Hill.

Fan, L. T., and Wang, C. S. (1964). *The Discrete Maximum Principle*, Wiley.

Freedman, D. (1967). Timid play is optimal, *Ann. Math. Statist.*, **38**, 1281–1284.

Gale, D. (1960). *The Theory of Linear Economic Models*, McGraw-Hill.

Gale, D. (1968). A mathematical theory of optimal economic development, *Bull. Amer. Math. Soc.*, **74**, 207–223.

Gilbert, J., and Mosteller, F. (1966). Recognizing the maximum of a sequence, *J. Amer. Statist. Soc.*, **61**, 35–73.

Gittins, J. C. (1979). Bandit processes and dynamic allocation indices, *J. Roy. Statist. Soc., B*, **41**, 148–164.

Gittins, J. C., and Jones, D. M. (1974). A dynamic allocation index for the sequential design of experiments. In *Progress in Statistics* (Ed. J. Gani), pp. 241–266, North Holland, Amsterdam.

Gittins, J. C., and Nash, P. (1977). Scheduling, queues and dynamic allocation indices, *Proc. EMS, Prague 1974*, pp. 191–202. Czechoslovak Academy of Sciences, Prague.

Glazebrook, K. D. (1976). A profitability index for alternative research projects, *Omega*, **4**, 79–83.

Glazebrook, K. D. (1982). On a sufficient condition for superprocesses due to Whittle, *J. Appl. Prob.*, **19**, 99–110.

Harrison, J. M. (1975a). A priority queue with discounted linear costs, *Operations Res.*, **23**, 260–269.

Harrison, J. M. (1975b). Dynamic scheduling of a multi-class queue: discount optimality, *Operations Res.*, **23**, 270–282.

Harrison, J. M. (1976). Dynamic scheduling of a two-class queue: small interest rates, *Siam J. Appl. Math.*, **31**, 51–61.

Holt, C., Modigliani, F., Muth, J. F., and Simon, H. A. (1960). *Planning, Production, Inventories and Work-force*, Prentice-Hall.

Howard, P., and Matheson, I. (1972). Risk-sensitive Markov decision processes, *Management Sci.*, **18**, 356–369.

Jacobson, D. H. (1973). Optimal stochastic linear systems with exponential performance criteria and their relation to deterministic differential games, *IEEE Trans. Automatic Control*, **AC-18**, 124–131.

Jacobson, D. H. (1977). *Extensions of Linear-Quadratic Control, Optimization and Matrix Theory*, Academic Press.

Kalman, R. E. (1960). A new approach to linear filtering and prediction problems, *Trans. ASME Ser. D, J. Basic Eng.*, **82**, 35–45.

Kalman, R. E. (1963). New methods in Wiener filtering theory. In *Proceedings of the Symp. on the Engineering Applic. of Random Function Theory and Probability*, (Eds. F. Kozin and J. L. Bogdanoff) Wiley, New York.

Kalman, R. E., and Bucy, R. S. (1961). New results in linear filtering and prediction theory, *Trans. ASME Ser. D, J. Basic Eng.*, **83**, 95–108.

Kemeny, J., and Snell, J. (1960). *Finite Markov Chains*, Van Nostrand.

Kiefer, J. (1953). Sequential minimax search for a maximum, *Proc. Ann. Math. Soc.*, **4**, 502–506.

Kumar, D. R., and Van Schuppen, J. H. (1981). On the optimal control of stochastic systems with an exponential-of-integral performance index, *J. Math. Anal. Appls.*, **80**, 312–332.

Lawden, D. F. (1963). *Optimal Trajectories for Space Navigation*, Butterworths, London.

Lehnigk, S. H. (1966). *Stability Theorems for Linear Motions*, Prentice-Hall.

Margenau, H., and Murphy, G. M. (1943). *The Mathematics of Physics and Chemistry*, Van Nostrand.

Morf, M., and Kailath, T. (1975). Square-root algorithms for least-squares estimation, *IEEE Trans. Autom. Control*, **AC-20**, 487–497.

Morton, R. M. (1968). On the dynamic programming approach to Pontryagin's maximal principle, *J. Appl. Prob.*, **5**, 679–692.

Nash, P., and Gittins, J. C. (1977). A hamiltonian approach to optimal stochastic resource allocation, *Adv. Appl. Prob.*, **9**, 55–68.

Newton, G. C. (1952). Compensation of feed-back control systems subject to saturation, *J. Franklin. Inst.*, **254**, 281–286; 391–413.

Newton, G. C., Gould, L. A., and Kaiser, J. F. (1957). *Analytical Design of Linear Feedback Controls*, Wiley.

Prabhu, N. U. (1965). *Stochastic Processes*, Macmillan.

Reid, D. W., and Teo, K. L. (1980). Optimal parameter selection of parabolic systems, *Maths. of Op. Res.*, **5**, 467–479.

Ross, S. H. (1969). A problem in optimal search and stop, *J. Op. Res. Soc. Am.*, **17**, 984–992.

Ross, S. M. (1974). Dynamic programming and gambling models, *Adv. Appl. Prob.*, **6**, 593–606.

Seber, G. (1977). *Linear Regression Analysis*, Wiley.

Serfozo, R. (1981). Optimal control of random walks, birth and death processes, and queues, *Adv. Appl. Prob.*, **13**, 61–83.

Silverman, L. (1976). Discrete Riccati equations, In *Control and Dynamic Systems*, (Ed. C. T. Leondes), vol. 12, pp. 313–386, Academic Press.

Speyer, J. L. (1976). An adaptive terminal guidance scheme on an exponential cost criterion with application to homing missile guidance, *IEEE Trans. Automat. Control*, **AC-21**, 371–375.

Speyer, J. L., Deyst, J., and Jacobson, D. H. (1974). Optimisation of stochastic linear systems with additive measurement and process noise using exponential performance criteria, *IEEE Trans. Automat. Control*, **AC-19**, 358–366.

Stewart, G. W. (1973). *Introduction to Matrix Computations*, Academic Press.

Weiss, G., and Pinedo, M. (1980). Scheduling tasks with exponential service times on non-identical processors to minimise various cost functions, *J. Appl. Prob.*, **17**, 187–202.

Whittle, P. (1963). *Prediction and Regulation by Linear Least-Square Methods*, English Universities Press, London.

Whittle, P. (1971). *Optimization under Constraints*, Wiley Interscience.

Whittle, P. (1975). An approximate characterisation of optimal stopping boundaries, *J. Appl. Prob.*, **10**, 158–165.

Whittle, P. (1980). Multi-armed bandits and the Gittins index, *J. Roy. Statist. Soc. Ser. B*, **42**, 143–149.
Whittle, P. (1981a). Arm-acquiring bandits, *Ann. Prob.*, **9**, 284–292.
Whittle, P. (1981b). Risk-sensitive linear/quadratic/Gaussian control, *J. Appl. Prob.*, **18**,
Whittle, P., and Gait, P. A. (1969). Reduction of a class of stochastic control problems, *J. Inst. Maths. Applics.*, **6**, 131–140.
Wiener, N. (1949). *Extrapolation, Interpolation and Smoothing of Stationary Time Series*, Wiley, New York.
Willems, I. L. (1970). *Stability of Dynamic Systems*, Nelson.
Wong, E. (1971). *Stochastic Processes in Information and Dynamical Systems*, McGraw-Hill.
Wonham, W. M. (1968). On the Separation Theorem of Stochastic Control, *SIAM Journal on Control*, **6**, 312–326.

Index to Volume I

Contents of Volume II

PART IV. INFINITE-HORIZON PROBLEMS

CONTENTS OF VOLUME II

PART V. CONTINUOUS TIME

PART VI INCOMPLETELY OBSERVED PROCESS STATE

ENVOI

FLEISS • Statistical Methods for Rates and Proportions, *Second Edition*
FRANKEN, KÖNIG, ARNDT, and SCHMIDT • Queues and Point Processes
GALAMBOS • The Asymptotic Theory of Extreme Order Statistics
GIBBONS, OLKIN, and SOBEL • Selecting and Ordering Populations: A New Statistical Methodology
GNANADESIKAN • Methods for Statistical Data Analysis of Multivariate Observations
GOLDBERGER • Econometric Theory
GOLDSTEIN and DILLON • Discrete Discriminant Analysis
GROSS and CLARK • Survival Distributions: Reliability Applications in the Biomedical Sciences
GROSS and HARRIS • Fundamentals of Queueing Theory
GUPTA and PANCHAPAKESAN • Multiple Decision Procedures: Theory and Methodology of Selecting and Ranking Populations
GUTTMAN, WILKS, and HUNTER • Introductory Engineering Statistics, *Third Edition*
HAHN and SHAPIRO • Statistical Models in Engineering
HALD • Statistical Tables and Formulas
HALD • Statistical Theory with Engineering Applications
HAND • Discrimination and Classification
HARTIGAN • Clustering Algorithms
HILDEBRAND, LAING, and ROSENTHAL • Prediction Analysis of Cross Classifications
HOEL • Elementary Statistics, *Fourth Edition*
HOEL and JESSEN • Basic Statistics for Business and Economics, *Third Edition*
HOLLANDER and WOLFE • Nonparametric Statistical Methods
JAGERS • Branching Processes with Biological Applications
JESSEN • Statistical Survey Techniques
JOHNSON and KOTZ • Distributions in Statistics
Discrete Distributions
Continuous Univariate Distributions—1
Continuous Univariate Distributions—2
Continuous Multivariate Distributions
JOHNSON and KOTZ • Urn Models and Their Application: An Approach to Modern Discrete Probability Theory
JOHNSON and LEONE • Statistics and Experimental Design in Engineering and the Physical Sciences, Volumes I and II, *Second Edition*
JUDGE, HILL, GRIFFITHS, LÜTKEPOHL and LEE • Introduction to the Theory and Practice of Econometrics
JUDGE, GRIFFITHS, HILL and LEE • The Theory and Practice of Econometrics
KALBFLEISCH and PRENTICE • The Statistical Analysis of Failure Time Data
KEENEY and RAIFFA • Decisions with Multiple Objectives
LAWLESS • Statistical Models and Methods for Lifetime Data
LEAMER • Specification Searches: Ad Hoc Inference with Nonexperimental Data
McNEIL • Interactive Data Analysis
MANN, SCHAFER and SINGPURWALLA • Methods for Statistical Analysis of Reliability and Life Data
MARTZ and WALLER • Bayesian Reliability Analysis
MEYER • Data Analysis for Scientists and Engineers
MILLER • Survival Analysis
MILLER, EFRON, BROWN, and MOSES • Biostatistics Casebook
MONTGOMERY and PECK • Introduction to Linear Regression
NELSON • Applied Life Data Analysis
OTNES and ENOCHSON • Applied Time Series Analysis: Volume I, Basic Techniques
OTNES and ENOCHSON • Digital Time Series Analysis
POLLOCK • The Algebra of Econometrics